The IMA Volumes in Mathematics and its Applications

Volume 141

Institute for Mathematics and its Applications (IMA)

The **Institute for Mathematics and its Applications** was established by a grant from the National Science Foundation to the University of Minnesota in 1982. The primary mission of the IMA is to foster research of a truly interdisciplinary nature, establishing links between mathematics of the highest caliber and important scientific and technological problems from other disciplines and industries. To this end, the IMA organizes a wide variety of programs, ranging from short intense workshops in areas of exceptional interest and opportunity to extensive thematic programs lasting a year. IMA Volumes are used to communicate results of these programs that we believe are of particular value to the broader scientific community.

The full list of IMA books can be found at the Web site of the Institute for Mathematics and its Applications:

http://www.ima.umn.edu/springer/volumes.html

Douglas N. Arnold, Director of the IMA

* * * * * * * * * * *

IMA ANNUAL PROGRAMS

1982–1983	Statistical and Continuum Approaches to Phase Transition
1983–1984	Mathematical Models for the Economics of Decentralized Resource Allocation
1984–1985	Continuum Physics and Partial Differential Equations
1985–1986	Stochastic Differential Equations and Their Applications
1986–1987	Scientific Computation
1987–1988	Applied Combinatorics
1988–1989	Nonlinear Waves
1989–1990	Dynamical Systems and Their Applications
1990–1991	Phase Transitions and Free Boundaries
1991–1992	Applied Linear Algebra
1992–1993	Control Theory and its Applications
1993–1994	Emerging Applications of Probability
1994–1995	Waves and Scattering
1995–1996	Mathematical Methods in Material Science
1996–1997	Mathematics of High Performance Computing
1997–1998	Emerging Applications of Dynamical Systems
1998–1999	Mathematics in Biology

Continued at the back

Maria-Carme T. Calderer Eugene M. Terentjev
Editors

Modeling of Soft Matter

With 20 Illustrations

Maria-Carme T. Calderer
School of Mathematics
University of Minnesota
http://www.math.umn.edu/%7Emcc/

Eugene M. Terentjev
Cavendish Laboratory
Cambridge University
http://www.poco.phy.cam.ac.uk/~emt1000

Series Editor:
Douglas N. Arnold
Arnd Scheel
Institute for Mathematics and its Applications
University of Minnesota
Minneapolis, MN 55455
http://www.ima.umn.edu/

Mathematics Subject Classification (2000): 76Axx, 76A15, 82Cxx

ISBN 978-1-4899-8758-7 ISBN 978-0-387-32153-0 (eBook)
DOI 10.1007/978-0-387-32153-0

Printed on acid-free paper.

(SBA)

Camera-ready copy provided by the IMA.

9 8 7 6 5 4 3 2 1

springeronline.com

FOREWORD

This IMA Volume in Mathematics and its Applications

MODELING OF SOFT MATTER

contains papers presented at a very successful workshop with the same title. The event, which was held on September 27–October 1, 2004, was an integral part of the 2004-2005 IMA Thematic Year on "Mathematics of Materials and Macromolecules: Multiple Scales, Disorder, and Singularities." We would like to thank Maria-Carme T. Calderer (School of Mathematics, University of Minnesota) and Eugene M. Terentjev (Cavendish Laboratory, University of Cambridge) for their superb role as workshop organizers and editors of the proceedings.

We take this opportunity to thank the National Science Foundation for its support of the IMA.

Series Editors

Douglas N. Arnold, Director of the IMA

Arnd Scheel, Deputy Director of the IMA

PREFACE

The physics of soft matter in particular, focusing on such materials as complex fluids, liquid crystals, elastomers, soft ferroelectrics, foams, gels and particulate systems is an area of intense interest and contemporary study. Soft matter plays a role in a wide variety of important processes and application, as well as in living systems. For example, gel swelling is an essential part of many biological processes such as motility mechanisms in bacteria and the transport and absorption of drugs. Ferroelectrics, liquid crystals, and elastomers are being used to design ever faster switching devices. Experiments of the last decade have provided a great deal of detailed information on structures and properties of soft matter. But the integration of mathematical modeling and analysis with experimental approaches promises to greatly increase our understanding of underlying principles and provide a predictive power. The articles presented in this volume share such an integrated approach and span several areas of applied and computational mathematics, continuum and statistical mechanics.

Several articles deal with interfacial phenomena in soft matter, from both, static and dynamic points of view, including a survey on evolution of interfaces in complex fluids, and on the role of surface forces in bulk ordering. A related article deals with the role of line tension in wetting. Modeling of nano-composite films of nematic liquids is the topic of one of the works, that also explores the effective conductivity properties of such a composites. There are two articles in the subject of elastomers with two distinctive thrusts: studies of stripe phenomena, and also swelling of gels made of liquid filled networks. Ferroelectricity in smectic C* liquid crystals is also presented pointing to the challenges of nonlocal electric phenomena due to the spontaneous polarization in the material. New models of nonrigid particulate systems is the subject of one of the articles, that addresses the problem of stress transmission and isostatic states of such systems. One of the articles is devoted to the non-equilibrium statistical mechanics of nematic liquids and the Fokker-Planck equation. Workers in areas of complex and non-Newtonian fluids may benefit from the article on constitutive equations involving the orientational order parameter. This work establishes comparisons among well known theories of non-Newtonian fluids.

The editors wish to thank all the contributors of the volume and also the speakers and participants of the IMA workshop on soft matter physics.

The goal of this volume is to motivate the applied mathematics and the soft matter physics communities to continue collaborative tasks of identifying beautiful and novel scientific problems and set the stage for further research.

The editors wish to thank the IMA staff for all their assistance, and especially Patricia V. Brick and Dzung N. Nguyen for their efforts in making this volume possible.

Maria-Carme T. Calderer
School of Mathematics
University of Minnesota
http://www.math.umn.edu/%7Emcc/

Eugene M. Terentjev
Cavendish Laboratory
Cambridge University
http://www.poco.phy.cam.ac.uk/~emt1000

CONTENTS

AN ENERGETIC VARIATIONAL FORMULATION WITH PHASE FIELD METHODS FOR INTERFACIAL DYNAMICS OF COMPLEX FLUIDS: ADVANTAGES AND CHALLENGES

JAMES J. FENG[§*], CHUN LIU[†], JIE SHEN[‡], AND PENGTAO YUE[§]

Abstract. The use of a phase field to describe interfacial phenomena has a long and fruitful tradition. There are two key ingredients to the method: the transformation of Lagrangian description of geometric motions to Eulerian description framework, and the employment of the energetic variational procedure to derive the coupled systems. Several groups have used this theoretical framework to approximate Navier-Stokes systems for two-phase flows. Recently, we have adapted the method to simulate interfacial dynamics in blends of microstructured complex fluids. This review has two objectives. The first is to give a more or less self-contained exposition of the method. We will briefly review the literature, present the governing equations and discuss a suitable numerical schemes, such as spectral methods. The second objective is to elucidate the subtleties of the model that need to be handled properly for certain applications. These points, rarely discussed in the literature, are essential for a realistic representation of the physics and a successful numerical implementation. The advantages and limitations of the method will be illustrated by numerical examples. We hope that this review will encourage readers whose applications may potentially benefit from a similar approach to explore it further.

Key words. Energetic variational formulation, phase field methods, Cahn-Hilliard equation, two-phase flows, complex fluids, free interfacial motions.

AMS(MOS) subject classifications. 76A02, 76A15, 76A05, 76M30, 76T20, 76T10, 76R99, 76M45,76M22, 76D45, 76B10, 76D05.

1. Introduction. Most complex fluids have complicated internal microstructures, whose conformation is coupled with the macroscopic dynamics of the material [1]. On the one hand, this coupling gives rise to novel flow behavior. On the other hand, it plays a central role in achieving desirable structure and property in advanced engineering materials. Complex fluids are often used in composites, of which polymer-dispersed liquid crystals and polymer blends are good examples [2, 3]. In these two-phase systems, the components are separated by myriad interfaces that move and deform with the flow; the interfacial morphology to a large extent determines the dynamics of the mixture.

A fluid-mechanical theory for two-phase mixtures of complex fluids has to contend with two difficulties: the moving internal boundaries (or internal transition regions) and the complex rheology of the components.

[*](jfeng@chml.ubc.ca).

[†]Department of Mathematics, Penn State University, University Park, PA 16802 (liu@math.psu.edu).

[‡]Department of Mathematics, Purdue University, West Lafayette, IN 47907 (shen@math.purdue.edu).

[§]Department of Chemical & Biological Engineering and Department of Mathematics, University of British Columbia, Vancouver, BC V6T 1Z4, CANADA.

The former is a well-known mathematical difficulty. The movement of the interfaces is naturally amenable to a Lagrangian description, while the bulk flow is conventionally solved in an Eulerian framework. A great deal of effort has gone to reconciling these two considerations in a numerical scheme [4]. The latter difficulty stems from the fact that the rheology of each component depends on the internal microstructure, which is coupled with the flow field [5, *e.g.*]. Thus, these materials feature dynamic coupling of three disparate length scales: molecular or supra-molecular conformation inside each component, mesoscopic interfacial morphology and macroscopic hydrodynamics. The complexity of such materials has for the large part prohibited theoretical and numerical analysis.

A conceptually straightforward way of handling the moving interfaces is to employ a moving mesh that has grid points on the interfaces and deforms according to the flow on both sides of the boundary. This has been implemented in boundary integral and boundary element methods [6–8], finite-element methods [9–11] and a finite-difference method [12, 13]. Besides the overhead in keeping track of the moving mesh, these methods break down when large displacement of internal domains causes mesh entanglement or when the interfaces undergo singular topological changes such as breakup and coalescence. Thus, these methods have been limited mostly to single drops undergoing relatively mild deformations. As an alternative, fixed-grid methods have been developed that regularized the interface [4]. These include the volume-of-fluid (VOF) method [14, 15], the front-tracking method [16, 17] and the level-set method [18–20]. All these approaches have the advantage of converting the Lagrangian description of a geometric motion into the Eulerian description. Instead of computing the flow of the two components with matching boundary conditions on the interface, these methods represent the interfacial tension as a body force or bulk stress spread over a narrow region covering the interface. Then a single set of governing equations can be written over the entire domain and solved on a fixed grid in a purely Eulerian framework.

The phase-field method is also a fixed-grid method; it differs from those aforementioned in that the interface is diffusive in a physical rather than numerical sense. Thus, it is also known as the diffuse-interface model. More precisely, the diffuse interface is introduced through an energetic variational procedure that results in a thermodynamic consistent coupling system. The basic idea was derived from the consideration that the two components, though nominally immiscible, does mix in reality within a narrow interfacial region. A phase-field variable ϕ is introduced, which can be thought of as the volume fraction, to demarcate the two species and indicate the location of the interface. A mixing energy is defined based on ϕ which, through a convection-diffusion equation, governs the evolution of the interfacial profile. The phase-field method can be viewed as a physically motivated level-set method, and Lowengrub and Truskinovsky [21] have argued for the advantage of using a physically determined ϕ profile instead of an artificial smoothing function for the interface. When the thickness of the interface approaches zero, the diffuse-interface model becomes asymptotically identical to a sharp-interface level-set formulation. It also reduces properly to the classical sharp-interface model in general.

The idea of diffuse interfaces can be traced back to van der Waals [22–25], and has since been developed for numerous applications, e.g., phase transition and critical phenomena [26, 27], solidification and dendritic growth in alloys [28, 29], interfacial tension theories [30], phase-separation [31, 32, 27] and two-phase flows [33–40]. Recently, we [41] have generalized the theoretical model to simulate interfacial dynamics in complex fluids. Taking advantage of the energy-based formulation, they are able to resolve the dual difficulties for complex fluid mixtures—moving interfaces and complex rheology—in a unified framework. So far, we have applied the method to a number of problems on drop dynamics of viscoelastic and liquid crystalline fluids [42–46]. In the following, we first give a brief but self-contained derivation of the theoretical model, and describe a numerical algorithm using spectral methods. Then we will illustrate the advantages and limitations of the model by several numerical examples. We hope to convince the reader that the diffuse-interface idea can be developed into a versatile CFD tool for multi-phase and multi-component complex fluids.

2. An energy-based phase-field theory. The phase-field model can be derived from the general procedure of Lagrangian mechanics [21, 37]. We write out the Lagrangian (action functional) of the system based on its free energy, and carry out variations with respect to the field variations (and the flow map). This amounts to following the "least-action principle" and various dynamical laws, and will lead to evolution equations for these variables (including the momentum equation — force balance equations). The dissipative portions of these equations need to be derived separately, for instance via irreversible thermodynamics [47]. The entire procedure has been demonstrated previously for Newtonian, viscoelastic and nematic liquid-crystalline fluids [37, 41, 48], and even for fluid-structure interactions (with the help of a Eulerian description of elasticity) [49]. In the following, we will use an example of a Newtonian-nematic blend with planar anchoring for illustration.

For an immiscible blend of a nematic liquid crystal and a Newtonian fluid, there are three types of free energies: mixing energy of the interface, bulk distortion energy of the nematic, and the anchoring energy of the liquid crystal molecules on the interface. We introduce a phase-field variable ϕ such that the concentration of the two components is $(1+\phi)/2$ and $(1-\phi)/2$, respectively. We express the mixing energy density in the Landau-Ginzburg form:

$$f_{mix}(\phi, \nabla\phi) = \frac{\lambda}{2}|\nabla\phi|^2 + \frac{\lambda}{4\epsilon^2}(\phi^2-1)^2, \tag{2.1}$$

where the parameter λ is the mixing energy density, and ϵ is a capillary width representative of the thickness of interface. The gradient energy term $\lambda|\nabla\phi|^2/2$ and the bulk energy term $f_0 = \lambda(\phi^2-1)^2/(4\epsilon^2)$ represent the "philic" and "phobic" tendencies between the species, their competition giving rise to the equilibrium ϕ profile. Note that f_{mix} is the diffuse-interface counterpart of the interfacial tension. In fact, one can relate the conventional interfacial tension σ to the parameters in the mixing energy. For instance, from an equilibrium hyperbolic-tangent ϕ profile that is the 1D energy minimizer, one obtains [34, 41]

$$\sigma = \frac{2\sqrt{2}}{3}\frac{\lambda}{\epsilon}. \tag{2.2}$$

The orientation of the nematic liquid crystal is described by the director field $\boldsymbol{n}(\boldsymbol{x})$. The Frank distortion energy expresses the energy penalty for distorting the orientation [50]:

$$f_{bulk} = K\left[\frac{1}{2}\nabla\boldsymbol{n} : (\nabla\boldsymbol{n})^{\mathrm{T}} + \frac{(|\boldsymbol{n}|^2-1)^2}{4\delta^2}\right], \tag{2.3}$$

where K is the elastic constant. The second term on the right-hand side regularized the original Frank energy to allow defects [51]. The nematic prefers to orient on the interface along an easy axis [50]; any deviation from it is penalized by an anchoring energy. Here we assume that the easy axis is any direction in the tangential plane, and write the anchoring energy as

$$f_{anch} = \frac{A}{2}(\boldsymbol{n}\cdot\nabla\phi)^2, \tag{2.4}$$

with the positive constant A representing the anchoring strength. This is the diffuse-interface counterpart of the Rapini-Popoular energy [52]. Unlike in the sharp-interface picture, both f_{mix} and f_{anch} are volumetric energy densities. Finally, the total free energy density for the two-phase material is written as:

$$f(\phi,\boldsymbol{n},\nabla\phi,\nabla\boldsymbol{n}) = f_{mix} + \frac{1+\phi}{2}f_{bulk} + f_{anch} \tag{2.5}$$

where $(1+\phi)/2$ is the volume fraction of the nematic component, and $\phi = 1$ in the purely nematic phase.

Variation of the system's action functional with respect to the phase-field variable ϕ, the nematic director $\boldsymbol{n}$ and the displacement leads to evolution equations for ϕ, $\boldsymbol{n}$ and the momentum equation. Augmented by the dissipative effects, the governing equations of the system are:

$$\frac{\partial \phi}{\partial t} + \boldsymbol{v} \cdot \nabla \phi = \gamma_1 \nabla^2 \frac{\delta F}{\delta \phi} \tag{2.6}$$

$$\frac{\partial \boldsymbol{n}}{\partial t} + \boldsymbol{v} \cdot \nabla \boldsymbol{n} = \gamma_2 \boldsymbol{h}, \tag{2.7}$$

$$\nabla \cdot \boldsymbol{v} = 0, \tag{2.8}$$

$$\rho \left(\frac{\partial \boldsymbol{v}}{\partial t} + \boldsymbol{v} \cdot \nabla \boldsymbol{v} \right) = -\nabla p + \nabla \cdot \left[\mu (\nabla \boldsymbol{v} + \nabla \boldsymbol{v}^T) + \boldsymbol{\sigma}^e \right], \tag{2.9}$$

where γ_1 is the interfacial mobility and γ_2 determines the relaxation time of $\boldsymbol{n}$. $F = \int f d\Omega$ is the total free energy of the system, whose variations produce

$$\frac{\delta F}{\delta \phi} = \lambda \left[-\nabla^2 \phi + \frac{\phi(\phi^2 - 1)}{\epsilon^2} \right] + \frac{1}{2} f_{bulk} - A \nabla \cdot \left[(\boldsymbol{n} \cdot \nabla \phi) \boldsymbol{n} \right], \tag{2.10}$$

and the molecular field

$$\begin{aligned} \boldsymbol{h} &= -\frac{\delta F}{\delta \boldsymbol{n}} \\ &= K \left[-\nabla \cdot \left(\frac{1+\phi}{2} \nabla \boldsymbol{n} \right) + \frac{1+\phi}{2} \frac{(\boldsymbol{n}^2 - 1)\boldsymbol{n}}{\delta^2} \right] + A (\boldsymbol{n} \cdot \nabla \phi) \nabla \phi. \end{aligned} \tag{2.11}$$

Note that the right-hand side of the dynamic equation (2.6) dictates the relaxation of the phase-field variable ϕ, with a relaxation time proportional to $1/\gamma_1$. In the limit of γ_1 approaching zero, we recover the kinematic condition for the interface. Moreover, as ϵ approaches zero, the dynamics of ϕ will preserve the profile of the transition (hyperbolic-tangent in this case), a key advantage of phase field approach. The last two terms in equation (2.10) represent coupling between the phase field and the Frank distortion energy and anchoring energy. When the interface is thin, f_{bulk} is dominated by the mixing energy near the interface and therefore negligible. The last term may have an effect on the interfacial ϕ profile for strong anchoring. But it is a higher order effect, negligible if the effects of interfacial tension and surface anchoring are assumed to be additive (cf. equation (2.12) below). Thus for simplicity, the last two terms on the right-hand-side of equation (2.10) are neglected hereafter. There are applications, e.g. [28], where the interface is relatively thick and the ϕ profile has physical consequences.

In the variation with respect to displacement, we have assumed equal density between the two species. A small density mismatch may be handled by the Boussinesq approximation [37]. In the more general situation, the mass-averaged mixture velocity becomes non-solenoidal within the interfacial region, and a theory for compressible mixtures can be constructed [21]. The pressure is a Lagrange multiplier for the constraint of incompressibility. The elastic stress tensor is derived as part of the variational procedure [41], and in this case can be written out as

$$\sigma^e = -\lambda(\nabla\phi\otimes\nabla\phi) - K\frac{1+\phi}{2}(\nabla \boldsymbol{n})\cdot(\nabla \boldsymbol{n})^{\mathrm{T}} - A(\boldsymbol{n}\cdot\nabla\phi)\boldsymbol{n}\otimes\nabla\phi. \tag{2.12}$$

3. Numerical scheme. While the coupled nonlinear system (2.6–2.9) are adequate mathematical models for the mixtures of complex fluids, it is a challenging task to construct a numerical scheme which is capable of correctly capturing, at a reasonable cost, the complex spatial and temporal features of these two-phase flows.

We propose to discretize the coupled nonlinear system (2.6–2.9) in time with a stabilized semi-implicit second-order scheme. The guiding principle here is that we only want to solve decoupled, constant-coefficient elliptic equations at each time step while preserving the overall second-order time accuracy and having a reasonably large stability region.

To simplify the presentation, we shall only describe our approach for the Cahn-Hilliard equation

$$\frac{\partial\phi}{\partial t} + \gamma\nabla^2\left(\nabla^2\phi - \frac{(|\phi|^2-1)\phi}{\epsilon^2}\right) = h_1, \tag{3.1}$$

and for the time-dependent Stokes equations

$$\begin{aligned}\frac{\partial \boldsymbol{v}}{\partial t} - \nu\Delta\boldsymbol{v} + \nabla p &= \boldsymbol{h_2},\\ \nabla\cdot\boldsymbol{v} &= 0,\end{aligned} \tag{3.2}$$

where the forcing functions h_1 and $\boldsymbol{h_2}$ would include all the extra nonlinear terms in (2.6–2.9) which will be treated explicitly to avoid solving nonlinear equations at each time step. The treatment for the nematic director equation 2.7) is very similar.

Let us consider first the Cahn-Hilliard equation (3.1). A main difficulty associated with the numerical approximation of (3.1) is that a standard semi-implicit scheme leads to a very stiff system (when $\epsilon \ll 1$) which dictates a very small time step. This difficulty can be alleviated by using the following shifted semi-implicit scheme:

$$\begin{aligned}\frac{3\phi^{k+1}-4\phi^k+\phi^{k-1}}{2\Delta t} + \gamma(\Delta^2 - \frac{C_s}{\epsilon^2}\Delta)\phi^{k+1} &= 2h_1^k - h_1^{k-1}\\ &+ \frac{\gamma}{\epsilon^2}\Delta[2(|\phi^k|^2-(1+C_s))\phi^k\\ &- (|\phi^{k-1}|^2-(1+C_s))\phi^{k-1}];\end{aligned} \tag{3.3}$$

where C_s is a stabilizing parameter typically in the range of $[1,5]$. Ample numerical results indicate that the above stabilized semi-implicit scheme allows much larger time step than the standard semi-implicit scheme does. We observe that (3.3) is a fourth-order equation for ϕ^{k+1} with constant coefficients.

Next, we describe our approach for solving the time-dependent Stokes problem (3.2).

- If the boundary conditions are periodic, the pressure in (3.2) can be easily eliminated using the divergence-free conditions so (3.2) can be efficiently solved by using a Fourier-spectral method [37].
- If the velocity satisfies a free-slip boundary condition (cf. [53]), then, the time discrete approximation of (3.2) can be split into a sequence of Poisson-type equations for the velocity and for the pressure.
- If the boundary conditions in all but one direction are periodic, (3.2) can be reduced into a sequence of one-dimensional fourth-order equations using a Fourier expansion in all but one direction [54].
- Finally, for the general cases, we shall use a projection type scheme (see the recent review paper [55]) to decouple the computation of the velocity from the pressure. For example, we may use the new consistent splitting scheme introduced in [56]. To be specific, we assume here that the velocity is subjected to a homogeneous Dirichlet boundary condition:

$$\begin{aligned} &\frac{3\boldsymbol{v}^{k+1}-4\boldsymbol{v}^{k}+\boldsymbol{v}^{k-1}}{2\Delta t} - \nu\Delta\boldsymbol{v}^{k+1}+\nabla(2p^{k}-p^{k-1}) = 2\boldsymbol{h_2}^{k}-\boldsymbol{h_2}^{k-1}, \\ &\boldsymbol{v}^{k+1}|_{\partial\Omega} = 0, \end{aligned} \tag{3.4}$$

$$(\nabla\psi^{k+1},\nabla q) = \left(\frac{3\boldsymbol{v}^{k+1}-4\boldsymbol{v}^{k}+\boldsymbol{v}^{k-1}}{2\Delta t},\nabla q\right),\ \forall q\in H^1(\Omega), \tag{3.5}$$

$$p^{k+1} = \psi^{k+1}+2p^{k}-p^{k-1}-\nu\nabla\cdot\boldsymbol{v}^{k+1}, \tag{3.6}$$

Note that (3.4) is a Poisson-type equation for $\boldsymbol{v}^{k+1}$ while (3.5) is a Poisson equation (with homogeneous Neumann boundary conditions) in the weak form for ψ^{k+1}.

Hence, after a time discretization to the coupled nonlinear system (2.6–2.9), we only need to solve, at each time step, a sequence of constant-coefficient elliptic equations which can be efficiently handled by one of the many existing numerical methods using finite difference, finite elements or spectral methods. Since we shall confine ourselves to simple geometries in this study, we choose to use the well-conditioned and fast spectral-Galerkin methods developed in [57, 54, 58] which are capable of solving constant-coefficient elliptic equations in simple geometries with quasi-optimal computational complexity, i.e., the number of operations per time step is of order $O(N\log N)$, N being the number of unknowns. The high resolution property of the spectral method and the efficiency of the fast spectral-Galerkin algorithms allow us to numerically solve the coupled nonlinear system (2.6–2.9) at a reasonable cost. For example, with a 750 MHz Sparc-v9 processor, the two-dimensional problems with a spatial resolution of 1024×1024 or 2048×1024 typically take about 1 minute of CPU time per time step. For all the simulations reported below, we have carried out grid and time-step refinements to ensure convergence. If we take 4.164ϵ to be a nominal interfacial thickness (cf. [41]), this layer typically requires 7–10 grids to resolve. Coarser grids will generate spurious oscillations in the solution, especially in the vicinity of the interface.

4. Advantages of the diffuse-interface model. Needless to say, the greatest payoff of adopting a diffuse-interface picture is the ease with which moving interfaces can be handled. Compared with the traditional sharp-interface view of internal boundaries, there is no longer a need to track the position of the interface, and to impose matching boundary conditions for solving the flow inside each component separately. As mentioned earlier, the interfacial tension is now represented by an elastic stress tensor concentrated in the interfacial region. The cost is the additional dynamics for ϕ; we have to deal with the physics of the convection-diffusion process as well as the numerical burdens of an additional equation. These will be discussed in the next section among the subtle issues that need special consideration.

The diffuse-interface formulation also brings about several "side benefits" that may be of great importance to the physical applications at hand. Here, we illustrate in some detail three of such benefits that we have noted in our simulations. These advantages reflect the fact that the phase-field idea transforms the Lagrangian description of a geometric motion into Eulerian coordinates, and easily represents the competition between various energy functionals for the multiphase material.

4.1. Short-range molecular forces during topological changes. For the same reason that the phase-field method handles moving interfaces easily, so it does singular topological changes such as breakup and coalescence. In the sharp-interface convention, such events require an *ad hoc* treatment. For filament breakup and drop coalescence, for example [59, 60], the thinning neck or film has to be artificially removed once its thickness reaches a prescribed threshold. In contrast, the diffuse-interface is represented by the contour of $\phi = 0$, which deforms and reconnects smoothly during flow. Thus, no artificial trigger is needed for drop breakup and coalescence. As an example, Fig. 1 illustrates the head-on collision and subsequent coalescence of two Newtonian drops in a Newtonian matrix. The draining film develops a "dimple" in the middle [61] and the rupture occurs toward the outside of the film, trapping some matrix fluid inside.

In reality, film rupture is effected by short-range forces such as van der Waals force [62]. Interestingly, the phase-field model is rooted in the physics of molecular interaction between the two species, and thus contains short-range molecular forces. To see this, consider the simple situation in Fig. 2, with a liquid film (F) of uniform thickness h sandwiched between semi-infinite domains of another fluid (A). For a thick film, the phase-field variable at the center approaches the bulk value, say $\phi_0 \rightarrow -1$, at the

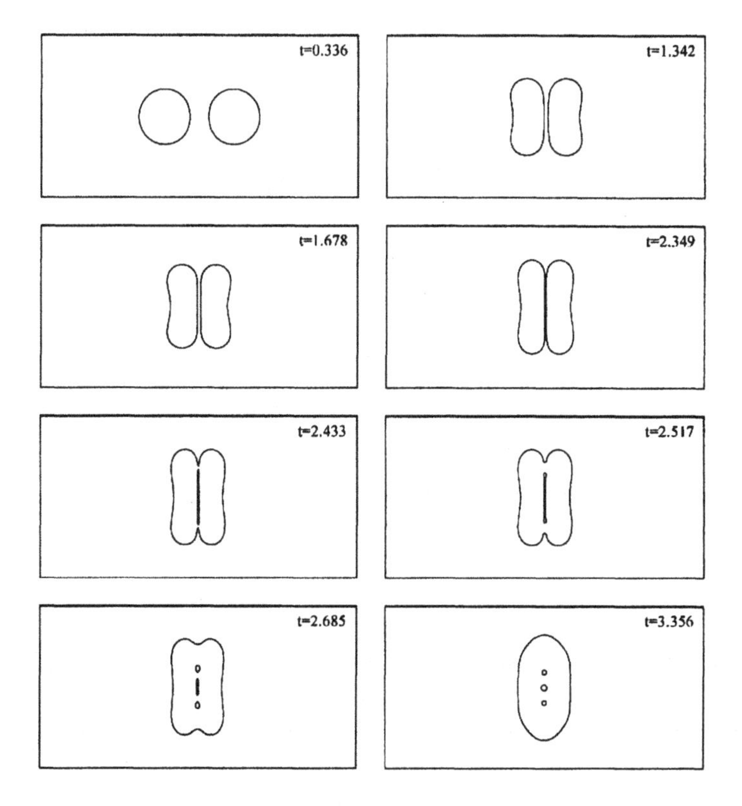

FIG. 1. *Collision and coalescence of two Newtonian drops in a Newtonian matrix. The Reynolds number, defined using D and U, is* $Re = 33.6$, *and the Weber number is* $We = 12$. *Other parameters are:* $\epsilon = 0.01$ *and* $\gamma = 3.365 \times 10^{-5}$ *(after Yue et al. [41], ©Cambridge University Press.)*

center. For a thin film, however, conceivably ϕ inside F will differ from the bulk value: $\phi_0 > -1$. From the elastic stress tensor due to the mixing energy (cf. [42]), one may calculate the disjoining pressure in the diffuse-interface model:

$$\Pi_\phi = -\lambda f_0 = -\frac{\lambda(\phi_0^2 - 1)^2}{4\epsilon^2}. \tag{4.1}$$

which implies an attractive force between the interfaces as with van der Waals force. If we estimate ϕ_0 based on a hyperbolic tangent ϕ-profile as in a one-dimensional equilibrium interface [41],

$$\phi_0 = -\tanh\left(\frac{h}{2\sqrt{2}\epsilon}\right), \tag{4.2}$$

Then the disjoining pressure in Eq. (4.1) can be shown to be of the same order of magnitude as the van der Waals force. As the film thickness

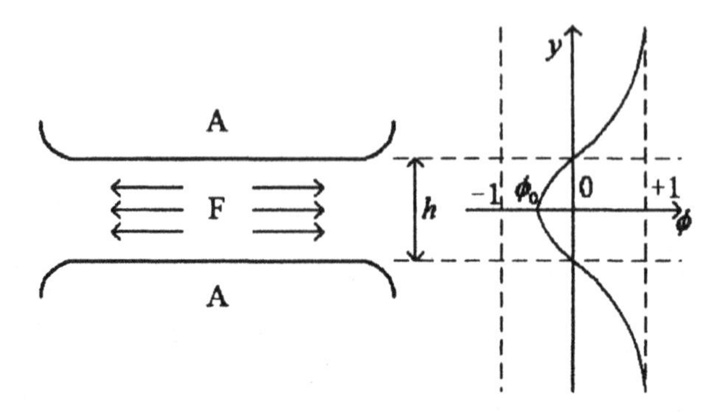

FIG. 2. *A cartoon for a draining film and the corresponding ϕ profile.*

approaches zero, however, the van der Waals force goes to infinity while Π_ϕ remains finite. A more detailed comparison can be found in Ref. [42]. On a fundamental level, the discrepancy between van der Waals force and Π_ϕ stems from the truncation of the Cahn-Hilliard free energy at the quadratic term $|\nabla\phi|^2$. An elegant explanation has been given by Pismen [63].

4.2. Complex rheology. Because of its energy-based formalism, our diffusive interface method incorporates complex rheology easily. The non-Newtonian rheology is typically due to microstructures whose conformation deviates from equilibrium under deformation. The conformation of the microstructure is often governed by a free energy, e.g., the Frank distortion energy for a liquid crystal or the free energy of a polymer chain. In Section 2, we showed how this microstructural energy can be added to the mixing energy to form the total free energy of the multi-phase system, which will give rise to the proper constitutive equation for the microstructured fluids in addition to the evolution equation of the phase field variable. Thus, interfacial dynamics and complex rheology are included in a *unified* theoretical framework.

This procedure is general in that various types of constitutive relations can be derived by the same procedure. As a second example, we consider here the important case of a viscoelastic polymer solution modeled as a suspension of Hookean dumbbells in a Newtonian solvent [64]. Instead of the least-action principle, we follow a formally different but essentially equivalent "virtual-work principle" [5]. For a single dumbbell with a connector $\boldsymbol{Q}$, its elastic energy is $\frac{1}{2}H\boldsymbol{Q}\cdot\boldsymbol{Q}$, where H is the elastic constant. For an ensemble of dumbbells with configuration distribution $\Psi(\boldsymbol{Q})$, the average energy can be written as

$$f_d = \int_{R^3}\left(kT\ln\Psi + \frac{1}{2}H\boldsymbol{Q}\cdot\boldsymbol{Q}\right)\Psi dQ, \tag{4.3}$$

where k is the Boltzmann constant and T is the temperature, and the integration is over all possible configurations of $\boldsymbol{Q}$. Now the total free energy density of the two-phase system is:

$$f = f_{mix} + \frac{1+\phi}{2} n f_d, \tag{4.4}$$

where n is the number density of the dumbbells. Since the stress tensor due to f_{mix} has been derived (cf. equation 2.12 and [41]), we will only consider the elastic stress due to the dumbbell energy f_d. We impose a virtual displacement $\delta \boldsymbol{x}$ on the material, which takes place instantaneously so that the dumbbells deform affinely with no slip between the bead and the surrounding fluid. The corresponding change in the distribution function Ψ can be obtained from the Fokker-Planck equation for Ψ [64]. Now we may calculate the resultant variation in the dumbbell free energy. Omitting the intermediate steps [42], we eventually arrive at:

$$\begin{aligned} \delta f_d &= \int_{R^3} \left(kT \ln \Psi + kT + \frac{H}{2} \boldsymbol{Q} : \boldsymbol{Q} \right) \delta \Psi d\boldsymbol{Q} \\ &= (-kT\boldsymbol{I} + H < \boldsymbol{QQ} >) : (\nabla \delta \boldsymbol{x})^{\mathrm{T}}, \end{aligned} \tag{4.5}$$

where $< \cdot >= \int_{R^3} \cdot \Psi d\boldsymbol{Q}$ and $\boldsymbol{I}$ is the identity tensor. Thus the dumbbell stress tensor is:

$$\boldsymbol{\tau}_d = -nkT\boldsymbol{I} + nH < \boldsymbol{QQ} >, \tag{4.6}$$

which obeys the Maxwell equation. This is exactly the Kramers expression for the polymer elastic stress tensor [64]. The same procedure can be followed for other microstructural free energies, such as the Marrucci-Greco nematic potential energy for liquid-crystalline polymers [65, 5].

4.3. Energy conservation. An additional advantage of the phase-field method over other interface-regularizing methods is its energy conservation: a solution to the governing equations in Section 2 obeys an energy law. For example, multiplying equation (2.9) by the velocity $\boldsymbol{v}$, equation (2.6) by the chemical potential $\delta F/\delta \phi$ and equation (2.7) by the molecular field $\delta F/\delta \boldsymbol{n}$, integrating over the entire domain and summing the results, we obtain:

$$\begin{aligned} &\frac{d}{dt} \int_\Omega \left(\frac{\rho}{2} |\boldsymbol{v}|^2 + f \right) d\Omega \\ &= - \int_\Omega \left(\mu \nabla \boldsymbol{v} : \nabla \boldsymbol{v}^{\mathrm{T}} + \gamma_1 \left| \nabla \frac{\delta F}{\delta \phi} \right|^2 + \gamma_2 \left| \frac{\delta F}{\delta \boldsymbol{n}} \right|^2 \right) d\Omega, \end{aligned} \tag{4.7}$$

where f is the system's potential energy density (cf. equation 2.5), and surface work has been omitted. Physically, the law states that the total energy of the system (excluding thermal energy) will decrease from internal dissipation. Based on such energy laws, Lin and Liu [66, 67] have established the existence of classical and weak solutions for Leslie-Ericksen fluids. In general, energy laws play an important role in the convergence of finite-dimensional approximations to partial differential equations, especially when the solution is not smooth [51]. This constitutes one of the advantages of our method over previous methods that do not maintain the system's total energy budget. In VOF simulations, density is the labeling function subjected to smoothing. The level-set method renormalizes the distance function. In either case, the conservation of energy cannot be maintained.

Note that the energy conservation holds exactly when all the coupling terms in equation (2.10) are kept. For numerical conveniences, we have omitted such terms in applications where the interface will remain thin and the coupling terms have at most a localized effect. This omission will violate the energy conservation. When the geometry is simple and the solution is smooth, non-conservation of energy usually does not compromise the quality of the solution. But difficulties may arise in the presence of rapid spatial variations, which are characteristic of microstructured fluids with internal boundaries and/or defects [1, 43].

5. Physical and numerical subtleties. Although the convergence of the phase-field model to the sharp-interface model has been established by asymptotic expansion for regular velocity fields [24, 25, 40, 33, 39, 21, 37], there are some subtle issues that merit further discussion. One such issue, for example, concerns incompressibility. While the phase-field formulation imposes incompressibility throughout the domain (hence also on the interface), the sharp-interface model satisfies this condition only weakly on the interface. In fact, the system would be over-determined with such a constraint on the interface. For phase-field models, we are allowed to impose $\nabla \cdot \boldsymbol{v} = 0$ everywhere thanks to the diffused transition layer. The same holds for VOF and level-set methods through the introduction of an artificial transition layer. Physically, one may consider the sharp interface and the diffuse interface different approximations of the real physical situation, the former by relaxing incompressibility on the interface and the latter by introducing the transition layer.

The phase-field method can be viewed from two complementary angles: as a representation of the microscopic physics on the interface or as a numerical device for simulating moving boundary problems without tracking the interface. Depending on the applications, one or the other viewpoint may be more appropriate. For applications such as solidification of alloys [28, 29] and near-critical systems [26, 33], it is essential to ensure that the phase-field equation captures the dynamics at the interface because the interfacial profile is of direct interest. On the other hand, the two-phase flow problems we have simulated involve "immiscible" components with interfacial thickness on the order of tens of nanometers. Beyond indicating the position and movement of the interface, the ϕ profile has little direct bearing on the macroscopic properties of interest. Thus, there is a degree of freedom or ambiguity in choosing the dynamics of the phase field and the parameter values. In particular, the interfacial thickness in the model can be much thicker than in reality; there is no need, nor perhaps the capability, to resolve the interface down to nanometer scales. From this an array of subtle issues arise, which must be handled with care for the model to be physically sound and numerically efficient.

5.1. Cahn-Hilliard and Allen-Cahn dynamics. As long as our physical problem conceptually consists of sharp interfaces, the diffusion dynamics of the phase-field variable is to a large extent fictitious. Thus, one can choose Cahn-Hilliard, Allen-Cahn or other types of dynamics. We can view all such choices as a relaxation or approximation of the kinematic transport equations. Based on similar considerations, we have neglected certain coupling terms in the Cahn-Hilliard equation due to presence of microstructures (cf. equation 2.10). One requirement on the diffusion dynamics is that they maintain the integrity of the interface. In other words, the "phobic" and "philic" tendencies should be balanced such that the transition layer neither smoothes out nor steepens into a shock wave.

The Cahn-Hilliard equation follows from the physical argument that the flux be proportional to the gradient of a generalized chemical potential. This differs from the conventional Fick's law, which leads to the Allen-Cahn dynamics. The advantage of the Cahn-Hilliard equation is the following conservation of total system "mass":

$$\frac{d}{dt}\int_{\Omega} \phi(x,t)\, dx = 0, \tag{5.1}$$

if the following no-flux boundary condition is imposed.

$$\frac{\partial}{\partial \boldsymbol{n}}\left(\frac{\delta F_{mix}}{\delta \phi}\right) = 0, \tag{5.2}$$

where $\boldsymbol{n}$ is the normal direction to the boundary.

A disadvantage of the Cahn-Hilliard equation is that its higher (4th) order causes numerical complications. Shen [54] and Yue *et al.* [41] used a procedure of splitting it into two second-order Helmholtz equations.

The Allen-Cahn equation is easier to handle numerically but does not automatically ensure conservation of mass; a Lagrange multiplier can be introduce to enforce it as a constraint [68]:

$$\frac{\partial \phi}{\partial t} + \boldsymbol{v} \cdot \nabla \phi = \gamma_1 \left(-\frac{\delta F}{\delta \phi} + \sigma \right), \tag{5.3}$$

with $\int_{\Omega} \phi(x,t)\, dx = \int_{\Omega} \phi(x,0)\, dx$.

Another possibility is the "advected field" method [69], which is a compromise between phase-field and level-set approaches. To impose mass conservation on the Allen-Cahn equation, an additional term proportional to the local curvature is added:

$$\frac{\partial \phi}{\partial t} + \boldsymbol{v} \cdot \nabla \phi = \gamma_1 \left[\frac{df_0}{d\phi} - (\Delta \phi + c|\nabla \phi|) \right], \tag{5.4}$$

where c is the curvature of interface. In the sharp-interface limit, the new term cancels the diffusion flux incurred by the Allen-Cahn dynamics, thus mass is conserved. On the downside, the *ad hoc* term prevents interfacial tension to be incorporated into the momentum equation via the phase field. Instead, it has to be added "by hand" through a spread-out delta function as in level-set and VOF methods.

Finally, we must point out that the phase-field dynamics do play a central role in a special class of two-phase flow problems where the interface undergoes topological changes such as breakup or coalescence[42]. The length scale of such critical processes approaches that of the interfacial thickness. In reality, these processes are dominated by short-range forces. As illustrated in the last section, the Cahn-Hilliard dynamics does contain a type of short-range force; it produces a disjoining pressure comparable to the van der Waals force. Then the question arises as to how closely this type of short-range force approximates reality in a particular experiment. The answer likely depends on the complex details of the experiment, as short-range forces from several sources can take part, typically imparting a stochasticity to the problem [70, 71].

5.2. Interfacial relaxation. Secondary to the ambiguity in interfacial dynamics is the determination of parameter values. For the diffuse interface to reproduce the macroscopic behavior of a sharp-interface, the model parameters must be judiciously chosen. In particular, the parameter γ_1 determines the rate of relaxation of the ϕ field. However, there is little experimental information on γ_1 for the thin-interface two-phase flows that we are interested in. Jacqmin[34] juxtaposed two considerations on this: "straining flows can thin or thicken an interface and this must be resisted by a high enough diffusion. On the other hand, too large a diffusion will overly damp the flow". We will discuss several manifestations of interfacial relaxation in the following.

One interesting effect of interfacial relaxation is the initial "contraction" of a drop in a quiescent fluid. As an initial condition, we impose the hyperbolic tangential ϕ profile at the interface (equation 4.2), with $\phi = \pm 1$ in the two bulk phases. On commencing the simulation, however, we notice a very small shift in ϕ such that the interface $\phi = 0$ shrinks slightly, and ϕ deviates from ± 1 slightly in the bulk (Fig. 3). The reason for this artificial shrinkage is that the initial ϕ field is not the equilibrium one that minimizes the total free energy in 2D. Thus, the interface tends to shrink to reduce the mixing energy. Since $\int_\Omega \phi \, d\Omega$ is conserved by the Cahn-Hilliard equation with the zero-flux boundary condition (equation 5.2), the shrinking interface causes the bulk ϕ value to change slightly, incurring an energy penalty

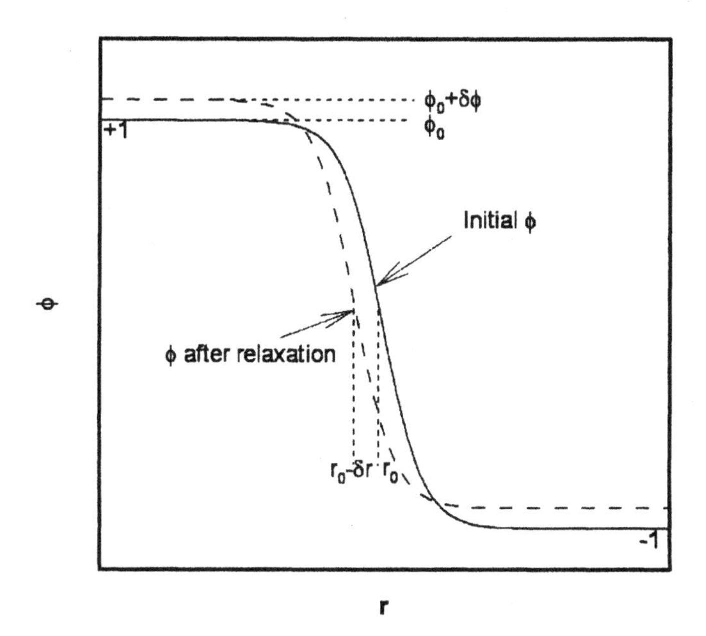

FIG. 3. *A diagram showing the initial contraction of a drop in a quiescent matrix fluid.*

in the bulk energy f_0. The competition between the bulk and interfacial energies results in a slightly relaxed ϕ field that has a lower energy than our initial condition. For a circular drop of radius r, one can calculate the shift in the bulk value of ϕ analytically if $\epsilon/r \ll 1$:

$$\delta\phi = \frac{\sqrt{2}\epsilon}{6r}. \tag{5.5}$$

In general, such a formula will not be available. But one may always choose a sufficiently small ϵ so that the initial shift is insignificant to the accuracy of the results.

Another important consequence of interfacial relaxation is the change in apparent interfacial tension[41, 43]. To simulate an experiment with two immiscible fluids, one chooses appropriate values for the mixing energy λ and capillary width ϵ so as to match λ/ϵ to the experimental interfacial tension σ according to a formula based on some equilibrium ϕ profile[34, 41]. As ϕ evolves during flow, the matching formula no longer holds. Yue *et al.* [41] have shown an example of drop deformation in shear flows, where the deviation of the ϕ profile from the equilibrium one increases the effective interfacial tension. As a result, the drop deformation

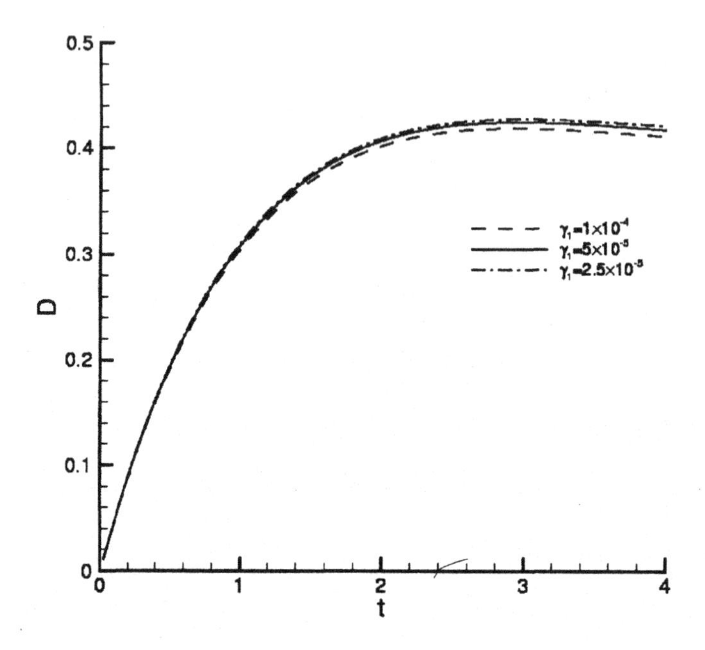

FIG. 4. *Effect of the mobility parameter γ_1 on the deformation of a drop after startup of a simple shear. The drop is Newtonian while the matrix is a viscoelastic Oldroyd-B fluid (after Yue et al. [41], ©Cambridge University Press.)*

is underpredicted. Since the rate of relaxation is controlled by γ_1, it has an effect on the drop deformation as well. In this case, Fig. 4 shows that a smaller γ_1 increases the drop deformation slightly.

5.3. Interfacial thickness. The capillary width ϵ is another parameter that needs to be chosen carefully. This is a well-recognized issue in phase-field models for alloy solidification [29]. In our simulations of two-phase flows, the interfacial thickness h, defined for example by 90% of the jump in ϕ, is typically on the order of 4ϵ. The smallest h that one can resolve depends on the macroscopic length scale and the computational capacity. But it is typically much thicker than the nano-scale real interfaces. Thus, it is a delicate task to pick an ϵ within one's computational reach that produces approximately the correct macroscopic behavior of a much thinner interface. As mentioned before, ϵ affects the effective interfacial tension, the relaxation of the interface and the short-range molecular forces. The philosophy behind choosing an appropriate value is perhaps best illustrated by a situation involving drastic topological changes.

Figures 5 and 6 show simulations with a larger or smaller ϵ than in Fig. 1 with all other parameters unchanged. The early stage of the simula-

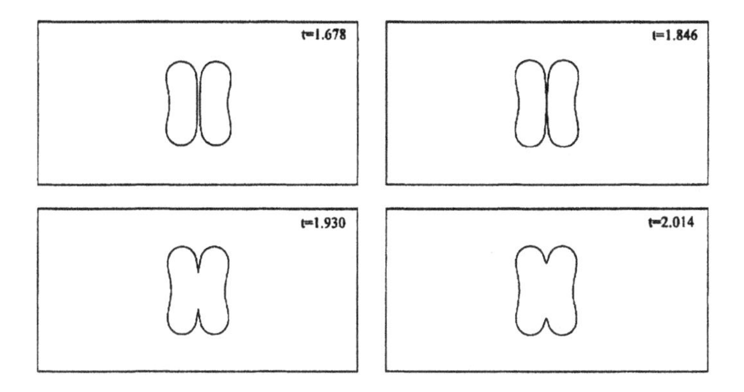

FIG. 5. *Collision and coalescence of two Newtonian drops in a Newtonian matrix with a thicker interface. The parameters are the same as Fig. 1 except for* $\epsilon = 0.02$ *(after Yue et al. [42], ©Elsevier.)*

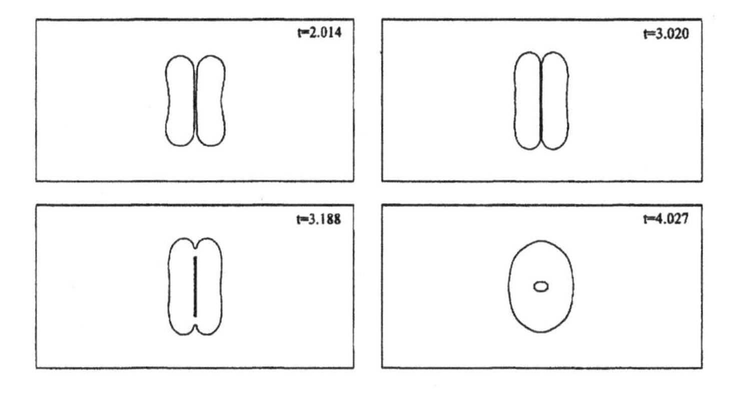

FIG. 6. *Collision and coalescence of two Newtonian drops in a Newtonian matrix with a thinner interface. The parameters are the same as Fig. 1 except for* $\epsilon = 0.005$ *(after Yue et al. [42], ©Elsevier.)*

tions, say for $t \leq 1.342$, is identical with Fig. 1. This is before the interfacial profiles of the two drops overlap. For a larger ϵ, the interfaces of the two drops overlap at an earlier time during their approach, and the ensuing coalescence occurs more readily (Fig. 5). Note that the interface does not have time to develop the dimpled shape, and no matrix fluid is trapped inside the drop. On the other hand, a smaller ϵ prolongs the coalescence process (Fig. 6). As compared with Fig. 1, the points of rupture are more toward the ends of the film. This produces a less pronounced waist in the resultant compound drop, and the entrapped matrix filament does not break up but retracts into a droplet. The optimal ϵ cannot be determined by an *a priori* criterion. Rather, it needs to reflect the range of the molecular forces at work in the particular experiment to be simulated. Owing to a degree of randomness in the short-range forces, the coalescence time in experiments often exhibits a Gaussian distribution [70, 71]. Obviously, such intricate details cannot be reproduced by the disjoining pressure in a phase-field formulation. Instead, one may hope to capture the macroscopic dynamics in some average sense by using optimal values for the model parameters.

Note that the effect of ϵ is not to be confused with numerical resolution of the interface. For each ϵ value tested here, mesh refinement has confirmed that the grid used is adequate for resolving the interface (see also [41]).

5.4. Adaptive mesh refinement. We argue that adaptive mesh refinement is capable of addressing all aforementioned issues. As has been established before, the diffuse-interface model will stay close to the sharp-interface model, with the conventional interfacial tension, when the interfacial thickness tends to zero [33, 37]. Note that the ϕ profile as a solution to equation (2.6) is "nontrivial" only within the interfacial layer, whose thickness scales with ϵ. Therefore, for sufficiently small transition thickness ϵ and elastic relaxation time γ, the effect of interfacial relaxation becomes negligible, and the difference between Cahn-Hilliard and Allen-Cahn dynamics becomes irrelevant. In fact, they represent two different regularizations of the kinematic transport of the phase field.

However, in some cases, such as those involving surfactant monolayers, the interfacial profile needs to be numerically resolved for accurate evaluation of the interfacial stress. The disparity between small ϵ and the global length scale implies the need for a locally refined grid inside the interfacial region.

Although procedures for dynamically adaptive meshing seem to be available [72, 73], they have not been used in a diffuse-interface framework as we are aware. So far, we have used spectral methods with structured grids; the resolution of the interface is the numerical bottleneck [41] that must be tackled before the method can be used for large-scale flow simulations in three dimensions. Such an adaptive meshing scheme seems to be most conveniently implemented within a finite-element formulation. In addition, moving-mesh schemes may serve the same purpose. Code development along both directions is underway, and will be reported in the near future.

5.5. Topological control. So far we have considered it an advantage that the phase-field method automatically handles topological changes such as merging and rupture of interfaces. This is the case when the nature of the short-range forces are understood and more or less adequately represented by the phase-field dynamics [42]. In certain applications, however, this may become a liability [74]. For instance, surface-active agents greatly modify the behavior of interfaces, stabilizing drops in emulsions and bubbles in foams against coarsening [1]. Membranes may prevent vesicles in contact from coalescing. If one chooses to use a phase-field model

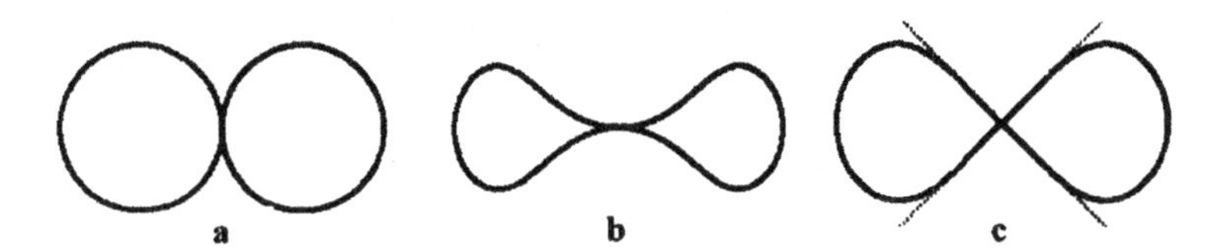

FIG. 7. *Singular cases in 2-D. The inner intersect angles are π, 0, $\pi/2$ for cases a, b and c respectively. The Euler number χ is 2, 1 and 1.5. The Euler-Poincaré index number η is always 2.*

in such situations, it is desirable to retain some control of the topological events within the model. This consists of retrieving topological information from the phase field formulation, monitoring the occurrence of topological events, and even using the information to design a criterion for prohibiting unphysical changes of topology.

Du *et al.* [75] have recently developed a method for topological control in a phase-field model via the Euler number. The idea, briefly outlined below, applies equally well to other simulation methods for free boundary and interface problems such as the level-set methods.

Given an oriented (regular) compact (i.e., without boundary) surface Γ, the well-known Gauss-Bonnet formula states that

$$\int_{\Gamma} K\, ds = 2\pi\chi, \tag{5.6}$$

where $K = k_1 k_2$ is the Gaussian curvature of the surface in R^3, ds is the area element and $\chi/2$ in 3D or χ in 2D is the Euler number [76]. The number χ is a commonly used topological quantity. For some frequently encountered surfaces, we have $\chi = 2$ for a sphere, $\chi = 0$ for a torus and $\chi = -2$ for a torus with 2 holes. For 2D curves, K is the curvature and $\chi = 1$ for a circle.

Such a concept can be generalized to the cases involving singularities, as illustrated in Figure 7. For instance, in 2 dimensional cases, we will have that:

$$2\pi\eta = \int_{\Gamma} K\, ds + \sum_{i=1}^{n}(\pi - \alpha_i) = 2\pi\chi + \sum_{i=1}^{n}(\pi - \alpha_i), \tag{5.7}$$

where α_i are the inner angle at each vertices. And η, the Euler-Poincaré index number, is the topological integer, the genus of the surface.

In [75], we derived a phase-field representation of χ. Let Γ be a smooth oriented compact surface of a domain Ω in $\mathbf{R}^3$ (note that Γ is allowed to have multiple disconnected pieces). Let p be a monotonically increasing function defined from $\mathbf{R}$ to $\mathbf{R}$ with $p(0) = 0$. We define the phase-field function as $\phi(x) = p(d(x))$ where the signed distance function $d(x) =$

$dist(x, \Gamma)$ is defined to be positive inside Ω and negative outside Ω. The level sets of ϕ are denoted by $\Gamma_\mu = \{x \in \Omega | \phi(x) = \mu\}$. In particular, we have $\Gamma = \Gamma_0$. We also define $\Omega' = \{x \in \Omega \,|\, b < \phi(x) < a\}$, which forms a banded (layered) neighborhood around the surface for $b < 0 < a$. Further define $\Lambda(M) = \lambda_1(M)\lambda_2(M) = \Lambda(\nabla^2 d(x))$ for a singular matrix M with λ_1, λ_2 being the two non-zero eigenvalues of $M = \nabla^2 d(x)$. Since we can view that k_1, k_2 remain close to constant along the normal directions in the thin layer region Ω', we have that

$$
\begin{aligned}
\frac{\chi}{2} &= \frac{1}{4\pi}\int_\Gamma k_1(x)k_2(x)\, ds \\
&= \frac{1}{4\pi(a-b)}\int_{p^{-1}(b)}^{p^{-1}(a)} p'(\tau)d\tau \int_\Gamma k_1(x)k_2(x)\, ds \\
&= \frac{1}{4\pi(a-b)}\int_{\Omega(a,b)} p'(d(x))k_1(x)k_2(x)\, dx \qquad (5.8)\\
&= \frac{1}{4\pi(a-b)}\int_{\Omega(a,b)} p'(d(x))\Lambda(\nabla^2 d(x))\, dx \\
&= \frac{1}{4\pi(a-b)}\int_{\Omega(a,b)} \frac{1}{p'(d(x))}\Lambda(\nabla^2\phi - p''\nabla_i d\nabla_j d)\, dx\,. \qquad (5.9)
\end{aligned}
$$

In practice, the function p and the constants a, b will be chosen such that p' is relatively small outside of the transition layer. Now, since $p(x)$ is monotone, hence we have that $p'(d(x)) = |\nabla\phi(x)|$ and $p''(d(x)) = \frac{\nabla|\nabla\phi|^2\cdot\nabla\phi}{2|\nabla\phi|^2}$. In the end we have the following theorem [75]:

THEOREM 5.1. *If $\phi = \phi(x)$ of Ω as $\phi(x) = p(d(x))$ where the* signed *distance function $d(x) = dist(x, \Gamma)$. For any monotone increasing function p, there exists $b < 0 < a$, such that the following matrix M, where*

$$
M(x)_{ij} = \frac{1}{2\sqrt{\pi(a-b)|\nabla\phi|}}\left(\nabla_i\nabla_j\phi - \frac{\nabla|\nabla\phi|^2\cdot\nabla\phi}{2|\nabla\phi|^4}\nabla_i\phi\nabla_j\phi\right), \qquad (5.10)
$$

is a singular matrix for $\forall x \in \Omega(a,b)$ in the sense that it always has a zero eigenvalue, and the Euler number of Γ can be obtained as:

$$
\frac{\chi}{2} = \int_{\Omega(a,b)} F(x)\, dx \qquad (5.11)
$$

where F denote the coefficient of the linear term of the characteristic polynomial of M.

Numerical simulations, such as that in Figure 8 and Figure 9, show that the Euler number thus computed indeed captures the occurrence of critical topological events[75].

Besides detecting the occurrence of critical topological events, this quantity also provides an important tool in designing a scheme to prevent

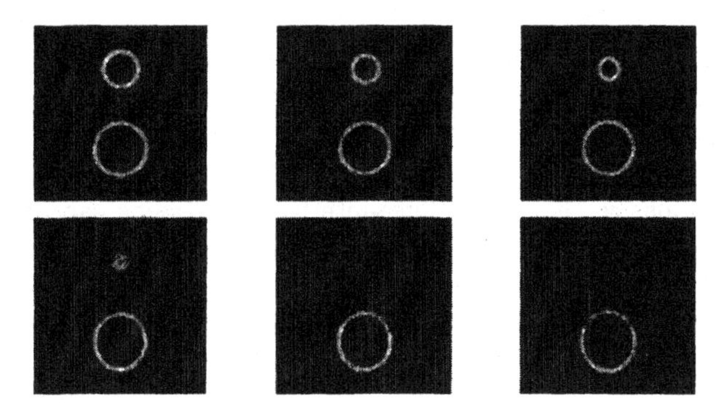

FIG. 8. *Coalescence of two bubbles in a Newtonian fluid with the time valued at 0.00, 0.10, 0.18, 0.22, 0.24, 0.28 (after Du et al. [75].)*

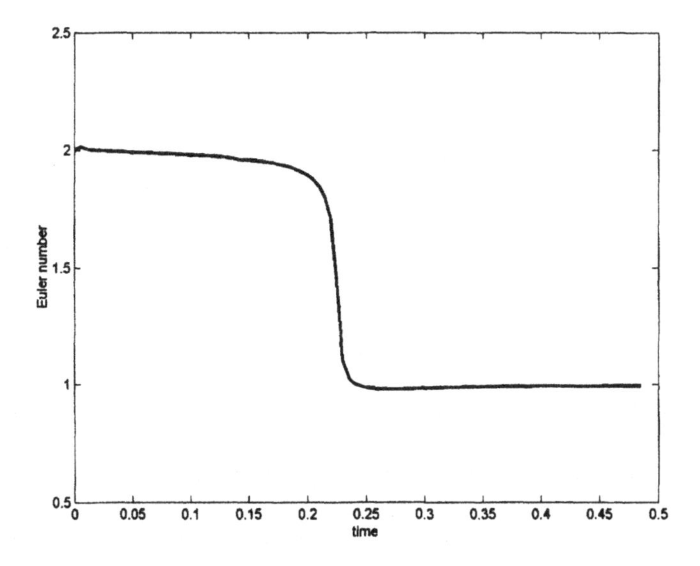

FIG. 9. *A plot of the Euler number in time with the annihilation of the small bubble (after Du et al. [75].)*

topological changes from happening. For instance, one may use a Lagrange multiplier to enforce the constancy of the Euler number over the entire domain. Since the constraint will involve a cost functional of high derivatives, more detailed analysis and numerical studies are needed in this area.

6. Concluding remarks. This article aims to introduce the energetic variation based phase-field approach to readers interested in the fluid dynamics of immiscible complex fluids. Although various versions of the model have been used in the past to great degrees of success, we highlight the generic advantages inherent in the formalism. More importantly, perhaps, we discuss several detailed key crucial issues (challenges) with the method whose proper treatment is prerequisite to a physically realistic and numerically practicable implementation of the model.

We emphasize that the diffuse-interface treatment can alternatively be seen as a physical model or a numerical device. It can be viewed as a physics motivated approximation (regularization) of the sharp interface models. The employment of the phase field method changes the Lagrangian description of the interface motion into Eulerian description. The energetic variational procedure ensures that the resulting coupling system will still preserve the overall energy law. The method seems to be more appropriate for the drop dynamics problems that we have simulated, although there are other applications where the opposite is true. As such, the interfacial dynamics and model parameters do not directly correspond to measurable quantities and their determination is a delicate matter. We have advocated the view that the criterion should be that the diffuse-interface model accurately predict the macroscopic dynamics of the two-phase system, including drastic changes of the interfacial morphology. Several numerical experiments are shown to illustrate these issues and how they can be resolved to a satisfactory degree of accuracy. The inherent ambiguity vanishes as the interfacial thickness shrinks. Thus, we suggest adaptive mesh refinement as the solution when a thin interface has to be resolved. It is also necessary for computing large-scale 3D flows of blends of rheologically complex fluids.

7. Acknowledgments. Acknowledgment is made to the Donors of The Petroleum Research Fund, administered by the American Chemical Society, for partial support of this research. J.J.F. was also supported by the NSERC and the Canada Research Chairs program, and the NNSF of China (No. 20490220). C.L. was also supported by NSF grant DMS-0405850. J.S. was partially supported by NSF grants DMS-0243191 and DMS-0311915. The authors also want to thank IMA for the support. This work was completed during the the workshop "Modeling of Soft Matter", for which the authors want to thank the organizers. The authors also want to thank Professors Q. Du, X. Feng and N. Walkington for many discussions over the years.

REFERENCES

[1] R.G. Larson. *The Structure and Rheology of Complex Fluids.* Oxford, New York, 1999.

[2] J.L. West. Polymer-dispersed liquid crystals. In R.A. Weiss and C.K. Ober, editors, *Liquid-Crystalline Polymers, , R.A. Weiss and C.K. Ober editors*, Volume 435 of *ACS Symp. Ser.*, Chapter 32, pp. 475–495. ACS, Washington, D.C., 1990.

[3] C.L. Tucker and P. Moldenaers. Microstructural evolution in polymer blends. *Ann. Rev. Fluid Mech.*, 34:177–210, 2002.

[4] J.A. Sethian and P. Smereka. Level set methods for fluid interfaces. *Ann. Rev. Fluid Mech.*, 35:341–372, 2003.

[5] J.Feng, G. Sgalari, and L.G. Leal. A theory for flowing nematic polymers with orientational distortion. *J. Rheol.*, 44:1085–1101, 2000.

[6] V. Cristini, J. Blawzdziewicz, and M. Loewenberg. Drop breakup in three-dimensional viscous flows. *Phys. Fluids*, 10:1781–1783, 1998.

[7] E.M. Toose, B.J. Geurts, and J.G.M. Kuerten. A boundary integral method for two-dimensional (non)-Newtonian drops in slow viscous flow. *J. Non-Newtonian Fluid Mech.*, 60:129–154, 1995.

[8] R.E. Khayat. Three-dimensional boundary-element analysis of drop deformation for Newtonian and viscoelastic systems. *Int. J. Num. Meth. Fluids*, 34:241–275, 2000.

[9] H.H. Hu, N.A. Patankar, and M.Y. Zhu. Direct numerical simulations of fluid-solid systems using the arbitrary lagrangian-eulerian technique. *J. Comput. Phys.*, 169:427–462, 2001.

[10] B. Ambravaneswaran, E.D. Wilkes, and O.A. Basaran. Drop formation from a capillary tube: Comparison of one-dimensional and two-dimensional analyses and occurrence of satellite drops. *Phys. Fluids*, 14:2606–2621, 2002.

[11] R.W. Hooper, V.F. de Almeida, C.W. Macosko, and J.J. Derby. Transient polymeric drop extension and retraction in uniaxial extensional flows. *J. Non-Newtonian Fluid Mech.*, 98:141–168, 2001.

[12] S. Ramaswamy and L.G. Leal. The deformation of a viscoelastic drop subjected to steady uniaxial extensional flow of a Newtonian fluid. *J. Non-Newtonian Fluid Mech.*, 85:127–163, 1999.

[13] S. Ramaswamy and L.G. Leal. The deformation of a Newtonian drop in the uniaxial extensional flow of a viscoelastic liquid. *J. Non-Newtonian Fluid Mech.*, 88:149–172, 1999.

[14] J. Li and Y. Renardy. Numerical study of flows of two immiscible liquids at low Reynolds number. *SIAM Review*, 42:417–439, 2000.

[15] Y. Renardy, M. Renardy, T. Chinyoka, D.B. Khismatullin, and J. Li. A viscoelastic vof-prost code for the study of drop deformation. *Proc. 2004 ASME Heat Transfer/Fluids Engineering Summer Conference, ASME*, HT-FED2004-56114, 2004.

[16] J. Glimm, C. Klingenberg, O. McBryan, B. Plohr, D. Sharp, and S. Yaniv. Front tracking and two-dimensional Riemann problems. *Adv. Appl. Math.*, 6:259–290, 1985.

[17] S.O. Unverdi and G. Tryggvason. A front-tracking method for viscous, incompressible, multi-fluid flows. *J. Comput. Phys.*, 100:25–37, 1992.

[18] Y.C. Chang, T.Y. Hou, B. Merriman, and S. Osher. A level set formulation of eulerian interface capturing methods for incompressible fluid flows. *J. Comput. Phys.*, 124:449–464, 1996.

[19] J.A. Sethian. *Level Set Methods and Fast Marching Methods: evolving interfaces in computational geometry, fluid mechanics, computer vision, and materials science, 2nd edition.* Cambridge University Press, New York, 1999.

[20] S. Osher and R. Fedkiw. Level set methods: An overview and some recent results. *J. Comput. Phys.*, 169:463–502, 2001.

[21] J. Lowengrub and L. Truskinovsky. Quasi-incompressible Cahn-Hilliard fluids and topological transitions. *Proc. Roy. Soc. Lond. A*, 454:2617–2654, 1998.

[22] J.D. van der Waals. The thermodynamic theory of capillarity under the hypothesis of a continuous variation of density. *Verhandel Konink. Akad. Weten. Amsterdam, (Sect. 1)*, 1:1–56, 1892.

[23] J.D. van der Waals. The thermodynamic theory of capillarity under the hypothesis of a continuous variation of density, translation by j. s. rowlingson. *J. Statist. Phys*, 20:197–244, 1979.

[24] J. Dunn and J. Serrin. On the thermomechanics of interstitial working. *Arch. Rational Mech. Anal.*, 88(2):95–133, 1985.

[25] D. Joseph. Fluid dynamics of two miscible liquids with diffusion and gradient stresses. *European J. Mech. B Fluids*, 9(6):565–596, 1990.

[26] P.C. Hohenberg and B.I. Halperin. Theory of dynamic critical phenomena. *Rev. Mod. Phys.*, 49:435–479, 1977.

[27] A.J. Bray. Theory of phase-ording kinetics. *Advances in Physics*, 51(2):481–587, 2002.

[28] J.A. Warren and W.J. Boettinger. Prediction of dendritic growth and microsegregation patterns in a binary alloy using the phase-field method. *Acta Metall. Mater.*, 43:689–703, 1995.

[29] A. Karma. Phase-field formulation for quantitative modeling of alloy solidification. *Phys. Rev. Lett.*, 87:115701, 2001.

[30] C.I. Poser and I.C. Sanchez. Interfacial tension theory of low and high molecular weight liquid mixtures. *Macromolecules*, 14:361–370, 1981.

[31] J.W. Cahn and J.E. Hilliard. Free energy of a nonuniform system. III. nucleation in a two-component incompressible fluid. *J. Chem. Phys.*, 31:688–699, 1959.

[32] A.M. Lape na, S.C. Glotzer, S.A. Langer, and A.J. Liu. Effect of ordering on spinodal decomposition of liquid-crystal/polymer mixtures. *Phys. Rev. E*, 60:R29–R32, 1999.

[33] D.M. Anderson, G.B. McFadden, and A.A. Wheeler. Diffuse-interface methods in fluid mechanics. *Ann. Rev. Fluid Mech.*, 30:139–165, 1998.

[34] D. Jacqmin. Calculation of two-phase Navier-Stokes flows using phase-field modelling. *J. Comput. Phys.*, 155:96–127, 1999.

[35] M. Verschueren, F.N. van de Vosse, and H.E.H. Meijer. Diffuse-interface modelling of thermocapillary flow instabilities in a hele-shaw cell. *J. Fluid Mech.*, 434:153–166, 2001.

[36] V.E. Badalassi, H.D. Ceniceros, and S. Banerjee. Computation of multiphase systems with phase field model. *J. Comput. Phys.*, 190:371–397, 2003.

[37] C. Liu and J. Shen. A phase field model for the mixture of two incompressible fluids and its approximation by a Fourier-spectral method. *Physica D*, 179:211–228, 2003.

[38] F. Boyer. A theoretical and numerical model for the study of incompressible mixture flows. *Comput. Fluids*, 31:41–68, 2002.

[39] T. Qian, X.-P. Wang, and P. Sheng. Generalized navier boundary condition for the moving contact line. *Commun. Math. Sci.*, 1(2):333–341, 2003.

[40] M. Gurtin, D. Polignone, and J. Vinals. Two-phase binary fluids and immiscible fluids described by an order parameter. *Math. Models Methods Appl. Sci.*, 6(6):815–831, 1996.

[41] P. Yue, J.J. Feng, C. Liu, and J. Shen. A diffuse-interface method for simulating two-phase flows of complex fluids. *J. Fluid Mech.*, 515:293–317, 2004.

[42] P. Yue, J.J. Feng, C. Liu, and J. Shen. Diffuse-interface simulations of drop coalescence and retraction in viscoelastic fluids. *J. Non-Newtonian Fluid Mech.*, accepted, 2005.

[43] P. Yue, J.J. Feng, C. Liu, and J. Shen. Interfacial force and Marangoni flow on a nematic drop retracting in an isotropic fluid. *J. Colloid Interface Sci.*, accepted, 2005.

[44] P. Yue, J.J. Feng, C. Liu, and J. Shen. Viscoelastic effects on drop deformation in steady shear. *J. Fluid Mech.*, submitted, 2004.

[45] P. Yue, J.J. Feng, C. Liu, and J. Shen. Transient drop deformation upon startup of shear in viscoelastic fluids. *Phys. Fluids*, submitte, 2005.

[46] P. Yue, J.J. Feng, C. Liu, and J. Shen. Heart-shaped bubbles rising in a nematic fluid. *in preparation*, 2005.

[47] S.R. de Groot and P. Mazur. *Nonequilibrium Thermodynamics*. North Holland, 1962.

[48] C. Liu, J. Shen, J.J. Feng, and P. Yue. Variational approach in two-phase flows of complex fluids: transport and induced elastic stress. In A. Miranville, editor, *Mathematical Models and Methods in Phase Transitions,*. Nova Pub-

lications, 2005.

[49] C. Liu and N.J. Walkington. An eulerian description of fluids containing visco-elastic particles. *Arch. Ration. Mech. Anal.*, 159(3):229–252, 2001.

[50] P.G. de Gennes and J. Prost. *The Physics of Liquid Crystals.* Oxford, New York, 1993.

[51] C. Liu and N.J. Walkington. Approximation of liquid crystal flows. *SIAM J. Numer. Anal.*, 37:725–741, 2000.

[52] A. Rapini and M. Popoular. Distortion d'une lamelle nematique sous champ magnetique conditions d'ancrage aux parois. *J. Phys. (Paris) C*, 30:54–56, 1969.

[53] C. Liu and J. Shen. On liquid crystal flows with free-slip boundary conditions. *Dis. Cont. Dyn. Sys.*, 7:307–318, 2001.

[54] J. Shen. Efficient spectral-Galerkin method. II. direct solvers of second and fourth order equations by using Chebyshev polynomials. *SIAM J. Sci. Comput.*, 16:74–87, 1995.

[55] J.L. Guermond, P. Minev, and J. Shen. An overview of projection methods for incompressible flows. *Comput. Methods Appl. Mech. Eng.*, submitted, 2005.

[56] J.L. Guermond and J. Shen. A class of truly consistent splitting schemes for incompressible flows. *J. Comput. Phys.*, 192:262–276, 2003.

[57] J. Shen. Efficient spectral-Galerkin method. I. direct solvers of second and fourth order equations by using Legendre polynomials. *SIAM J. Sci. Comput.*, 15:1489–1505, 1994.

[58] J. Shen. Efficient spectral-Galerkin method. III. polar and cylindrical geometries. *SIAM J. Sci. Comput.*, 18:1583–160, 1997.

[59] X. Zhang, R.S. Padgett, and O.A. Basaran. Nonlinear deformation and breakup of stretching liquid bridges. *J. Fluid Mech.*, 329:207–245, 1996.

[60] M.R. Nobari, Y.-J. Jan, and G. Tryggvason. Head-on collision of drops—a numerical investigation. *Phys. Fluids*, 8:29–42, 1996.

[61] A.N. Zdravkov, G.W.M. Peters, and H.E.H. Meijer. Film drainage between two captive drops: PEO-water in silicon oil. *J. Colloid Interface Sci.*, 266:195–201, 2003.

[62] A. Bhakta and E. Ruckenstein. Decay of standing foams: drainage, coalescence and collapse. *Adv. Colloid Interface Sci.*, 70:1–124, 1997.

[63] L.M. Pismen. Nonlocal diffuse interface theory of thin films and the moving contact line. *Phys. Rev. E*, 64:021603, 2001.

[64] R.B. Bird, D.F. Curtiss, R.C. Armstrong, and O. Hassager. *Dynamics of Polymeric Liquids, Vol. 2. Kinetic Theory.* Wiley, New York, 1987.

[65] G. Marrucci and F. Greco. The elastic constants of Maier-Saupe rodlike molecule nematics. *Mol. Cryst. Liq. Cryst.*, 206:17–30, 1991.

[66] F.H. Lin and C. Liu. Nonparabolic dissipative systems, modeling the flow of liquid crystals. *Comm. Pure Appl. Math.*, 48:501–537, 1995.

[67] F.H. Lin and C. Liu. Existence of solutions for the Ericksen-Leslie system. *Arch. Rat. Mech. Anal.*, 154:135–156, 2000.

[68] X. Yang, J.J. Feng, C. Liu, and J. Shen. Contraction and pinch-off phenomena of a liquid filament. *J. Comput. Phys.*, submitted, 2005.

[69] T. Biben, C. Misbah, A. Leyrat, and C. Verdier. An advected-field approach to the dynamics of fluid interfaces. *Europhys. Lett.*, 63:623–629, 2003.

[70] G.E. Charles and S.G. Mason. The coalescence of liquid drops with flat liquid-liquid interfaces. *J. Colloid Sci.*, 15:236–267, 1960.

[71] P. Ghosh and V.A. Juvekar. Analysis of the drop rest phenomenon. *Chem. Eng. Res. Design*, 80:715–728, 2002.

[72] C.F. Ollivier-Gooch. Coarsening unstructured meshes by edge contraction. *Int. J. Numer. Methods Eng.*, 57:391–414, 2003.

[73] V. Cristini and Y.C. Tan. Theory and numerical simulation of droplet dynamics in complex flows - a review. *Lab Chip*, 4:257–264, 2004.

[74] M. van Sint Annaland, N.G. Deen, and J.A.M. Kuipers. Numerical simulation of gas bubbles behaviour using a three-dimensional volume of fluid method.

Chem. Eng. Sci., 60:2999–3011, 2005.

[75] Q. Du, C. Liu, and X. Wang. Retrieving topological information for phase field models. *SIAM J. Appl. Math.*, accepted for publication, 2005.

[76] M.P. do Carmo. *Differential Geometry of Curves and Surfaces*. Prentice-Hall, Englewood Cliffs, NJ, 1976.

NON-EQUILIBRIUM STATISTICAL MECHANICS OF NEMATIC LIQUIDS*

CHII J. CHAN* AND EUGENE M. TERENTJEV†

Abstract. The rotational diffusion of a general-shape object (a molecule) in a flow of uniaxial nematic liquid crystal is considered in the molecular field approximation. The full corresponding Fokker-Planck equation is derived, and then reduced to the limit of diffusion of orientational coordinates in a field of uniaxial nematic potential and the flow gradient. The spectrum of orientational relaxation times follows from this analysis. As a second main theme of this work, we derive a complete form of microscopic stress tensor for this molecule from the first principles of momentum balance. Averaging this microscopic stress with the non-equilibrium probability distribution of orientational coordinates produces the anisotropic part of the continuum Leslie-Ericksen viscous stress tensor and the set of viscous coefficients, expressed in terms of molecular parameters, nematic order and temperature. The axially-symmetric limits of long-rod and thin-disk molecular shapes allow comparisons with existing theories and experiments on discotic viscosity. The article concludes with more complicated aspects of non-linear constitutive equations, microscopic theory of rotational friction and the case of non-uniform flow and director gradients.

Key words. Rotational diffusion, Microscopic stress tensor, Nematic liquid crystals, Constitutive equations, Leslie coefficients, Orientational relaxation.

AMS(MOS) subject classifications. 76A15 Liquid crystals; 74A25 Molecular, statistical, and kinetic theories; 60J60 Diffusion processes.

1. Introduction. On the macroscopic continuum level, the dissipation of energy in a liquid flow is determined by the viscous stress tensor, which in the linear regime is proportional to the flow gradients with a factor of viscous coefficient. Kinetic theory of fluid viscosity has the aim of deriving this stress tensor, and the viscous coefficients, from the molecular parameters, interaction forces and temperature, thereby relating the kinetic linear response coefficient with the thermal fluctuations in the medium. Kinetic theory of viscosity of classical isotropic liquids is based on a complicated and delicate analysis of pair correlation functions out of equilibrium; it has a famous history of successful developments [1–3] although by far not everything is understood in this field.

In this article we describe a non-equilibrium statistical theory of the hydrodynamics of nematic liquid crystals, the liquids with a spontaneously broken orientational symmetry due to the anisotropic pair interactions between constituent particles (molecules). In developing the *nematodynamics* we aim to justify prevalent phenomenological theories and determine the underlying principles governing the orientational dynamics of the molecules

*This work has been partially funded by the Institute for Mathematics and its Applications (IMA), University of Minnesota, and the Cambridge Commonwealth Trust.

†Cavendish Laboratory, University of Cambridge, Madingley Road, Cambridge CB3 0HE, U.K. (emt1000@cam.ac.uk).

under simple shear flow. From a fundamental perspective, such studies reveal important physical insights which may help to answer some of the most important questions in rheological studies of nematic liquid crystals: To what extent does shear flow affect the molecular alignments? What is the microscopic basis for nematic liquid crystals displaying flow-induced transitions into an ordered or unstable state? Such questions represent typical phenomena abundant in physics for which a simple physical analysis often reveals deep underlying principles, yet a detailed and rigorous solution is necessary to confirm the analysis.

From a more practical perspective, the inherent nature of nematic liquid crystals to acquire a preferred orientation of anisotropic molecules, and preserve it in the presence of flow, provides a natural advantage to these materials to be used as precursors for the manufacturing of high performance fibres. The preferred orientation and degree of alignment of the molecules are found to have a predominant effect on the mechanical and thermal properties of the materials, and the optimization and control of preferred alignment is of paramount importance. Unfortunately a fundamental understanding of the factors affecting the development of preferred alignment is still lacking, which may hinder their further development.

In comparison to thermotropic nematics, i.e. dense liquids of anisotropic molecules, dilute suspensions of non-spherical particles are reasonably well understood [4]. The intrinsic viscosities of suspensions of oblate and prolate spheroids have been calculated allowing low volume fraction viscosity measurements to be used to estimate particle aspect ratio. Studies have been completed which extend the observations to the interactions of several particles. Models that include the influence of Brownian motion have also been developed [6]. However the majority of these theoretical studies have focused solely on rod-shaped nematic molecules, as opposed to disk-shaped particles or a more general case of spheroidal molecules with uniaxial symmetry. As the concentration of particles increases, the particles no longer rotate freely, but their motion is limited through excluded volume interactions as well as long range inter-particle and hydrodynamic forces. For particles with a large length/thickness ratio the effective excluded volume is much larger than their actual volume. As a result, their relative motion will be geometrically constrained, and the physics becomes non-trivial since many-particle correlations have to be considered. This situation will be applicable to a thermotropic nematic liquid as well.

In general the orientation of the director in a flowing nematic is determined by four external influences which tend to compete with, and in the steady state balance one another. The first effect is the influence of flow alignment; in the case of simple planar shear this tends to rotate the director until it lies almost, though not quite, in the direction in which the fluid is moving. Secondly there is the influence exerted by applied fields such as the magnetic fields. Thirdly there is the influence exerted by the solid surfaces which contain the liquid which affects the dynamics of thin layers (the strong anchoring effect). Finally the director alignments may be influenced by the curvature elasticity of the nematic itself. In this work, we will not consider the effects of external magnetic fields and surface anchoring effects, since we are primarily interested in the bulk property of the system subject to shear flow without imposing external fields.

There are traditionally two approaches towards studying the rheological behavior of liquid crystalline materials: the *top-down* macroscopic theory based on classical mechanics such as the Leslie-Ericksen model [5, 7] or the time-dependent Ginzburg-Landau theory, and the *bottom-top* molecular theory employing statistical approach that aims to derive fundamental constitutive equations governing the dynamics of the variables we are interested in. The macroscopic models assume the system being close to equilibrium and consider the dynamics of the *slow variables* such as the director or the order parameter tensor, while a complete microscopic theory allows us to consider even nematic systems driven far from equilibrium. Such is the case for a *tumbling* nematic that are often observed in high shear flow regime. A phenomenological explanation had for a long time failed to account for this phenomenon but as we shall see later, this phenomenon can be understood from a microscopic perspective.

Some of the earliest attempts on microscopic approach include works by Diogo and Martins [8]. They consider the viscosity coefficients to be proportional to the characteristic relaxation time which is related to the probability of overcoming the nematic potential barrier during molecular reorientation. Although such consideration does give microscopic expressions for the Leslie coefficients, their model was not constructed as a self-consistent statistical theory, and contains too many free parameters. Therefore, a more elaborate statistical theory was required. In 1983 Kuzuu and Doi [9] proposed the first non-equilibrium statistical model that describes the hydrodynamics of nematic liquid crystals made of ellipsoidal molecules using the concept of averaging the microscopic stress tensor over the non-equilibrium distribution function. However their expression for the microscopic stress tensor was not accurately derived and gave only the symmetric part of it. To introduce antisymmetric elements to the stress tensor, they invoke an external magnetic field in an ad hoc manner which makes it hard to reconcile with intrinsic antisymmetric viscous stress in liquid crystals in the absence of external field. Osipov and Terentjev suggested another approach [10] which assumes that the overall nonequilibrium distribution function should consist of an original equilibrium part and an additional non-equilibrium part due to the flow gradients, but their derivation of microscopic stress tensor is questionable and their derived Leslie coefficients are not always consistent with flow alignment experiments.

All these approaches either suffer from some theoretical shortcomings or they are confined to specific nematic systems composed only of long rod-like molecules. Although the later analysis is highly relevant to the rheology of liquid crystalline polymers [11], a more elaborate microscopic theory on the nematodynamics of spheroidal molecules will serve a greater purpose to a wider class of nematic systems. We generally follow the approach of [10], improving on several shortcomings of their treatment. Our work is also motivated by recent interests in the studies of *discotic* nematic liquid crystals. To our present knowledge, there have been little theoretical studies for the case of discotic nematic liquid crystalline phases in shear flow, though lately there has been a revival in experimental and theoretical interests in these materials due to their applications in high performance fibres (eg. mesophase pitch-based carbon fibres, Kevlar) [12, 13]. Another example that highlights many important technological applications in these materials is kaolin clay suspensions (plate-shaped particles) which have seen limited rheological characterization [14]. The work outlined in this article should assist in characterizing some of the main microstructure features and textures developed in these materials under flow.

The outline of this paper is as follows. In Section 2 we discuss theoretical concepts of non-equilibrium statistical physics and hydrodynamics which allow us to derive the kinetic equation governing the evolution of the orientation distribution function of the molecules. We also attempt to solve the kinetic equation which gives us the dominant orientation relaxation time. In Section 3 we demonstrate how the microscopic stress tensor can be derived using classical equation of motion for fluids. In Section 4 we put the results of kinetic modelling and the microscopic dynamics together to derive the average macroscopic stress tensor. Its coefficients are a complete set of the Leslie's coefficients, expressed in terms of molecular and order parameters. We discuss their validity, followed by a brief discussion on the unusual non-linear effects which exist in discotic nematics only. Section 5 outlines some attempts to derive the rotational frictional constant from a microscopic description, focusing primarily on discotic nematics. Finally in Section 6 we consider more realistic situations when spatial inhomogeneities and domain structures (such as those with point defects or disclinations) exist in nematic liquid crystals, and construct a new molecular theory to account for the Ericksen stress in the complete Leslie-Ericksen theory.

2. Kinetic equation. *In this section we discuss some of the concepts of rotational diffusion and Brownian motion. We demonstrate that the dynamical evolution of a general ellipsoidal-shaped molecule in rotational motion in a nematic potential can be described essentially by a multi-phase variable Fokker-Planck equation. The solutions of the kinetic equation in the weak flow limit suggest a rich spectrum of relaxation times. The dominant relaxation time is found to depend linearly on the rotational friction constant and exhibits an Arrhenius activation exponential dependence on the inter-molecular coupling strength.*

2.1. Rotational Brownian motion and hydrodynamics. A nematic fluid contains many anisotropic molecules, all of similar size in a dense phase. On a mean-field level, each molecule can be considered to be immersed in a thermodynamic bath which acts as a source of background fluctuations. We can therefore consider each molecule to undergo rotational Brownian motion since it experiences a constant flux of stochastic torques. The problem of an arbitrary-shaped rigid body executing rotational Brownian dynamics is however a technically complicated one. The reason for this is at least three-fold: (1) rotations about different axes do not commute, (2) the range of relevant variables, the angles specifying the body's orientation, is finite. This introduces the peculiar nature of the topological constraints to the system. (3) Relation between angular velocity and angular momentum is tensorial, not vectorial, as in translational motion.

Despite the above complications, the rotational Brownian motion in a mean-field potential is thoroughly described within the framework of Smoluchowski equation. Jefferey and Hinch *et. al.* [15, 6] had solved similar problems for a dilute suspension of ellipsoids in a flow. We note that in addition to the rotational Brownian motion the molecules also execute translational random motions which we will not deal with in this article. The orientational degree of freedom is described by the dynamical variable $\boldsymbol{u}$, which is the unit vector defining the direction of principal axis (parallel to the long axis for rod-like molecules and perpendicular to the plane of a disk-like molecule).

The rotational Brownian motion can be best visualised as the trajectory of $\boldsymbol{u}(t)$ on the surface of the sphere defined by $|\boldsymbol{u}| = 1$. The movement of $\boldsymbol{u}(t)$ can be considered as random steps due to random stochastic force and external potential (see Fig. 1). The hydrodynamics of rotational motion can be addressed by first considering a general spheroid immersed in a stationary viscous liquid. We consider the molecule rotating with an angular velocity $\boldsymbol{\omega}$ by the influence of a torque $\boldsymbol{\Gamma}$ exerted by an external field $U(\boldsymbol{u})$. Consider a small rotation $\delta\boldsymbol{\varphi}$ of the molecule that changes $\boldsymbol{u}$ to $\boldsymbol{u} + \delta\boldsymbol{\varphi} \times \boldsymbol{u}$. The work needed for this change is $-\boldsymbol{\Gamma} \cdot \delta\boldsymbol{\varphi}$, which must be equal to the change in U, i.e.,

$$-\boldsymbol{\Gamma} \cdot \delta\boldsymbol{\varphi} = U(\boldsymbol{u}+\delta\boldsymbol{\varphi} \times \boldsymbol{u}) - U(\boldsymbol{u}) = (\delta\boldsymbol{\varphi} \times \boldsymbol{u}) \cdot \frac{\partial U}{\partial \boldsymbol{u}} = \delta\boldsymbol{\varphi} \cdot \left(\boldsymbol{u} \times \frac{\partial U}{\partial \boldsymbol{u}}\right). \tag{2.1}$$

Hence

$$\Gamma_\beta = -\partial_\beta U \,, \qquad \text{where } \partial_\beta = \left(\boldsymbol{u} \times \frac{\partial}{\partial \boldsymbol{u}}\right)_\beta . \tag{2.2}$$

The operator ∂_β is called the rotational operator that plays the role analogous to the gradient operator ∇ in translational motion. Now if the molecules are immersed in a flowing medium, there will be a *residue* angular

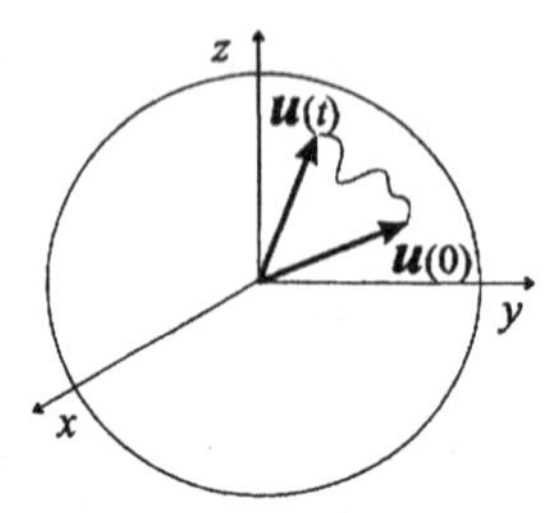

FIG. 1. *Rotational Brownian diffusion by the unit vector $\boldsymbol{u}$ along the molecular axis, which explores the space on the surface of unit sphere.*

velocity for a spheroid with aspect ratio $p = a/b$ given by [15]:

$$
\begin{aligned}
\boldsymbol{\omega}^{\text{res}} &= \boldsymbol{u} \times \left\{ \frac{p^2}{p^2+1} \boldsymbol{g} \cdot \boldsymbol{u} - \frac{1}{p^2+1} \boldsymbol{g}^{\mathsf{T}} \cdot \boldsymbol{u} \right\} \\
&= \boldsymbol{u} \times \left\{ \frac{1}{2} \left(\frac{p^2}{p^2+1} \right) (\boldsymbol{g}^s + \boldsymbol{g}^a) \cdot \boldsymbol{u} - \frac{1}{2} \frac{1}{p^2+1} (\boldsymbol{g}^s + \boldsymbol{g}^a)^{\mathsf{T}} \cdot \boldsymbol{u} \right\} \\
&= \frac{1}{2} \frac{p^2-1}{p^2+1} (\boldsymbol{u} \times \boldsymbol{g}^s \cdot \boldsymbol{u}) + \frac{1}{2} \ (\boldsymbol{u} \times \boldsymbol{g}^a \cdot \boldsymbol{u}) \\
&= \frac{1}{2} \frac{p^2-1}{p^2+1} (\boldsymbol{u} \times \boldsymbol{g}^s \cdot \boldsymbol{u}) + \frac{1}{2} \nabla \times \boldsymbol{v} - \frac{1}{2} \ (\boldsymbol{u} \cdot \nabla \times \boldsymbol{v}) \, \boldsymbol{u}
\end{aligned} \tag{2.3}
$$

where $g_{\alpha\beta}$ is the velocity gradient $\nabla_\beta v_\alpha$, and $g^s_{\alpha\beta}$ and $g^a_{\alpha\beta}$ are the symmetric and asymmetric part of this velocity gradient, respectively.

2.2. Langevin equation. The stochastic effects on the particle's rotational motion in viscous medium can be considered in a coherent fashion using the method of Langevin stochastic equation and the Fokker-Planck kinetic equation. The latter allows us to find explicit dynamical evolution of the distribution function in terms of the orientation of the director. To see how this can be applied to anisotropic fluid motion we first consider the dynamical equations of motion for the particle's angular velocity.

We first note that we can always diagonalize the moment of inertia tensor in the spheroid's principal-axis frames,

$$I_{\alpha\beta} = I_\perp \delta_{\alpha\beta} + (I_\parallel - I_\perp) u_\alpha u_\beta \tag{2.4}$$

where $\boldsymbol{u}$ is the unit vector of the molecule's axis. The instantaneous angular velocity of the molecule is

$$\dot{\boldsymbol{\Psi}} = \dot{\psi} \boldsymbol{u} + \boldsymbol{\omega} \tag{2.5}$$

where the first term on the right hand side denotes angular velocity about the molecular axis, while $\boldsymbol{\omega}$ is the transverse angular velocity due to large rotational motion perpendicular to the molecular axis, i.e. $\boldsymbol{\omega} \perp \boldsymbol{u}$.

For a molecule moving with an instantaneous angular momentum $\mathbf{L}$, we can immediately write down its expression,

$$L_\alpha = I_{\alpha\beta}\dot{\Psi}_\beta = I_\parallel \dot{\psi} u_\alpha + I_\perp \omega_\alpha. \tag{2.6}$$

The rotational motion thus obeys the equation of motion:

$$\dot{L}_\alpha = I_{\alpha\beta}\dot{\Psi}_\beta = I_\parallel \ddot{\psi} u_\alpha + I_\perp \dot{\omega}_\alpha + I_\parallel \dot{\psi}\dot{u}_\alpha. \tag{2.7}$$

The first two terms on the right hand side are the expected rotational torques about $I_\parallel$ and $I_\perp$ respectively, while the last term represents the *gyroscopic effect.* This term vanishes for the case of an infinitely long and thin rod ($I_\parallel \ll I_\perp$) but may be large for disk-like molecules. We will soon see that this term gives rise to non-trivial modifications to the kinetic equation and the stress tensor.

At this stage we have to be careful about the meaning of $\boldsymbol{\omega}$. To evaluate all physical observables in the laboratory frame we have to make a coordinate transformation from the body's frame to the laboratory frame. Therefore for a molecule rotating with an instantaneous angular velocity $\dot{\boldsymbol{\Psi}}$, the transverse angular velocity $\boldsymbol{\omega}$ in the body frame is transformed in the following manner:

$$\left.\frac{d\omega}{dt}\right|_{lab} = \dot{\boldsymbol{\Psi}} \times \boldsymbol{\omega} + \left.\frac{d\omega}{dt}\right|_{body} = -\dot{\psi}\dot{\boldsymbol{u}} + \boldsymbol{u} \times \ddot{\boldsymbol{u}}. \tag{2.8}$$

Having obtained the general equation of motion in equation (2.7), we can write down the Langevin equation in terms of a vector stochastic torque $\boldsymbol{\xi}$ and a possible external torque $\boldsymbol{\Gamma}$,

$$\begin{aligned} \dot{L}_\alpha &= I_\parallel \ddot{\psi} u_\alpha + I_\perp \dot{\omega}_\alpha + I_\parallel \dot{\psi}\dot{u}_\alpha \\ &= -\lambda_{\alpha\beta}(\omega_\beta + \dot{\psi} u_\beta - \omega_\beta^{\text{res}}) + \Gamma_\alpha + \xi_\alpha \end{aligned} \tag{2.9}$$

where $\lambda_{\alpha\beta}$ is the frictional constant tensor and the vector $(\omega_\beta + \dot{\psi} u_\beta - \omega_\beta^{\text{res}})$ is the net angular velocity of the molecule relative to the reservoir.

To get the equation of motion for the dynamical variable $\dot{\psi}$, we can multiply the above equation by u_α to eliminate the gyroscopic term,

$$I_\parallel \ddot{\psi} = -u_\alpha \lambda_{\alpha\beta} \Omega_\beta + \Gamma_\alpha u_\alpha + \xi_\alpha u_\alpha \tag{2.10}$$

where we replace $(\omega_\beta + \dot{\psi} u_\beta - \omega_\beta^{\text{res}})$ with Ω_β. Equation (2.10) is the equation of motion for the dynamical variable $\dot{\psi}$ dictating the angular rotation about the molecular axis.

Substituting (2.10) into Equation (2.9) we obtain a similar equation of motion for the transverse angular velocity $\boldsymbol{\omega}$,

$$\begin{aligned} I_\perp \dot{\omega}_\alpha = &-(\delta_{\alpha\beta} - u_\alpha u_\beta)\lambda_{\beta l}\Omega_l + (\delta_{\alpha\beta} - u_\alpha u_\beta)\Gamma_\beta \\ &+(\delta_{\alpha\beta} - u_\alpha u_\beta)\xi_\beta - I_\parallel \dot{\psi}(\boldsymbol{\omega} \times \boldsymbol{u})_\alpha \end{aligned} \tag{2.11}$$

2.3. Generalized Fokker-Planck equation. The rotational diffusion of the anisotropic molecules is captured by the Fokker-Planck equation which describes the dynamical evolution in time of the system's phase space distribution function $W(\boldsymbol{\omega}, \dot{\psi}, \boldsymbol{u}, t)$ [16]:

$$\begin{aligned}\frac{\partial W}{\partial t} = &-\frac{\partial}{\partial u_\alpha}(\dot{u}_\alpha W) - \frac{\partial}{\partial \omega_\alpha}(\dot{\omega}_\alpha W) - \frac{\partial}{\partial \dot{\psi}}(\ddot{\psi} W) \\ &+ \frac{1}{2}\frac{\partial^2}{\partial \omega_\alpha \partial \omega_\beta}\left[\frac{1}{I_\perp^2}\langle(\xi_\alpha - u_\alpha \xi_\gamma u_\gamma)(\xi_\beta - u_\beta \xi_\mu u_\mu)\rangle W\right] \\ &+ \frac{1}{2}\frac{\partial^2}{\partial \dot{\psi}^2}\left[\frac{1}{I_\parallel^2} u_\alpha\, \Xi^{\psi}_{\alpha\beta}\, u_\beta W\right]\end{aligned} \tag{2.12}$$

where $\Xi^{\psi}_{\alpha\beta}(t-t') = \langle \xi_\alpha(t)\xi_\beta(t')\rangle$ is the correlation function between the vector stochastic torque ξ that perturbs $\dot{\psi}$. It can be shown directly that the following form for $\Xi^{\psi}_{\alpha\beta}$ indeed satisfies the fluctuation-dissipation theorem,

$$\Xi^{\psi}_{\alpha\beta} = \Xi_\perp \delta_{\alpha\beta} + (\Xi_\parallel - \Xi_\perp) u_\alpha u_\beta. \tag{2.13}$$

2.3.1. Reduced Fokker-Planck equation. We next consider obtaining the coordinate dependence of the distribution function. We note that there are intrinsically two time-scales of interest, the fast relaxation time after which the system reaches the Maxwell equilibrium velocity distribution,

$$\tau_\omega = \frac{I_\perp}{\lambda_\perp} \tag{2.14}$$

and the slow relaxation time

$$\tau_u = \frac{(\Delta\theta)^2}{2D_r} \tag{2.15}$$

after which the system reaches the equilibrium Boltzmann distribution of angular coordinates. $\Delta\theta$ is the free angular volume the molecule rotates in the diffusion limit and D_r is the rotational diffusion constant related to the microscopic friction constants via the fluctuation-dissipation theorem. This is the characteristic time for the relaxation of fluctuations of the system back to equilibrium under Brownian forces. Their ratio we define as a small parameter:

$$\alpha^2 = \frac{\tau_\omega}{\tau_u} = \frac{2kTI_\perp}{\lambda_\perp^2} \ll 1 \tag{2.16}$$

where we omit the dimensionless term $(\Delta\theta)^2$. The smallness of α is not obvious at this stage, but will become apparent later. Substituting Equations

(2.10) and (2.11) into Equation (2.12), and introducing the dimensionless variables:

$$\tau = \frac{\lambda_\perp}{I_\perp}t, \quad \omega' = \sqrt{\frac{I_\perp}{kT}}\omega, \quad \dot{\psi}' = \sqrt{\frac{I_\parallel}{kT}}\dot{\psi}\,, \tag{2.17}$$

we obtain the dimensionless form for Equation (2.12),

$$\begin{aligned} &\frac{\partial W}{\partial \tau} + \alpha\partial_\beta(\omega_\beta W) + \alpha\frac{\partial}{\partial\omega_\beta}(\delta_{\alpha\beta} - u_\alpha u_\beta)\left(\frac{\Gamma_\beta}{kT}W\right) \\ &= \frac{\partial}{\partial\omega_\alpha}\left[\omega_\alpha - \Omega_\alpha + \frac{1}{2}\frac{\partial}{\partial\omega_\beta}(\delta_{\alpha\beta} - u_\alpha u_\beta) + \alpha A\dot{\psi}(\boldsymbol{\omega}\times\boldsymbol{u})_\alpha\right]W \\ &\quad + \Delta A^2\frac{\partial}{\partial\dot{\psi}}\left(\dot{\psi} - \phi + \frac{1}{2}\frac{\partial}{\partial\dot{\psi}}\right)W \end{aligned} \tag{2.18}$$

with the notations

$$A = \sqrt{\frac{I_\parallel}{I_\perp}}, \quad \Delta = \frac{\lambda_\parallel}{\lambda_\perp}, \quad \phi = \frac{1}{2}\boldsymbol{u}\cdot\nabla\times\boldsymbol{v}\,. \tag{2.19}$$

2.3.2. Elimination of fast variables. We now describe in a qualitative fashion the meanings of the two time-scales introduced in the previous section. The situation where the variables describing a phenomenon can be divided into two sets, one evolving on a rapid time scale and one evolving on a slow time scale, is of frequent occurrence in physics. It is often desirable to eliminate or average over the rapid variables in order to study the dynamics of the slow variables. Such coarse-graining is done by assuming that the velocity distribution of the Brownian particle rapidly thermalizes while the coordinate distribution remains far from equilibrium for a much longer time. This means that the velocity distribution is close to a Maxwell distribution while the position distribution still has not evolved too far from the initial distribution. The equation, obtained after integrating out the fast variables by estimating the phase distribution function to be the product of the reduced distribution function in terms of the slow variable and a Maxwell distribution for fast variables, is formally known as the Smoluchowski equation [17]. The basic assumption that thermalization occurs on a time scale short with respect to the time for appreciable changes in the positional distribution is almost always satisfied in the high friction (overdamped) limit.

We can apply the above concepts to the case of a nematic to obtain the coordinate-only Smoluchowski equation. The fast variables in this case are the angular velocity both along and perpendicular to the director axis ($\boldsymbol{\omega}$ and $\dot{\psi}$) while the slow variable is the angular orientation $\boldsymbol{u}(t)$. Assuming the quasi-equilibrium state when the longitudinal angular velocity distribution function has thermalized, we can approximate:

$$W(\boldsymbol{\omega}, \dot{\psi}, \boldsymbol{u}, t) = \exp\left\{-\frac{1}{2}(\dot{\psi} - \phi)^2\right\}W'(\boldsymbol{\omega}, \boldsymbol{u}, t). \tag{2.20}$$

Substituting this into Equation (2.18) and integrating over $\dot{\psi}$ eventually gives the angular velocity dependence of $W'(\boldsymbol{\omega}, \boldsymbol{u}, t)$:

$$\begin{aligned}\dot{W}' &= \alpha\partial_\beta(\omega_\beta W') + \alpha\frac{\partial}{\partial\omega_\beta}(\delta_{\alpha\beta} - u_\alpha u_\beta)\frac{\Gamma_\beta}{kT}W' \\ &= \alpha A\phi(\boldsymbol{u})\frac{\partial}{\partial\omega_\beta}\left[(\boldsymbol{\omega}\times\boldsymbol{u})_\beta W'\right] \\ &\quad + \frac{\partial}{\partial\omega_\alpha}\left[\omega_\alpha - \Omega_\alpha + (\delta_{\alpha\beta} - u_\alpha u_\beta)\frac{\partial}{\partial\omega_\beta}\right]W'.\end{aligned}$$

Introducing a relative velocity $\xi_\alpha = \omega_\alpha - \Omega_\alpha$, we may naively proceed with integration over the remaining fast variable $\boldsymbol{\omega}$ using

$$W'(\boldsymbol{\omega}, \boldsymbol{u}, t) = e^{-\frac{1}{2}\xi^2}w(\boldsymbol{u}, t). \tag{2.21}$$

This however gives the trivial equation

$$\dot{w} + \alpha\partial_\beta(\Omega_\beta w) = 0 \tag{2.22}$$

with the diffusion term missing from the equation. We conclude that the non-trivial Smoluchowski equation with the diffusion term must come from **adding small corrections to the distribution function** that contains the last bits of non-relaxed Maxwell distribution. Hence we suggest that,

$$W'(\boldsymbol{\omega}, \boldsymbol{u}, t) = e^{-\frac{1}{2}\xi^2}\left[w(\boldsymbol{u}, t) + \alpha y(\boldsymbol{\xi}, \boldsymbol{u})\right] \tag{2.23}$$

where the smallness is controlled by the natural parameter – the ratio of relaxation times $\alpha \ll 1$, and the form of the correction term $y(\boldsymbol{\xi}, \boldsymbol{u})$ is to be determined self-consistently. Substituting (2.23) into (2.21), and neglecting terms of second orders in α, the equation transforms into:

$$\begin{aligned}\dot{w}(\boldsymbol{u}, t) + \alpha\Omega_\beta\partial_\beta w + \alpha\xi_\beta\left[\partial_\beta w + \Omega_\alpha\partial_\alpha\Omega_\beta w - \frac{\Gamma_\beta}{kT}w + A\phi(\boldsymbol{\Omega}\times\boldsymbol{u})_\beta w\right] \\ + \alpha\xi_\alpha\xi_\beta\partial_\alpha\Omega_\beta w = \alpha(\delta_{\alpha\beta} - u_\alpha u_\beta)\left[\frac{\partial^2 y}{\partial\xi_\alpha\partial\xi_\beta} - \xi_\alpha\frac{\partial y}{\partial\xi_\beta}\right].\end{aligned}$$

Assuming $y = a + b_i\xi_i + c_{ij}\xi_i\xi_j$ we determine uniquely the coefficients b_i and c_{ij}

$$c_{ij} = -\frac{1}{2}w\partial_i\Omega_j \tag{2.24}$$

$$b_j = -\partial_j w - \omega_i(\partial_i\Omega_j)w + \frac{\Gamma_j}{kT}w - A\phi(\boldsymbol{u})(\boldsymbol{\Omega}\times\boldsymbol{u})_j w. \tag{2.25}$$

Integrating over the fast variable ξ_i, we finally have the desired dimensionless Smoluchowski's equation for the coordinate-only distribution function

$w(\boldsymbol{u},t)$, which we will call here the *orientational distribution function*:

$$\begin{aligned}\dot{w} + \alpha\partial_\beta(\Omega_\beta w) = {} & \alpha^2\partial_\beta(\partial_\beta w - \frac{\Gamma_\beta}{kT}w) + \alpha^2\partial_\beta(\Omega_\alpha\partial_\alpha\Omega_\beta w)\\ & + \frac{A}{2}\alpha^2\partial_\beta\left[(\boldsymbol{u}\cdot\nabla\times\boldsymbol{v})(\boldsymbol{\Omega}\times\boldsymbol{u})_\beta w\right]\\ & + \alpha^2\partial_\beta\left[\Omega_\alpha(\partial_\alpha\Omega_\beta)w\right].\end{aligned} \tag{2.26}$$

The right-hand side now contains small but non-vanishing terms proportional to α^2 (compare with Equation (2.22) where this was missing in the leading order in α). The first term on the right hand side gives the diffusional term in a non-equilibrium system with external potential, which describes rotational diffusion mechanism. The term $\alpha\partial_\beta(\Omega_\beta w)$ incorporates the linear effects of perturbation due to external flow.

2.3.3. Non-linear effects. The last two terms in Equation (2.26) deserve further discussion. We note that these terms have not been shown in previous work [9, 10], but their presence necessarily describe novel non-linear effects due to higher flow and intrinsic geometrical shape of the molecules. The term $\frac{1}{2}\alpha^2 A\partial_\beta\left[(\boldsymbol{u}\cdot\nabla\times\boldsymbol{v})(\boldsymbol{\Omega}\times\boldsymbol{u})_\beta w\right]$ reveals the gyroscopic motion of the molecules due to the non-vanishing moment of inertia along the molecular axis. This term is commonly neglected for thin rods with $A = \sqrt{I_\parallel/I_\perp} \ll 1$. This however is not the case for a discotic system, when $I_\parallel$ and $I_\perp$ are comparable. One expects that this gyroscopic effect will contribute essentially to the viscous torques and the antisymmetric stress tensor, and modify the 'shape' of the equilibrium distribution function. This conjecture will be pursued and verified in a quantitative fashion in Section 4.

On the other hand the second term $\alpha^2\partial_\beta\left[\Omega_\alpha(\partial_\alpha\Omega_\beta)w\right]$ arises as a result of algebra. This term vanishes in the weak-flow limit (small $\boldsymbol{\Omega}$) and will be present in both isotropic and anisotropic liquids. It therefore constitutes trivial higher order corrections to the overall stress tensor due to stronger external flow. It may explain the changes in the linear viscosity for a general spheroidal nematic before tumbling sets in, where the whole physical basis of the linear model breaks down, but it does not introduce any new symmetries into the effect.

2.4. Solving the kinetic equation. There is an intrinsic time-scale that may be related to the typical relaxation times of the orientational distribution function which may be obtained via solving the kinetic equation. In fact, as we will see shortly, the solutions give rise to a spectrum of relaxation times that relate to the relaxation of the various normal modes of angular rotations. This relaxation can be observed macroscopically in the relaxation spectrum of the order parameter [18].

The standard way to solve the nonlinear integral kinetic equation in the angular space is to expand the distribution function in spherical harmonics and solve the resulting equations for the expansion coefficients sometimes numerically. Although this method is always available, we can gain some insights by solving it analytically using a simple eigenfunction expansion. For the sake of simplicity, we rewrite the Smoluchowski equation (2.26) in *zero flow*, in the following form:

$$\frac{\partial w}{\partial t} = -\Lambda(\boldsymbol{u})w(\boldsymbol{u},t) \tag{2.27}$$

where $\Lambda(\boldsymbol{u})$ is a linear differential operator:

$$\Lambda(\boldsymbol{u})w(\boldsymbol{u},t) = -\alpha^2\partial_k\left(\partial_k w + \frac{\partial_k U}{kT}\, w\right). \tag{2.28}$$

Let w_n be the eigenfunctions:

$$\Lambda(\boldsymbol{u})w_n = \lambda_n\, w_n. \tag{2.29}$$

Expanding the distribution function in terms of the complete orthogonal set of eigenfunctions:

$$w(\boldsymbol{u},t) = \sum_n a_n(t)\, w_n(\boldsymbol{u}) \ , \tag{2.30}$$

we obtain the time dependence of coefficients $a_n(t)$, $a_n(t) = a_n(0)e^{-\lambda_n t}$. The equilibrium distribution function $w_{eq}(\boldsymbol{u})$ is an eigenfunction which by definition has infinite relaxation time. The eigenvalue being the inverse of the relaxation time therefore is 0, corresponding to the eigenfunction w_0 with $n = 0$. Therefore the full solution takes the form:

$$w(\boldsymbol{u},t) = w_{eq}(\boldsymbol{u}) + \sum_{n=1} a_n(0)e^{-\lambda_n t}\, w_n(\boldsymbol{u}) \tag{2.31}$$

where $a_0 = 1$ by normalization. Since in statistical equilibrium $\dot{w}_{eq} = 0$, substituting Equation (2.31) into (2.27) gives:

$$\partial_k\left[\partial_k w_n + \frac{\partial_k U(\boldsymbol{u}\cdot\boldsymbol{n})}{kT} w_n\right] = -\frac{\lambda_n}{\alpha^2} w_n \tag{2.32}$$

where $U(\boldsymbol{u}\cdot\boldsymbol{n})$ is the mean-field potential, which depends on the polar angle θ only. Equation (2.32) is very similar to solving the Schrodinger's equation in quantum mechanics. In this case the external potential has to be modified. This mapping is formally known as the 'Darboux transformation' or 'supersymmetry' [19]. The operator of course has to be made Hermitian but this can be achieved through a simple transformation [20].

Expanding the rotational operator ∂_k in spherical coordinates and writing the eigenfunction of (2.32) as $w_n = f_n(\theta)w_{eq}$, where $w_{eq} \equiv \exp(-U(\theta)/kT)$, we finally obtain the following differential equation:

$$\frac{\partial^2 f_n}{\partial\theta^2} + \left(\cot\theta - \frac{1}{kT}\frac{\partial U}{\partial\theta}\right)\frac{\partial f_n}{\partial\theta} = -\frac{\lambda_n}{\alpha^2} f_n\ . \tag{2.33}$$

f_n depends on θ since, for rotations of a spheroid, the general solution of Equation (2.33) must be an eigenfunction expansion in terms of the Legendre polynomials $P_n(\cos\theta)$. Equation (2.33) can be rearranged to give

$$\frac{e^{U(\theta)/kT}}{\sin\theta}\frac{\partial}{\partial\theta}\left[e^{U(\theta)/kT}\sin\theta\frac{\partial f_n}{\partial\theta}\right] = -\frac{\lambda_n}{\alpha^2}f_n. \tag{2.34}$$

Rearranging the equation further and taking care of the constants of integration, we finally obtain the following self-consistent integral equation:

$$f_n(\theta) = C - \frac{\lambda_n}{\alpha^2}\int_0^\theta \frac{e^{U(x)/kT}}{\sin x}dx \int_0^x w_n(z)\sin z\,dz. \tag{2.35}$$

Multiplying $e^{-U(\theta)/kT}$ on both sides of the equation,

$$w_n(\theta) = e^{-U(\theta)/kT}\left[C - \frac{\lambda_n}{\alpha^2}\int_0^\theta \frac{e^{U(x)/kT}}{\sin x}dx \int_0^x w_n(z)\sin z\,dz\right]. \tag{2.36}$$

At this stage we introduce the method of iterations [19]:

$$w_n(\theta) = w_0 + \frac{\lambda_1}{\alpha^2}w_1 + \left(\frac{\lambda_1}{\alpha^2}\right)^2 w_2 + \ldots \tag{2.37}$$

where λ_1 is the smallest non-vanishing coefficient corresponding to the first coefficient w_1. This method will be justified later, when the perturbation coefficient λ_1/α^2 is shown to be small.

Substituting the solution with only the leading terms in λ_1 and comparing the terms explicitly, we have the relation:

$$\begin{aligned} w_n(\theta) &= e^{-U(\theta)/kT}e^{-U(0)/kT}w_0(0) \\ &\times\left[1 - \frac{\lambda_1}{\alpha^2}\int_0^\theta \frac{e^{U(x)/kT}}{\sin x}dx \int_0^x e^{-U(z)/kT}\sin z\,dz\right]. \end{aligned} \tag{2.38}$$

A conceivable boundary condition for any eigenfunction is that the distribution function must vanish at $\theta = \pi$, i.e., $w_n(\pi) = 0$, as it must do since $w_n(\theta)$ is a single-valued function. This boundary condition gives

$$1 - \frac{\lambda_1}{\alpha^2}\int_0^\pi \frac{e^{U(x)/kT}}{\sin x}dx \int_0^x e^{-U(z)/kT}\sin z\,dz = 0. \tag{2.39}$$

The integrals can be evaluated using the saddle-point approximation. In Maier-Saupe mean-field approximation, $U(\theta) = -JS_2\cos^2(\theta)$, where J denotes the mean-field coupling strength (an explicit form for the energy constant J will be discussed in Section 5) and S_2 is the principle scalar order parameter of uniaxial nematic phase, discussed in much greater detail in Section 4. The ratio $q = JS_2/kT \simeq 4.5$ at the nematic transition, hence justifying the method of saddle-point approximation where $JS_2/kT \gg 1$. Recovering the full dimensional form finally gives the following value for λ_1:

$$\lambda_1 = \frac{4}{\pi} q^{\frac{3}{2}}\, e^{-q}\, D_r = \frac{4}{\pi\lambda_\perp} \frac{(JS_2)^{3/2}}{(kT)^{1/2}}\, e^{-JS_2/kT} \tag{2.40}$$

where $D_r = kT/\lambda_\perp$ is the rotational diffusion constant and $\lambda_\perp$ is the friction constant for the molecular rotation about any axis parallel to the plane of the disk. We now return to justifying the perturbation in terms of the small parameter λ_1/α^2. In dimensional form we have

$$\frac{\lambda_1}{\alpha^2} \longrightarrow \frac{\lambda_1 \lambda_\perp}{kT} = \frac{4}{\pi} q^{\frac{3}{2}} e^{-q}. \tag{2.41}$$

This is indeed small in the limit of large q and justifies the perturbation expansion. The inverse of λ_1 gives the dominant (longest) relaxation time[1]

$$\tau_1 = \frac{\pi}{4q^{3/2} D_r} e^{q} = \frac{\pi\lambda_\perp}{4} \frac{(kT)^{1/2}}{(JS_2)^{3/2}} e^{JS_2/kT} \tag{2.42}$$

which gives a dependence similar to the relaxation time for the molecular director correlation function $\langle \boldsymbol{u}(t)\boldsymbol{u}(0)\rangle = e^{-t/\tau_r}$, where $\tau_r = 1/2D_r$ is the rotational correlation time [11]. For a typical nematic liquid, $D_r \simeq 10^8$ sec^{-1}, and $\tau_1 \simeq 10^{-7}$ s. This result agrees well with typical molecular relaxation times for the principal tumbling motion [21]. Also, this time-scale is usually small compared to the typical flow rate hence justifying the validity of the continuum Leslie-Ericksen description for nematics in flow (see Section 3).

The fact that the rotational diffusion of a nematic liquid crystal is associated with a rich spectrum of relaxation times is due to higher-order modes of rotational motion contained in (2.37), involving spherical harmonics in azimuthal and polar coordinates. It could also be attributed to the generic non-spherical shape of the molecule and the anisotropic rotatory diffusion tensor. The various relaxation modes and times correspond to the non-collective relaxations around different symmetry axes of the molecules. Our results agree with Diogo's conclusion [8] that the relaxation times for the flipping motions of the molecules obey the Arrhenius law. The exponential factor accounts for the probability that the reorienting molecule has enough energy to overcome the potential barrier due to intermolecular nematic potential. In reality, however, we may need to consider the free

[1] In a passing remark we note that this problem can also be solved in a simpler way, with inspiration from Kramers problem on a particle's passage over a potential barrier [19]. In other words, the relaxation mechanisms for rotational motion in liquid crystals are similar to the overcoming of the potential barriers imposed by the average medium in a mean-field.

volume effects which exist even in the absence of nematic potential. This may give rise to the *Volger-Fulcher* type of glassy relaxation [22]. On the other hand, the explicit dependence of the relaxation time on the rotational friction constant is expected due to slow decay in the presence of high friction. A typical application of the rotational diffusion problem is observed in the dielectric relaxation of nematics in the presence of an external electric field [21, 18], where more than one Debye relaxation times are found corresponding to rotations around the long or short molecular axis. Similar phenomena are also observed via NMR of a nucleus inside the nematic liquid [21].

3. Viscous stress tensor. *In this section we discuss the non-equilibrium transport phenomena in a nematic liquid. We briefly review the classical Leslie-Ericksen theory and then construct a microscopic stress tensor using classical kinetic theory of simple fluids, and relate the macroscopic stress tensor in terms of microscopic parameters.*

3.1. Hydrodynamics of a uniaxial fluid. A nematic liquid crystal flows easily like a conventional liquid consisting of similar small molecules. The state of alignment however turns out to be rather complicated. In the first place, the flow depends on the angles the director makes with the flow direction and with the velocity gradient. Secondly, the translational motions are coupled to inner, orientational motions of the molecules. Consequently, in most cases the flow disturbs the alignments and causes the director to rotate. From the theoretical point of view the coupling between orientation and flow is a delicate matter.

The hydrodynamics for an isotropic classical fluid is well studied [3, 2]. The approach is to treat the fluid as a continuous medium and any small volume element is always assumed to be so large that it still contains a very great number of molecules. The dynamical situation is specified by by the fluid velocity field $\boldsymbol{v}(\boldsymbol{r},t)$, and by any two thermodynamic quantities pertaining to the fluid, for instance the pressure $p(\boldsymbol{r},t)$ and the density $\rho(\boldsymbol{r},t)$. The condition of incompressibility is always assumed, $\nabla \cdot \boldsymbol{v} = 0$. The equation of motion is then given by the linear Navier-Stokes form:

$$\rho\left[\frac{\partial \boldsymbol{v}}{\partial t} + (\boldsymbol{v}\cdot\nabla)\boldsymbol{v}\right] = -\nabla p + \nabla \boldsymbol{\sigma}_{\text{visc}} \tag{3.1}$$

where the right hand side denotes the total force, which comes from two contributions: the net pressure gradient and the viscous stress term. We have neglected the presence of additional external forces such as the potential term. The classical viscous stress is given by

$$\sigma_{\alpha\beta} = \eta\left(\frac{\partial v_\beta}{\partial x_\alpha} + \frac{\partial v_\alpha}{\partial x_\beta}\right) = 2\eta g^s_{\alpha\beta} \tag{3.2}$$

where η is the viscosity coefficient.

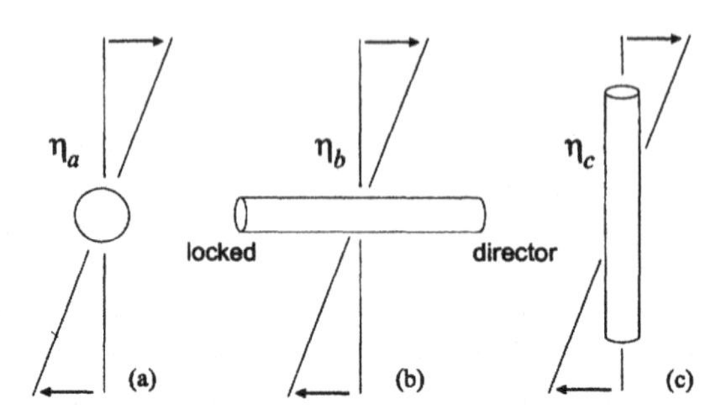

FIG. 2. *The three geometries of simple shear.*

For the case of a simple shear flow in a nematic liquid, the measured viscosity coefficient depends on the orientation of the director $\boldsymbol{n}$. The direction of $\boldsymbol{n}$ can be specified by the angles ϕ and θ. If the orientation of the director is fixed by external forces (for instance by a strong magnetic field), we can define three geometries of simple shear (see Fig. 2) as η_a : $\phi = 90^o$ $\theta = 90^o$ for director normal to shear plane; η_b : $\phi = 0^o$ $\theta = 0^o$ for director parallel to flow direction; η_c : $\phi = 0^o$ $\theta = 90^o$ for director parallel to velocity gradient. The three coefficients η_a, η_b and η_c are often called the Miesowicz coefficients.

3.2. Leslie-Ericksen theory. So far we have been concerned with the motion of a nematic liquid in which the orientation of the director is fixed. If we lift this restriction, we will have to consider an extra degree of freedom associated with the orientation of the director $\boldsymbol{n}(\boldsymbol{r}, t)$, which may introduce local unbalanced torques in the system. The phenomenological linear hydrodynamics of nematics is adequately described in the context of Leslie-Ericksen (LE) theory, by considering the entropy sources, due to all friction processes in the fluid. In short, and keeping the notation close to the de Gennes' monograph [21], the LE approach describes the dissipation due to a decrease in the stored energy,

$$T\dot{S} = \int \{\sigma^s_{\alpha\beta} g^s_{\alpha\beta} + h_\alpha N_\alpha\} d^3\boldsymbol{r} \tag{3.3}$$

where $g^s_{\alpha\beta}$ denotes the symmetric velocity gradient and h_α is the molecular field. Also, the corotational derivative

$$\boldsymbol{N} = \dot{\boldsymbol{n}} - \boldsymbol{\nu} \times \boldsymbol{n} \tag{3.4}$$

represents the rate of change of the director with respect to the flow background, and $\boldsymbol{\nu} = \frac{1}{2}\nabla \times \boldsymbol{v}$ is the flow rotation angular velocity.[2]

In irreversible processes, it is customary to write the entropy source as the product of 'flux' by the conjugate 'force' [25]. Choosing $\sigma^s_{\alpha\beta}$ as the

[2]Another approach, proposed by the Harvard group [23], assumes that the velocity field is sufficient to specify the state, and the orientation of the director is deduced from

force conjugate to $g^s_{\alpha\beta}$ and h_α as the force conjugate to N_α, we can write, in the limit of weak flux, the following linear functions of the fluxes for the forces, which satisfy the symmetry properties of uniaxial nematics:

$$\begin{aligned}\sigma^s_{\alpha\beta} &= \rho_1\delta_{\alpha\beta}g^s_{\mu\mu} + \rho_2 n_\alpha n_\beta + \rho_3\gamma_{\alpha\beta}n_\gamma n_\mu g^s_{\gamma\mu} \\ &\quad + \alpha_1 n_\alpha n_\beta n_\mu n_\rho g^s_{\mu\rho} + \alpha_4 g^s_{\alpha\beta} \\ &\quad + \frac{1}{2}(\alpha_5+\alpha_6)(n_\alpha n_\mu g^s_{\mu\beta} + n_\beta n_m u g^s_{\mu\alpha}) + \frac{1}{2}\gamma_2(n_\alpha N_\beta + n_\beta N_\alpha)\end{aligned} \tag{3.5}$$

$$h_\mu = \gamma_2' n_\alpha g^s_{\alpha\mu} + \gamma_1 N_\mu. \tag{3.6}$$

Note that all the coefficients ρ, α, γ have the dimensionality of viscosity, and the Onsager's symmetry of kinetic coefficients [25] implies that $\gamma_2' = \gamma_2$.

If the liquid is incompressible ($g^s_{\mu\mu} = 0$), we arrive at the Leslie-Ericksen theory where the total viscous stress tensor reads:

$$\begin{aligned}\sigma^{LE}_{\alpha\beta} &= \alpha_1 n_\alpha n_\beta n_\rho n_\mu g^s_{\mu\rho} + \alpha_4 g^s_{\alpha\beta} + \alpha_5 n_\alpha n_\mu g^s_{\mu\beta} + \alpha_6 n_\beta n_\mu g^s_{\mu\alpha} \\ &\quad + \alpha_2 n_\alpha N_\beta + \alpha_3 n_\beta N_\alpha\ ,\end{aligned} \tag{3.7}$$

where the viscosity constants $\alpha_1, ..., \alpha_6$ are called the Leslie coefficients. In the isotropic phase all of them vanish except α_4, which becomes the isotropic shear viscosity coefficient η. They have to fulfill the Onsager reciprocity, which for a nematic is known as the Parodi relation [26], $\alpha_2 + \alpha_3 = \alpha_6 - \alpha_5$. So effectively there are only five independent coefficients. Three of them are connected with the symmetric part of the stress tensor and the other two with the anti-symmetric part

$$\sigma^a_{\alpha\beta} = \frac{\gamma_1}{2}(n_\beta N_\alpha - n_\alpha N_\beta) + \frac{\gamma_2}{2}(n_\beta n_\mu g^s_{\mu\alpha} - n_\alpha n_\mu g^s_{\mu\beta}) \tag{3.8}$$

with

$$\gamma_1 = \alpha_3 - \alpha_2 \quad \text{and} \quad \gamma_2 = \alpha_2 + \alpha_3 \equiv \alpha_6 - \alpha_5. \tag{3.9}$$

The coefficients γ_1 and γ_2 determine the viscous torque acting on the molecule: γ_1 is characteristic of pure director rotations and γ_2 describes the contribution due to a shear flow. The equation of motion of the director reads:

$$\boldsymbol{n} \times (\gamma_1 \boldsymbol{N} + \gamma_2 \boldsymbol{n} \cdot g^s) = 0. \tag{3.10}$$

If we assume undeformed director field, the conservation law for angular momentum can be neglected. If one would like to consider the case

---

the gradients of $\boldsymbol{v}$. In this picture, a rotation of the director can only occur in the presence of a non-uniform flow. There is however experimental evidence to show that this choice of state variable is not sufficient to describe a nematic, while the EL choice is adequate [24].

of a deformed system, the stress tensor and the conservation of angular momentum have to be modified, and Equations (3.8) and (3.10) should be extended to a more general forms containing the additional elastic stress. Elastic stress induced by spatial inhomogeneities will be the subject of interest in Section 6. The status of the LE equation as a constitutive equation for nematics is therefore analogous to that of the Newtonian constitutive equation as a description for ordinary liquid.

3.3. Microscopic stress tensor. In general the transport coefficients can be obtained within the framework of classical kinetic theory [3, 2]. In this context, the macroscopic stress tensor can be defined as an ensemble average of σ_{ij}^m, the corresponding microscopic stress tensor, over the non-equilibrium distribution function $w\{\boldsymbol{x}_i\}$, where $\boldsymbol{x}_i$ are the relevant phase space variables. In fact the microscopic stress tensor describes the evolution of the microscopic momentum density $\mathbf{p}(\boldsymbol{R})$ according to the local conservation law:

$$\frac{d\mathbf{p}(\boldsymbol{R})}{dt} = \nabla \cdot \sigma^m(\boldsymbol{R}). \tag{3.11}$$

The general expression for the microscopic stress tensor can be obtained with the help of the microscopic equations of motion for individual molecules. For a nematic fluid composed of rigid elongated particles, an approximate expression for the microscopic stress tensor had been derived in the literature [9, 10]. Here we give a rigorous derivation for the microscopic stress tensor for a general spheroidal molecule with arbitrary shape.

A molecule can be considered as a rigid body made up of bounded points of mass m_k. Then the total momentum in the system of many such particles is:

$$\mathbf{p}(\boldsymbol{R}) = \sum_i \sum_k m_k \left[\boldsymbol{v}_i + (\boldsymbol{\omega}_i \times \boldsymbol{r}_{ik})\right] \delta(\boldsymbol{R} - \boldsymbol{r}_i - \boldsymbol{r}_{ik}) \tag{3.12}$$

where the index i indicates a molecule and k a point inside the molecule, see Fig. 3. $\boldsymbol{\omega}_i$ is the angular velocity of molecular rotation and $\boldsymbol{v}_i$ the velocity of its center of mass (COM). $\boldsymbol{r}_i$ is the position of the COM in the laboratory frame, while $\boldsymbol{r}_{ik}$ is the position of the point k in the molecular frame so that the velocity of a point k of the i-th molecule in the laboratory frame is $\boldsymbol{v}_{ik} = \boldsymbol{v}_i + \boldsymbol{\omega}_i \times \boldsymbol{r}_{ik}$.

Formally expanding the delta-function in powers of $\boldsymbol{r}_{ik}$, we have

$$\delta(\boldsymbol{R} - \boldsymbol{r}_i - \boldsymbol{r}_{ik}) = \delta(\boldsymbol{R} - \boldsymbol{r}_i) - \boldsymbol{r}_{ik} \cdot \nabla_R \delta(\boldsymbol{R} - \boldsymbol{r}_i) + f(\nabla^2\delta) + \ldots \tag{3.13}$$

Taking the time derivative in Equation (3.12) and substituting Equation (3.13) into (3.11), while working in the linear flow regime where higher order terms $\nabla^2\delta$ can be neglected, we find that the microscopic stress tensor can be separated into the translational (a function of $\boldsymbol{r}_i$ and its derivatives)

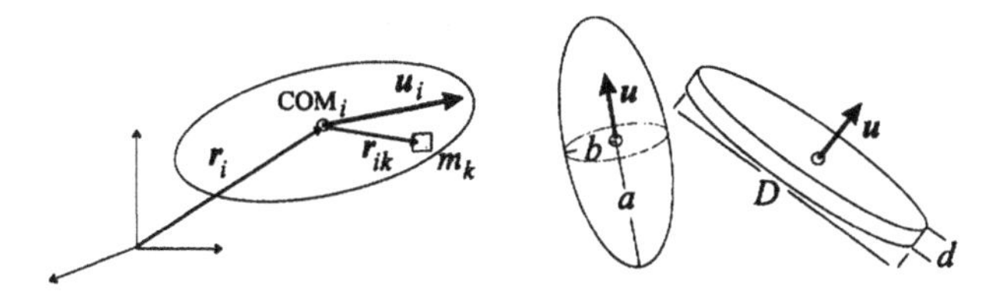

FIG. 3. *The molecule i (arbitrarily represented here as an ellipsoid, without loss of generality) has its center of mass coordinate r_i in the laboratory frame. In the frame of its COM, the position of a given mass element m_k is r_{ik}. The unit vector u_i represents the principal axis of the tensor of inertia moments of this molecule. For a uniaxial body this tensor is equal to $I_{\alpha\beta} = I_\perp \delta_{\alpha\beta} + (I_\parallel - I_\perp)u_\alpha u_\beta$.*

and orientational parts. Comparing these terms with the $\nabla \cdot \sigma^m(\boldsymbol{R})$ on the right hand side of definition (3.11) we obtain the orientational part of the microscopic stress tensor:

$$\begin{aligned}\sigma^{\text{or}}_{\alpha\beta} &= \sum_i \sum_k m_k \left[\boldsymbol{\omega}_i \times (\boldsymbol{\omega}_i \times \boldsymbol{r}_{ik}) + \dot{\boldsymbol{\omega}}_i \times \boldsymbol{r}_{ik}\right]_\alpha (\boldsymbol{r}_{ik})_\beta \, \delta(\boldsymbol{R} - \boldsymbol{r}_i) \\ &+ \sum_i \sum_k m_k (\boldsymbol{\omega}_i \times \boldsymbol{r}_{ik})_\alpha (\boldsymbol{\omega}_i \times \boldsymbol{r}_{ik})_\beta \delta(\boldsymbol{R} - \boldsymbol{r}_i).\end{aligned} \tag{3.14}$$

The translational part of the microscopic stress would determine the isotropic viscosity, arising from non-equilibrium pair correlations in liquid. Its contribution will remain in the nematic phase as well, adding a significant constant to the Leslie coefficient α_4, a fact often overlooked in molecular theories of nematic viscosity.

Expanding the tensors in (3.14) and grouping together the expressions for the inertial tensor of a body rotating about its COM, defined as

$$I_{\alpha\beta} = \sum_k m_k (r^2 \delta_{\alpha\beta} - r_\alpha r_\beta) \ , \tag{3.15}$$

we can rewrite the orientational part of microscopic stress tensor as a sum over all rigid molecules i:

$$\begin{aligned}\sigma^{\text{or}}_{\alpha\beta} &= -\sum_i \left[I_{\alpha\delta}(I^{-1})_{\nu l}\Gamma^l \epsilon_{\beta\nu\delta} + I_{mk}\epsilon_{\alpha lm}\omega_l \epsilon_{\beta jk}\omega_j + I_{\alpha\nu}\omega_\beta\omega_\nu - I_{\alpha\beta}\omega^2\right] \\ &\qquad \times \delta(\boldsymbol{R} - \boldsymbol{r}_i) \\ &+ \sum_i \epsilon_{\beta\nu\alpha}(I^{-1})_{\nu l}\Gamma^l \left(\frac{1}{2} Tr(I_{\alpha\beta})\right) \delta(\boldsymbol{R} - \boldsymbol{r}_i).\end{aligned} \tag{3.16}$$

Here Γ_i is the total moment of the force acting on the i-th molecule, arising from the dynamical relation $\Gamma_i = I_{ij}\dot{\omega}_j$. For a rigid uniaxial molecule, we should define the principal molecular frame in which the inertial tensor is diagonal with components $I_\perp$ and $I_\parallel$ (see Fig. 3):

$$I_{\alpha\beta} = I_\perp \delta_{\alpha\beta} + (I_\parallel - I_\perp) u_\alpha u_\beta. \tag{3.17}$$

The torque acting on the molecule i from all its neighbors is given by the rotational gradient of the pair potential,

$$\Gamma_\alpha(r^i) = -\sum_j \epsilon_{\alpha\beta\gamma} u^i_\beta \frac{\partial U(\boldsymbol{u}_i, \boldsymbol{u}_j, \boldsymbol{r}_{ij})}{\partial u^i_\gamma}, \tag{3.18}$$

where $U(\boldsymbol{u}_i, \boldsymbol{u}_j, \boldsymbol{r}_{ij})$ is the interaction potential for molecules i and j. Since all variables in (3.18) are related to the particle i, summing over the rest of the particles gives, by definition, the molecular field (often called the mean-field potential)

$$U(\boldsymbol{u}_i, \boldsymbol{r}_i) = \sum_j U(\boldsymbol{u}_i, \boldsymbol{u}_j, \boldsymbol{r}_{ij}). \tag{3.19}$$

Section 5.3.1 gives more detail to these concepts. Substituting Equations (3.17) and (3.18) into (3.16), we finally have

$$\begin{aligned} \sigma^{\mathrm{or}}_{\alpha\beta} = \sum_i & \left[\left(1 - \frac{I_\parallel}{2I_\perp}\right) u_\alpha \frac{\partial U}{\partial u_\beta} - \frac{I_\parallel}{2I_\perp} u_\beta \frac{\partial U}{\partial u_\alpha} + \left(\frac{I_\parallel - I_\perp}{I_\perp}\right) u_\alpha u_\beta u_m \frac{\partial U}{\partial u_m}\right] \\ & \times \delta(\boldsymbol{R} - \boldsymbol{r}_i) \\ & - \sum_i (I_\parallel - I_\perp) \left[(\boldsymbol{\omega} \times \boldsymbol{u})_\alpha (\boldsymbol{\omega} \times \boldsymbol{u})_\beta + u_\alpha u_\nu \omega_\nu \omega_\beta - \omega^2 u_\alpha u_\beta\right] \\ & \times \delta(\boldsymbol{R} - \boldsymbol{r}_i). \end{aligned} \tag{3.20}$$

For an ellipsoid, with semi-axes a and width b (see Fig. 3) the moments of inertia along and perpendicular to the director are $I_\parallel = \frac{2}{5}Mb^2$ and $I_\perp = \frac{1}{5}M(a^2 + b^2)$, with M the total mass of the molecule. For a long thin "rod-like" particle $p = a/b \gg 1$ and $I_\parallel \ll I_\perp$; for an oblate ellipsoid with $b \gg a$ they are of the same order of magnitude.[3] Substituting the I-values for an ellipsoid into Equation (3.20) gives the final form:

$$\begin{aligned} \sigma^{\mathrm{or}}_{\alpha\beta} = \sum_i & \left[\frac{p^2}{p^2+1} u_\alpha \frac{\partial U}{\partial u_\beta} - \frac{1}{(p^2+1)} u_\beta \frac{\partial U}{\partial u_\alpha} - \frac{p^2-1}{p^2+1} u_\alpha u_\beta u_m \frac{\partial U}{\partial u_m}\right] \\ & \times \delta(\boldsymbol{R} - \boldsymbol{r}_i) \\ & + \sum_i \frac{p^2-1}{p^2+1} I_\perp \left[(\boldsymbol{\omega} \times \boldsymbol{u})_\alpha (\boldsymbol{\omega} \times \boldsymbol{u})_\beta + u_\alpha u_\nu \omega_\nu \omega_\beta - \omega^2 u_\alpha u_\beta\right] \delta(\boldsymbol{R} - \boldsymbol{r}_i). \end{aligned} \tag{3.21}$$

The macroscopic continuum stress tensor is obtained by statistical averaging of (3.21) which implies the integration over the angles ($\boldsymbol{u}$) and angular velocity ($\boldsymbol{\omega}$) with a proper distribution function. The averaging over the velocity can be easily performed since it is determined by the one-particle

[3] Later in this text we shall be dealing with thin flat disks, with thickness d and diameter $D \gg d$, which have $I_\parallel = \frac{1}{8}MD^2$ and $I_\perp = \frac{1}{4}M(\frac{1}{3}d^2 + \frac{1}{4}D^2)$, also of the same order of magnitude.

local Maxwell distribution function $= \exp[-I_\perp(\omega - \omega_{\text{res}})^2/2kT]$, where ω_{res} is the background angular velocity due to flow. The second term on the right hand side of Equation (3.21) therefore gives, after averaging, the 'kinetic' part of the stress tensor.

The stress tensor in terms of microscopic orientational variables, but not molecular velocities (which have just been averaged out as fast variables), takes the form

$$\begin{aligned}\sigma^{\text{or}}_{\alpha\beta} = \sum_i \Big\{ & 3kT\tilde{p}\left(u_\alpha u_\beta - \frac{1}{3}\delta_{\alpha\beta}\right) + \frac{p^2}{p^2+1} u_\alpha \frac{\partial U}{\partial u_\beta} \\ & - \frac{1}{p^2+1} u_\beta \frac{\partial U}{\partial u_\alpha} - \tilde{p}\, u_\alpha u_\beta u_m \frac{\partial U}{\partial u_m} \Big\} \delta(\boldsymbol{R} - \boldsymbol{r}_i)\end{aligned} \tag{3.22}$$

where $\tilde{p} = (p^2 - 1)(p^2 + 1)$ is often called the form factor of the molecules. Note that the assumption made about ellipsoidal shape of the anisotropic molecule, leading to the particular expressions for $I_\parallel$ and $I_\perp$ and the resulting form of (3.22), was not necessary at all. The theory of microscopic stress at the level of (3.20) or (3.16) is totally general.

The separation of the orientational part of stress tensor into two parts, the kinetic and the potential, has an important physical significance. The kinetic part, proportional to $3kT$, represents the momentum flux due to the translation of individual molecules, while the second, potential, part represents the flux arising from intermolecular forces. Both are referred to a coordinate system moving with the local fluid velocity $\boldsymbol{v}$. In a dilute gas of molecules, the kinetic part gives the dominant contribution [11], while in a dense fluid, the orientational motion is inhibited and the potential part gives the dominant contribution. In the following sections, we will assume that the system has uniform density and the summation over the delta-functions is replaced by a constant number density $\rho(\boldsymbol{R})$.

3.4. Preliminary discussion points. It is obvious that in the limit $p \to \infty$, Equation (3.22) reduces to the familiar results for long rods system obtained previously [9, 10]. For the disk-like molecules the result is of special interest. Since in this case the form factor $\tilde{p}$ is negative, one may expect a change in sign of certain viscosity coefficients. One can speculate that more drastic differences in viscosity coefficients will arise from consideration of more precise mean-field potential.

It is interesting to compare our expressions with classical results of Kuzuu and Doi [9]. In their approach, the elastic stress tensor is obtained by relating changes in free energy to the elastic stress and virtual deformation [11]. They implicitly assumed that such free energy can be defined even in non-equilibrium state since the system behaves as an elastic material for instantaneous deformation. By making this approximation, they obtained the stress tensor:

$$\sigma_{\alpha\beta} = \tilde{p}\left[3\rho kT\left\langle u_\alpha u_\beta - \frac{1}{3}\delta_{\alpha\beta}\right\rangle - \rho\langle u_\alpha(\boldsymbol{u}\times\boldsymbol{\partial}U)_\beta\rangle\right]. \tag{3.23}$$

Our results therefore agree in the kinetic part of the stress tensor, but not in the potential-dependent part. In fact if one uses the free-energy approach, following Kuzuu and Doi, one finds that the expression they had derived contains only the symmetric part of our complete microscopic stress tensor. As a result, they had to introduce arbitrary magnetic field to generate asymmetric torque contribution, which however exists in nematics even in the absence of magnetic field. In this respect our results give a more accurate description since Equation (3.22) can be antisymmetrized, without imposing external conditions to the system.

A cautionary remark has to be made. We have assumed uniform liquid concentration throughout the sample. In reality a phase separation may occur between the isotropic and nematic phases with different concentrations, as often is the case in lyotropic systems. The kinetic equation will describe the internal dynamics in each of the phases, which is thermodynamically stable. It is however not sufficient to describe the hydrodynamics of the nematics near phase separation boundary. This fact must be borne in mind while comparing the theory with experiments in the bi-phase region.

4. Microscopic viscosity coefficients. *In this section we put together results from sections 2 and 3 to derive a set of the microscopic expressions for the Leslie coefficients, which are found to depend explicitly on the molecular form factor, the order parameters and the rotational friction constants. We also investigate the effects of non-linear corrections due to gyroscopic motions in discotic nematics. These effects give no corrections to the Miesowicz viscosities but generate a non-linear rotational viscosity γ' that depends on the aspect ratio p and the longitudinal moment of inertia $I_{\parallel}$.*

The motivation for finding the microscopic expressions for the Leslie's coefficients relies on the concept that the macroscopic continuum stress tensor is a result of averaging its microscopic equivalent σ^m over the appropriate *non-equilibrium* distribution function. The underlying assumption is that the nematic liquid crystal performs rotational Brownian motion in a mean-field potential and whose orientational distribution function satisfies the kinetic equation (Section 2). However, we note that the solution to the kinetic equation is non-trivial, even if one neglects the non-linear terms (though of course it can be done via eigenfunction expansion method when the flow term is neglected). Instead, we demonstrate how, by following the approach used by Kuzuu and others [9, 10], one can separate the macroscopic stress tensor into the symmetric and anti-symmetric parts, the microscopic viscosity coefficients can be obtained in a more elegant fashion.

4.1. Symmetric stress tensor. From Equation (3.7), the symmetric stress tensor of the LE phenomenological theory can be written as:

$$\begin{aligned}\sigma^s_{\alpha\beta} &= \alpha_1 n_\alpha n_\beta n_\rho n_\mu g^s_{\mu\rho} + \alpha_4 g^s_{\alpha\beta} \\ &+ \frac{1}{2}(\alpha_5+\alpha_6)(n_\alpha n_\mu g^s_{\mu\beta} + n_\beta n_\mu g^s_{\mu\alpha}) \\ &+ \frac{1}{2}(\alpha_2+\alpha_3)(n_\alpha N_\beta + n_\beta N_\alpha)\end{aligned} \tag{4.1}$$

where $g^s_{\alpha\beta}$ is the symmetric velocity gradient and $\boldsymbol{N}$ is the rate of angular rotation (3.4). Our aim is to derive a microscopic expression of the Leslie's viscosity coefficients from microscopic variables through a series of coarse-graining. The symmetric stress tensor can be obtained by averaging the microscopic stress tensor in Equation (3.22) over the non-equilibrium distribution function,

$$\sigma^s_{ij} = \rho\left\langle 3kT\tilde{p}\left(u_iu_j - \frac{1}{3}\delta_{ij}\right) + \frac{1}{2}\tilde{p}\left(u_i\partial_jU + u_j\partial_iU - 2u_iu_ju_m\partial_mU\right)\right\rangle \tag{4.2}$$

where ρ is the number density of the nematic liquid crystal. $\langle ... \rangle$ denotes the average over the non-equilibrium single-particle orientation distribution function $\langle ... \rangle = \int w(\boldsymbol{u},t)...d\boldsymbol{u}$. Obviously, averaging with $w_{eq}(\boldsymbol{u})$ alone will return zero.

We next use a trick, in this context often attributed to Doi [9]. We consider the kinetic equation, obtained in Section 2 and neglect higher order non-linear terms,

$$\dot{w} + \alpha\partial_k(\Omega_k w) = \alpha^2\partial_k\left(\partial_k w + \frac{\partial_k U}{kT}w\right). \tag{4.3}$$

Multiplying this equation by a factor $\left(u_iu_j - \frac{1}{3}\delta_{ij}\right)$ and integrating over the director orientation making use of the orientational version of integration by parts: $\int d\boldsymbol{u}A(\boldsymbol{u})\boldsymbol{\partial}B(\boldsymbol{u}) = -\int d\boldsymbol{u}\,[\boldsymbol{\partial}A(\boldsymbol{u})]\,B(\boldsymbol{u})$, we derive the following expressions for the four terms in (4.3):

$$\int \dot{w}\left(u_iu_j - \frac{1}{3}\delta_{ij}\right)d\boldsymbol{u} = \frac{\partial}{\partial t}\left\langle u_iu_j - \frac{1}{3}\delta_{ij}\right\rangle \tag{4.4}$$

$$\begin{aligned}\int \partial_k(\Omega_k w)\left(u_iu_j - \frac{1}{3}\delta_{ij}\right)d\boldsymbol{u} &= -\frac{1}{2}\left[\,g^a_{i\alpha}\langle u_\alpha u_j\rangle - g^a_{\alpha j}\langle u_\alpha u_i\rangle\right] \\ &+ \frac{\tilde{p}}{2}\left[2g^s_{\alpha\beta}\langle u_\alpha u_\beta u_i u_j\rangle\right. \\ &\left. - g^s_{\gamma i}\langle u_\gamma u_j\rangle - g^s_{\gamma j}\langle u_\gamma u_i\rangle\right]\end{aligned} \tag{4.5}$$

$$\int \partial_k(\partial_k w)\left(u_iu_j - \frac{1}{3}\delta_{ij}\right)d\boldsymbol{u} = -6\left\langle u_iu_j - \frac{1}{3}\delta_{ij}\right\rangle \tag{4.6}$$

$$\int \partial_k \frac{(w\partial_k U)}{kT}\left(u_i u_j - \frac{1}{3}\delta_{ij}\right) du = -\frac{1}{kT}\left\langle u_i\partial_j U + u_j \partial_i U - 2u_i u_j u_m \partial_m U\right\rangle . \tag{4.7}$$

Combining these results, Equation (4.3) after averaging gives,

$$\frac{\partial Q_{ij}}{\partial t} = F_{ij} + G_{ij} \tag{4.8}$$

where

$$Q_{ij} = \left\langle u_i u_j - \frac{1}{3}\delta_{ij}\right\rangle \tag{4.9}$$

$$F_{ij} = -6\alpha^2 \left\langle u_i u_j - \frac{1}{3}\delta_{ij}\right\rangle - \frac{\alpha^2}{kT}\left\langle u_i\partial_j U + u_j \partial_i U - 2u_i u_j u_m \partial_m U\right\rangle \tag{4.10}$$

$$G_{ij} = -\frac{1}{2}\alpha\tilde{p}\left[2g^s_{\alpha\beta}\langle u_\alpha u_\beta u_i u_j\rangle - g^s_{\gamma i}\langle u_\gamma u_j\rangle - \langle u_\gamma u_i\rangle g^s_{\gamma j}\right] + \frac{\alpha}{2}\left[g^a_{i\alpha}\langle u_\alpha u_j\rangle - \langle u_\alpha u_i\rangle g^a_{\alpha j}\right]. \tag{4.11}$$

Following this, the symmetric part of the macroscopic stress tensor can be written as:

$$\begin{aligned}
\sigma^s_{ij} &= \rho\left\langle 3kT\tilde{p}\left(u_i u_j - \frac{1}{3}\delta_{ij}\right) + \frac{\tilde{p}}{2}\left(u_i\frac{\partial U}{\partial u_j} + u_j\frac{\partial U}{\partial u_i} - 2u_i u_j u_m \frac{\partial U}{\partial u_m}\right)\right\rangle \\
&\equiv -\rho\frac{kT}{2\alpha^2}\tilde{p}F_{ij} = -\rho\frac{kT}{2\alpha^2}\tilde{p}\left[\frac{\partial Q_{ij}}{\partial t} - G_{ij}\right] \\
&= \rho\frac{kT\tilde{p}^2}{4\alpha}\left[-2g^s_{\alpha\beta}\langle u_\alpha u_\beta u_i u_j\rangle + \langle u_\gamma u_j\rangle g^s_{\gamma i} + \langle u_\gamma u_i\rangle g^s_{\gamma j}\right] \\
&\quad + \rho\frac{kT\tilde{p}}{4\alpha}\left[\langle u_\alpha u_j\rangle g^a_{i\alpha} - \langle u_\alpha u_i\rangle g^a_{\alpha j}\right] - \rho\frac{kT\tilde{p}}{2\alpha^2}\frac{\partial}{\partial t}\langle u_i u_j\rangle .
\end{aligned} \tag{4.12}$$

The various moments of orientational distribution function can be expressed generally in terms of the macroscopic average director field $\boldsymbol{n}$ and the delta-functions which must obey the directors symmetries that $\boldsymbol{n}$ and $-\boldsymbol{n}$ are equivalent. The derivation of the various moments is straightforward and we simply quote the results:

$$\langle u_i u_j \rangle = S_2 n_i n_j + \frac{1}{3}(1 - S_2)\delta_{ij} \tag{4.13}$$

$$\begin{aligned}\langle u_\alpha u_\beta u_i u_j \rangle &= S_4 n_\alpha n_\beta n_i n_j \\ &+ \frac{1}{15}\left(1 - \frac{10}{7}S_2 + \frac{3}{7}S_4\right)(\delta_{\alpha\beta}\delta_{ij} + \delta_{\alpha i}\delta_{\beta j} + \delta_{\alpha j}\delta_{\beta i}) \\ &+ \frac{1}{7}(S_2 - S_4)(n_\alpha n_\beta \delta_{ij} + n_i n_j \delta_{\alpha\beta} + n_i n_\alpha \delta_{j\beta} \\ &+ n_i n_\beta \delta_{\alpha j} + n_j n_\alpha \delta_{\beta i} + n_j n_\beta \delta_{i\alpha})\end{aligned} \tag{4.14}$$

where S_2 and S_4 are the scalar order parameters corresponding to the averaged second and fourth Legendre's polynomials of molecular orientation. The main scalar order parameter can be derived from the order parameter tensor S_{ij}:

$$S_{ij} = \left\langle u_i u_j - \frac{1}{3}\delta_{ij} \right\rangle = S_2 \left(n_i n_j - \frac{1}{3}\delta_{ij} \right). \tag{4.15}$$

Multiplying $n_i n_j$ to Equation (4.15) gives $S_2 = \frac{3}{2}\left\langle (\boldsymbol{n} \cdot \boldsymbol{u})^2 - \frac{1}{3} \right\rangle$. Thus S_2 is a scalar measure of how perfectly the molecules are oriented along $\boldsymbol{n}$. The expression for S_4 is derived in the same way.

Substituting the average moments we eventually obtain the desired expression of σ^s_{ij} in terms of velocity gradient and the directors,

$$\begin{aligned}\sigma^s_{ij} &= \frac{kT\tilde{p}^2}{4\alpha}\rho \left[-2S_4 n_\alpha n_\beta n_i n_j g^s_{\alpha\beta} + \frac{2}{35}(7 - 5S_2 - 2S_4) g^s_{ij} \right. \\ &\left. + \frac{1}{7}(3S_2 + 4S_4)(n_i n_\alpha g^s_{\alpha j} + n_j n_\beta g^s_{\beta j}) - \frac{2}{7}(S_2 - S_4) n_\alpha n_\beta g^s_{\alpha\beta}\delta_{ij} \right] \\ &- \frac{kT\tilde{p}S_2}{4\alpha}\rho \left[n_i(\frac{\dot{n}_j}{\alpha} - g^a_{j\alpha} n_\alpha) + n_j(\frac{\dot{n}_i}{\alpha} - g^a_{i\alpha} n_\alpha) \right].\end{aligned} \tag{4.16}$$

The term $n_\alpha n_\beta g^s_{\alpha\beta}\delta_{ij}$ contributes to the scalar pressure, which therefore does not appear in the LE stress tensor. Comparing with Equation (4.1), we find the corresponding Leslie's viscosity coefficients, after restoring to dimensional forms:

$$\alpha_1 = -\rho\lambda_\perp \tilde{p}^2 S_4 \tag{4.17}$$

$$\alpha_2 + \alpha_3 = -\rho\lambda_\perp \tilde{p} S_2 \tag{4.18}$$

$$\alpha_4 = \frac{\rho\lambda_\perp}{35}\tilde{p}^2 (7 - 5S_2 - 2S_4) \tag{4.19}$$

$$\alpha_5 + \alpha_6 = \frac{\rho\lambda_\perp}{7}\tilde{p}^2 (3S_2 + 4S_4) \tag{4.20}$$

4.2. Anti-symmetric stress tensor. For an isotropic liquid the stress tensor must be a symmetric function due to the demand on the local balance of torques. For anisotropic nematics, we expect the anisotropic part of the stress tensor to be non-vanishing due to the orientational torques of the director. The existence of a viscous stress in the fluid has to be a result of averaging over the non-equilibrium distribution function. We can write the non-equilibrium distribution function as $w = w_0(1 + h[\boldsymbol{u}])$ where h represents the deviation from the equilibrium distribution function w_0 (or w_{eq}) which in turn can be written in a very general form that reflects the symmetries of the terms in LE theory. The macroscopic antisymmetric stress tensor then takes the form:

$$\sigma^a_{\alpha\beta} = \frac{\rho}{2} \int \left(u_\alpha \frac{\partial U}{\partial u_\beta} - u_\beta \frac{\partial U}{\partial u_\alpha} \right) w_0[\boldsymbol{u}]\; h[\boldsymbol{u}]\; d\boldsymbol{u} \tag{4.21}$$

where the antisymmetric microscopic stress tensor follows from taking the antisymmetric part of (3.22). The antisymmetric stress tensor obtained this way has to be equivalent to the one obtained in the phenomenological LE formalism.

In this case there is no straightforward trick to solve $\sigma^a_{\alpha\beta}$, as was previously done for $\sigma^s_{\alpha\beta}$. Instead we have to solve the kinetic equation (4.3) to determine $h[\boldsymbol{u}]$ uniquely. The phenomenological antisymmetric stress tensor is given by Equation (3.8), which suggests that γ_1 is related to the rotation of the director $\dot{\boldsymbol{n}}$ and the flow vorticity $(\nabla \times \boldsymbol{v})$. Therefore we have the freedom to choose the nematic system in zero flow ($\boldsymbol{\Omega} = 0$) such that the solution of the kinetic equation gives h which is flow-independent and can be equated to the γ_1 term. The kinetic equation becomes:

$$\dot{w} = \alpha^2 \partial_\beta \left[\partial_\beta w + \frac{\partial_\beta U}{kT} w \right] = \alpha^2 w_0 \left[\partial^2 h - \frac{\partial_\beta U}{kT} \partial_\beta h \right]. \tag{4.22}$$

Assuming the mean field potential to be of Maier-Saupe form [21],

$$U(\theta) = -JS_2 \left[\frac{3}{2} (\boldsymbol{n} \cdot \boldsymbol{u})^2 - \frac{1}{2} \right], \tag{4.23}$$

and the equilibrium orientational distribution $w_0 \propto \exp[-U/kT]$, Equation (4.22) becomes

$$\partial^2 h - \frac{\partial_\beta U}{kT} \partial_\beta h \simeq \kappa (\boldsymbol{n} \cdot \boldsymbol{u})(\dot{\boldsymbol{n}} \cdot \boldsymbol{u})(1 + h) \tag{4.24}$$

with

$$\kappa = \frac{3 J_o S_2}{\alpha^2 kT}. \tag{4.25}$$

As designed, the only source of deviation from equilibrium here is the time dependence of the uniformly rotating director. We have assumed that the term $\boldsymbol{n} \cdot \dot{\boldsymbol{u}}$ is negligibly small for a nematic system approaching quasi-static state such that all molecules are almost aligned parallel to the averaged director $\boldsymbol{n}$. An equivalent argument is that $\boldsymbol{n} \cdot \dot{\boldsymbol{u}}$ is proportional to the angular velocity which is a fast dynamical variable that had been previously integrated out to yield the kinetic equation (4.3) in terms of $w(\boldsymbol{u}, t)$.

Symmetry of the problem suggests the following expression for the linear non-equilibrium correction h,

$$h = h_o(\boldsymbol{n} \cdot \boldsymbol{u})(\dot{\boldsymbol{n}} \cdot \boldsymbol{u}) \tag{4.26}$$

where h_o is a constant to be determined self-consistently. In this respect we can neglect h on the right hand side of Equation (4.24) since it produces non-linear terms. Substituting (4.26) into Equation (4.24) determines h_o (in its dimensional form):

$$h_o = -\frac{\lambda_\perp}{kT}\frac{q}{2+q} \tag{4.27}$$

where $q = JS_2/kT$ denotes the strength of the nematic order and $\lambda_\perp$ is the rotational friction constant. Substituting this result into Equation (4.21) and manipulating in spherical coordinates, we finally obtain the averaged antisymmetric stress tensor with the explicit coefficient in front

$$\sigma^a_{\alpha\beta} = \frac{1}{70}\frac{q^2}{2+q}\lambda_\perp \rho\,(7 + 5S_2 - 12S_4)\;(n_\alpha \dot{n}_\beta - \dot{n}_\alpha n_\beta). \tag{4.28}$$

Comparing with the continuum theory definition in Equation (3.7) we identify that:

$$\gamma_1 = \alpha_3 - \alpha_2 = \frac{1}{35}\frac{(JS_2/kT)^2}{2 + (JS_2/kT)}\,\lambda_\perp\,\rho(7 + 5S_2 - 12S_4), \tag{4.29}$$

which has the desired property of vanishing when S_2 goes to zero. Making use of Equations (4.17–4.20), we have the following microscopic expressions for the remaining Leslie coefficients:

$$\alpha_2 = -\frac{1}{2}(\rho\lambda_\perp\tilde{p}S_2 + \gamma_1) \tag{4.30}$$

$$\alpha_3 = -\frac{1}{2}(\rho\lambda_\perp\tilde{p}S_2 - \gamma_1) \tag{4.31}$$

$$\alpha_5 = \frac{1}{2}\rho\lambda_\perp\tilde{p}\left[S_2 + \frac{\tilde{p}}{7}(3S_2 + 4S_4)\right] \tag{4.32}$$

$$\alpha_6 = \frac{1}{2}\rho\lambda_\perp\tilde{p}\left[-S_2 + \frac{\tilde{p}}{7}(3S_2 + 4S_4)\right] \tag{4.33}$$

4.3. Theoretical predictions. 1) The microscopic expressions for the viscosity coefficients depend strongly on the order parameters and on the alignment of the director in the fluid. They also depend explicitly on the geometrical shape of the spheroid which manifest itself in the form factor $\tilde{p}$. Since the order parameters are the averaged property, the Leslie's coefficients do not depend explicitly on the exact form of the nematic potential U. Instead, the intermolecular potential determines uniquely the rotational friction constant $\lambda_\perp$ (see Section 5).

2) In the above analysis, we have paid particular attention to the fact that the general symmetric part of the microscopic stress tensor for a spheroid must be enriched with the form factor $\frac{p^2-1}{p^2+1}$, in contrast to previous works which treated only long-rods nematics [9, 10]. This allows us to take p to be asymptotically zero for the case of a discotic nematic, in which case the form factor $\tilde{p}$ goes to -1. This implies a change in the signs of certain viscosity coefficients, like α_5 and α_6. In the limit of small γ_1, both α_2 and α_3 are large and positive for a discotic nematic, but are negative for rod-like nematics. This is in accordance with earlier theoretical predictions [27, 28].

3) The geometrical shape appears to have no effects on the value of α_4. In the LE formalism, this term accounts for the Newtonian-like behavior which is present in isotropic liquid too. It accounts phenomenologically for contributions to momentum transport other than those due to rotational motions. For spherically symmetric molecules, this will be the sole contribution towards the viscosity of the liquid, mostly determined by the translational molecular degrees of freedom, which we haven't considered here at all. In the nematic case, according to Equation (4.20), the orientational part of α_4 vanishes in the limit of strong order when S_2 and S_4 tend to be 1. This corresponds to the fact that as the liquid approaches its full nematic alignment, its isotropic counterpart, independent of the shape of the anisotropic molecules, ceases to exist and so is α_4. This suggests that α_4 consists of two independent contributions: the isotropic α_4^{iso} which denotes contributions from momentum transport, in the style of classical works of Kirkwood and others [1–3], and the additional contribution α_4^{nem}, which we derived above.

4) From Equation (4.29), we see that as the intermolecular coupling strength q increases, the rotational viscosity γ_1 increases significantly, leading to large energy dissipation for uniform director rotation with respect to the matrix. This suggests that γ_1 characterizes director rotation that is associated with overcoming the potential barrier which is dictated by the order parameter S_2. A strong nematic potential therefore makes local rotational motions difficult and this increases the viscous loss.

5) There is an alternative approach towards evaluating the antisymmetric stress tensor, by taking the steady state solution $\dot{w} = 0$ in the kinetic equation, instead of the zero flow condition $\boldsymbol{\Omega} = 0$ as we have done. In

this case we are looking for the flow-dependent terms of $\sigma^a_{\alpha\beta}$ which will eventually give us values of γ_1 and γ_2 [29, 10]. This method bypasses some of the approximations in the calculation above, but gives similar expression of γ_1. We do not include such an alternative derivation here.

4.4. Reactive coefficient and director tumbling. We next discuss how an understanding of γ_1 and γ_2 leads to a description of non-trivial dynamics such as director tumbling. The ratio of the negative of the irrotational torque coefficient γ_2 over the rotational viscosity γ_1 is often known as the reactive coefficient[4] or *tumbling parameter* [21]. It represents the competing effects of strain to vorticity torques acting on the director $\boldsymbol{n}$. Our results from previous section give for this parameter, which is defined as the negative ratio of the two rotational viscosities:

$$-\frac{\gamma_2}{\gamma_1} = \frac{1+\alpha_3/\alpha_2}{1-\alpha_3/\alpha_2} = \tilde{p}\frac{35S_2}{7+5S_2-12S_4}\frac{2+JS_2/kT}{(JS_2/kT)^2}. \tag{4.34}$$

The form factor contributes to a sign inversion for γ_2/γ_1 between discotic and rod-like nematics. For long rods, $\tilde{p}$ goes to 1, $\alpha_3/\alpha_2 < 1$, and $-\gamma_2/\gamma_1 > 1$. For disk-like molecules, $\tilde{p}$ goes to -1, $\alpha_3/\alpha_2 > 1$, and $-\gamma_2/\gamma_1 < -1$. Our results agree well with the analytic solutions obtained via Poisson Bracket formalism of Volovik [28].

In a more quantitative manner, we can consider the time evolution of the director $\boldsymbol{n}$. This can be obtained from the conservation of angular momentum in the LE theory [30, 5]:

$$\frac{\partial n_i}{\partial t} = (\boldsymbol{\nu}\times\boldsymbol{n})_i - \frac{\gamma_2}{\gamma_1}\left[(g^s\cdot\boldsymbol{n})_i - (\boldsymbol{n}\cdot g^s\cdot\boldsymbol{n})\, n_i\right] \tag{4.35}$$

where as before, g^s is the symmetric velocity gradient, $\boldsymbol{\nu}$ is the vorticity defined in Equation (3.4) and the reactive coefficient is a factor in the second term. From (4.35) it is apparent that for $|\gamma_2/\gamma_1| > 1$, the straining motion dominates and the director tends towards a steady-state orientation angle θ relative to the stream lines, when the hydrodynamic torque $\boldsymbol{\Gamma}$ vanishes [27, 4]:

$$\tan\theta = \sqrt{\frac{\gamma_2+\gamma_1}{\gamma_2-\gamma_1}} = \sqrt{\frac{\alpha_3}{\alpha_2}}. \tag{4.36}$$

θ is called the **flow alignment angle**, defined as the angle between the director axis and the flow in the state of balanced nematic and viscous torques. We see that the straining term can be interpreted as the $\rho\lambda_\perp S_2$ term which is dictated by the rotational friction and the order parameter strength, while γ_1 dictates the vorticity effects. Steady-state alignments

[4]Here "reactive" means that the term is reversible, i.e. no sign changes in time-reversal and non-dissipative hence producing no dissipation either going 'forwards' or 'backwards' in time.

occur when shearing rotates the molecules until they are almost parallel or perpendicular to the flow direction and at this orientation they cease to rotate. As we had seen, for rod-like nematic, $\alpha_3/\alpha_2 < 1$, Equation (4.36) states that $\theta < 45^o$. In fact the rods align their axes almost parallel to the flow direction. For the discotic nematic, we have the opposite situation when $\alpha_3/\alpha_2 \gg 1$, and $\theta \approx 90^0$. The disks therefore tend to align with their axes perpendicular to the flow direction, with one of the long axes of the disks being parallel to the flow direction. The other long axis seems to point in the gradient direction, so that the director orients in the vorticity direction. Such behavior has indeed been observed in scattering studies [12], and agrees with earlier predictions [27].

Equation (4.35) also suggests that when $| \gamma_2/\gamma_1 | < 1$, the vorticity term dominates over the straining motion and the director can no longer find a steady-state orientation. This is reflected in $\gamma_1 \gg \rho\lambda_\perp S_2$ and $\alpha_3/\alpha_2 < 0$. As a result no alignment angle can be established. In this situation, the molecules deviate significantly from the average orientation, and even if the director is almost aligned with the flow field, there is a net torque on molecules that are not perfectly aligned with the field. Due to the anisotropic shape and the pair potential, the torque on any one molecule is transmitted to the others and the whole assembly of molecules continues rotating even at the instant the average direction is parallel to the flow. The nematics therefore do not have a preferred alignment angle, and at any orientation angle, the director experiences a viscous torque tending to rotate it. This leads to the tumbling phenomenon. The steady shear-flow properties of tumbling nematics are very different from those of flow-aligning nematics [31, 32], and the effects of tumbling and its arrest are believed to lead to observed transitions of normal stress differences from positive to negative values [30].

In this section we discussed several predictions of the LE theory pertaining to the tumbling of the director. However we note that these results do not immediately apply to real nematic liquids, confined within vessels when strong anchoring at the wall produces gradients in the director field. These gradients or distortions in the director field lead to elastic stresses known as Ericksen stresses. In passing we also note that both the flow-aligning and tumbling nematics are seen to produce large number of disclination lines under high rates of shear [33], and this observation can only be reconciled with the existence of elastic stress in the nematic medium. We will postpone this discussion to Section 6, when a spatially-varying director field in the presence of flow will be considered.

4.5. Non-linear effects. We return to the investigation of the non-linear effects that are present in the kinetic equation. As was briefly mentioned in Section 2, the non-linear effects were manifested in the following additional terms in the kinetic (Smoluchowski) equation (2.26). Their effects can be analyzed by considering their corrections to the symmetric and antisymmetric parts of the stress tensor separately.

4.5.1. Corrections to symmetric stress tensor. We first consider the effects of the gyroscopic term $\frac{1}{2}\alpha^2 A\partial_\beta\left[(\boldsymbol{u}\cdot\nabla\times\boldsymbol{v})(\boldsymbol{\Omega}\times\boldsymbol{u})_\beta w\right]$ on the symmetric stress tensor. Using the same method of averaging as outlined in Section 4.1, we obtain:

$$\begin{aligned}&\int \partial_\beta\left[(\boldsymbol{u}\cdot\nabla\times\boldsymbol{v})(\boldsymbol{\Omega}\times\boldsymbol{u})_\beta w\right]\left(u_iu_j-\frac{1}{3}\delta_{ij}\right)d\boldsymbol{u}\\ &\quad= -\int \partial_\beta\left(u_iu_j-\frac{1}{3}\delta_{ij}\right)(\boldsymbol{u}\cdot\nabla\times\boldsymbol{v})(\boldsymbol{\Omega}\times\boldsymbol{u})_\beta w\,d\boldsymbol{u}.\end{aligned} \tag{4.37}$$

Expanding $\boldsymbol{\Omega} = \frac{1}{2}\tilde{p}(\boldsymbol{u}\times g^s\cdot\boldsymbol{u})+\frac{1}{2}(\boldsymbol{u}\times g^a\cdot\boldsymbol{u})$, the integral (4.37) can be evaluated to give the correction that is added to the symmetric stress tensor in the original equation (4.8):

$$\sigma^s_{ij} = -\frac{kT}{2\alpha^2}\tilde{p}\left[\frac{\partial Q_{ij}}{\partial t}-G_{ij}-M_{ij}\right] \tag{4.38}$$

giving the non-linear addition $\sigma^{\rm NL}_{ij} = -\frac{kT}{2\alpha^2}\tilde{p}M_{ij}$. The next step involves expanding all moments of $\boldsymbol{u}$. The tensor M_{ij}, after manipulations, takes the form

$$\begin{aligned}M_{ij} &= \alpha^2\frac{A}{2}\left[-\tilde{p}\,\epsilon_{\alpha\beta\gamma}g^s_{\gamma l}(\nabla\times\boldsymbol{v})_p\langle u_\alpha u_\beta u_l u_i u_j u_p\rangle\right]\\
&+\alpha^2\frac{A}{2}\Bigg\{\frac{1}{2}\tilde{p}\,S_4\,(\boldsymbol{n}\cdot\nabla\times\boldsymbol{v})\left[n_i(\boldsymbol{n}\times g^s\cdot\boldsymbol{n})_j+n_j(\boldsymbol{n}\times g^s\cdot\boldsymbol{n})_i\right]\\
&+\frac{1}{2}\tilde{p}\,A'(g^a_{im}g^s_{mj}-g^s_{im}g^a_{mj})\\
&+\frac{1}{2}\tilde{p}\,B\Big[(\nabla\times\boldsymbol{v})_i(\boldsymbol{n}\times g^s\cdot\boldsymbol{n})_j+n_i\left(\boldsymbol{n}\times g^s(\nabla\times\boldsymbol{v})\right)_j\\
&+n_i\left((\nabla\times\boldsymbol{v})\times g^s\cdot\boldsymbol{n}\right)_j+(\boldsymbol{n}\cdot\nabla\times\boldsymbol{v})g^s_{mi}n_l\epsilon_{jlm}\\
&+(i\leftrightarrow j\ \text{terms})\Big]\\
&+\frac{1}{2}S_4(\boldsymbol{n}\cdot\nabla\times\boldsymbol{v})\left[n_i(\boldsymbol{n}\times g^a\cdot\boldsymbol{n})_j+n_j(\boldsymbol{n}\times g^a\cdot\boldsymbol{n})_i\right]\\
&+A'\left[2(\nabla\times\boldsymbol{v})_i(\nabla\times\boldsymbol{v})_j+g^a_{im}g^a_{mj}\right]\\
&+\frac{1}{2}B\Big[2(\boldsymbol{n}\cdot\nabla\times\boldsymbol{v})n_i(\nabla\times\boldsymbol{v})_j+(\nabla\times\boldsymbol{v})_i(\boldsymbol{n}\times g^a\cdot\boldsymbol{n})_j\\
&+n_i\left(\boldsymbol{n}\times g^a\cdot(\nabla\times\boldsymbol{v})\right)_j+n_i\left((\nabla\times\boldsymbol{v})\times g^a\cdot\boldsymbol{n}\right)_j\\
&+(\boldsymbol{n}\cdot\nabla\times\boldsymbol{v})g^a_{mi}n_l\epsilon_{jlm}+(i\leftrightarrow j\ \text{terms})\Big]\Bigg\}.\end{aligned} \tag{4.39}$$

Equation (4.39) shows that all terms are indeed second order in velocity gradient, with their coefficients expressed by the appropriate order parameters and obeying certain symmetry patterns. At this stage we are motivated by the fact that experimentally it is not easy to measure all of the Leslie's coefficients. What is often measured is the Miesovicz viscosity defined in the beginning of Section 3:

$$\eta_a = \frac{1}{2}\alpha_4 \quad \eta_b = \frac{1}{2}(\alpha_3 + \alpha_4 + \alpha_6) \quad \eta_c = \frac{1}{2}(-\alpha_2 + \alpha_4 + \alpha_5). \tag{4.40}$$

When each flow configuration is considered, we discover that all terms in Equation (4.39) vanish in one way or another. It seems to suggest that the effects of non-linear corrections only manifest themselves in some non-trivial flow configurations which involve the director and flow couplings. On the other hand some insights can be gained from consideration of the effects of non-linearity on the antisymmetric stress tensor.

4.5.2. Corrections to antisymmetric stress tensor. The non-linear corrections pertaining to the gyroscopic effects manifest itself strongly in the antisymmetric stress tensor, since it relates to the energy or entropy dissipation via rotation of the director axis in shear flow. The gyroscopic term changes the 'shape' of the distribution function which corresponds to energy loss via torques.

In steady state, the kinetic equation (2.26) becomes

$$\begin{aligned} \alpha\partial_\beta(\Omega_\beta w) &= \alpha^2\partial_\beta(\partial_\beta w - \frac{\Gamma_\beta}{kT}w) + \alpha^2\partial_\beta(\Omega_\alpha\partial_\alpha\Omega_\beta w) \\ &+ \frac{A}{2}\alpha^2\partial_\beta\left[(\boldsymbol{u}\cdot\nabla\times\boldsymbol{v})(\boldsymbol{\Omega}\times\boldsymbol{u})_\beta w\right] + \alpha\partial_\beta\left(\Omega_\alpha\partial_\alpha\Omega_\beta w\right) \end{aligned} \tag{4.41}$$

and we write the non-equilibrium distribution function with corrections:

$$w = w_0\left(1 + h^{(1)}[\boldsymbol{u}] + h^{(2)}[\boldsymbol{u}]\right) \tag{4.42}$$

where $h^{(1)}$ is the correction to the equilibrium distribution function w_0 that we had discussed before which is proportional to linear velocity gradient. $h^{(2)}[\boldsymbol{u}]$ is introduced to represent corrections that are second order in velocity gradients, and is relevant in this section since we are primarily interested in seeking non-linear corrections to the antisymmetric stress tensor due to the gyroscopic term $\alpha^2\frac{A}{2}\partial_\beta\left[(\boldsymbol{u}\cdot\nabla\times\boldsymbol{v})\,(\boldsymbol{\Omega}\times\boldsymbol{u})_\beta\, w\right]$. As such we can, for the moment, neglect the term $\alpha\partial_\beta\left(\Omega_\alpha\partial_\alpha\Omega_\beta w\right)$, and consider only the reduced equation:

$$\begin{aligned} &\alpha\left(\partial^2 h^{(2)} - \frac{\partial_\beta U}{kT}\partial_\beta h^{(2)}\right) - h^{(1)}\partial_k\Omega_k + \frac{\partial_k U}{kT}\Omega_k h^{(1)} - \Omega_k\partial_k h^{(1)} \\ &\qquad = \alpha\frac{A}{2}\partial_\beta\left[(\boldsymbol{u}\cdot\nabla\times\boldsymbol{v})\,(\boldsymbol{\Omega}\times\boldsymbol{u})_\beta\, w\right] \end{aligned} \tag{4.43}$$

which we obtain after substituting (4.42) into (4.41). We have deliberately left the gyroscopic term on the right hand side. The various terms can be evaluated explicitly in spherical coordinates, $\boldsymbol{u} = \boldsymbol{n}\cos\theta + \boldsymbol{e}\sin\theta$. The right hand side of Equation (4.43) yields

$$\begin{aligned}
&\frac{A}{2}\partial_\beta \left[(\boldsymbol{u}\cdot\nabla\times\boldsymbol{v})(\boldsymbol{\Omega}\times\boldsymbol{u})_\beta w\right] \\
&\quad = \frac{A}{2}\left\{\frac{1}{2}\tilde{p}\left[(\boldsymbol{u}\cdot\nabla\times\boldsymbol{v})(\boldsymbol{n}\times\boldsymbol{e})\cdot g^s\cdot\boldsymbol{u}\frac{1}{kT}\frac{\partial U}{\partial\theta} - (g^a\cdot\boldsymbol{u})(g^s\cdot\boldsymbol{u})\right]\right. \\
&\qquad + \frac{1}{2}\left[(\boldsymbol{u}\cdot\nabla\times\boldsymbol{v})(\boldsymbol{n}\times\boldsymbol{e})\cdot g^a\cdot\boldsymbol{u}\frac{1}{kT}\frac{\partial U}{\partial\theta} - (g^a\cdot\boldsymbol{u})^2\right] \\
&\qquad \left. - (\boldsymbol{u}\cdot\nabla\times\boldsymbol{v})^2\right\}.
\end{aligned} \tag{4.44}$$

We seek the appropriate general expression of $h^{(2)}$ that corresponds to Equation (4.44). The corrections due to $h^{(1)}$ becomes irrelevant since no matching of the velocity gradient terms can be found. An appropriate expression for $h^{(2)}$ will be $h^{(2)} = h_1 + h_2 + h_3$ where

$$\begin{aligned}
h_1 &= h_{p1} g^s_{\alpha\beta} g^a_{ij}\epsilon_{\gamma ij}(\boldsymbol{n}\times\boldsymbol{e})_\alpha n_\gamma n_\beta \\
&\quad + h_{p2} g^s_{\alpha\beta} g^a_{ij}\epsilon_{\gamma ij}(\boldsymbol{n}\times\boldsymbol{e})_\alpha n_\gamma e_\beta \\
&\quad + h_{p3} g^s_{\alpha\beta} g^a_{ij}\epsilon_{\gamma ij}(\boldsymbol{n}\times\boldsymbol{e})_\alpha e_\gamma e_\beta \\
&\quad + h_{p4} g^s_{\alpha\beta} g^a_{ij}\epsilon_{\gamma ij}(\boldsymbol{n}\times\boldsymbol{e})_\alpha e_\gamma n_\beta \\
&\quad + h_{p5} g^s_{m\alpha} g^a_{m\beta} n_\alpha n_\beta + h_{p6} g^s_{m\alpha} g^a_{m\beta} e_\alpha e_\beta \\
&\quad + h_{p7} g^s_{m\alpha} g^a_{m\beta} n_\alpha e_\beta + h_{p8} g^s_{m\alpha} g^a_{m\beta} e_\alpha n_\beta
\end{aligned} \tag{4.45}$$

where the correction coefficients must depend on the two angles $h = h(\theta, \phi)$.

The term h_2 has exactly the same general expansions as h_1 above except for every symmetric velocity gradient $g^s_{\alpha\beta}$ in these equations, we replace it by $g^a_{\alpha\beta}$. Finally we have

$$\begin{aligned}
h_3 &= h_a\epsilon_{\alpha jk}\epsilon_{\beta mn} g^a_{jk} g^a_{mn} n_\alpha n_\beta + h_b\epsilon_{\alpha jk}\epsilon_{\beta mn} g^a_{jk} g^a_{mn} e_\alpha e_\beta \\
&\quad + h_c\epsilon_{\alpha jk}\epsilon_{\beta mn} g^a_{jk} g^a_{mn} n_\alpha e_\beta + h_d\epsilon_{\alpha jk}\epsilon_{\beta mn} g^a_{jk} g^a_{mn} e_\alpha n_\beta.
\end{aligned} \tag{4.46}$$

The left hand side of Equation (4.43) can be evaluated explicitly in spherical polar coordinates to give

$$\frac{\partial^2 h^{(2)}}{\partial\theta^2} + \left(\cot\theta - \frac{1}{kT}\frac{\partial U}{\partial\theta}\right)\frac{\partial h^{(2)}}{\partial\theta} + \frac{1}{\sin^2\theta}\frac{\partial^2 h^{(2)}}{\partial\phi^2}. \tag{4.47}$$

Substituting h_1, h_2 and h_3 into Equation (4.47), and after a series of tedious algebra, we arrive at a form where explicit comparison on both sides can be made. For coefficients corresponding to h_1, we obtain the following relations

$$\frac{\partial^2 h_{p1}}{\partial\theta^2} + \left(\cot\theta - \frac{1}{kT}\frac{\partial U}{\partial\theta}\right)\frac{\partial h_{p1}}{\partial\theta} - \frac{h_{p1}}{\sin^2\theta} = \frac{A}{8}\tilde{p}\cos^2\theta\frac{1}{kT}\frac{\partial U}{\partial\theta} \tag{4.48}$$

$$\frac{\partial^2 h_{p2}}{\partial\theta^2} + \left(\cot\theta - \frac{1}{kT}\frac{\partial U}{\partial\theta}\right)\frac{\partial h_{p2}}{\partial\theta} - \frac{2h_{p2}}{\sin^2\theta} = \frac{A}{8}\tilde{p}\cos\theta\sin\theta\frac{1}{kT}\frac{\partial U}{\partial\theta} \tag{4.49}$$

$$\frac{\partial^2 h_{p3}}{\partial\theta^2} + \left(\cot\theta - \frac{1}{kT}\frac{\partial U}{\partial\theta}\right)\frac{\partial h_{p3}}{\partial\theta} - \frac{3h_{p3}}{\sin^2\theta} = \frac{A}{8}\tilde{p}\sin^2\theta\frac{1}{kT}\frac{\partial U}{\partial\theta} \tag{4.50}$$

$$\frac{\partial^2 h_{p5}}{\partial\theta^2} + \left(\cot\theta - \frac{1}{kT}\frac{\partial U}{\partial\theta}\right)\frac{\partial h_{p5}}{\partial\theta} - \frac{2h_{p6}}{\sin^2\theta} = -\frac{A}{4}\tilde{p}\cos^2\theta \tag{4.51}$$

$$\frac{\partial^2 h_{p6}}{\partial\theta^2} + \left(\cot\theta - \frac{1}{kT}\frac{\partial U}{\partial\theta}\right)\frac{\partial h_{p6}}{\partial\theta} - \frac{4h_{p6}}{\sin^2\theta} = -\frac{A}{4}\tilde{p}\sin^2\theta \tag{4.52}$$

$$\frac{\partial^2 h_{p7}}{\partial\theta^2} + \left(\cot\theta - \frac{1}{kT}\frac{\partial U}{\partial\theta}\right)\frac{\partial h_{p7}}{\partial\theta} - \frac{h_{p7}}{\sin^2\theta} = -\frac{A}{4}\tilde{p}\cos\theta\sin\theta. \tag{4.53}$$

For coefficients corresponding to h_2, we arrive at the same equations as h_1 but without the form factor $\tilde{p}$. For coefficients corresponding to h_3,

$$\frac{\partial^2 h_a}{\partial\theta^2} + \left(\cot\theta - \frac{1}{kT}\frac{\partial U}{\partial\theta}\right)\frac{\partial h_a}{\partial\theta} = -\frac{A}{8}\cos^2\theta \tag{4.54}$$

$$\frac{\partial^2 h_b}{\partial\theta^2} + \left(\cot\theta - \frac{1}{kT}\frac{\partial U}{\partial\theta}\right)\frac{\partial h_b}{\partial\theta} - 2\frac{h_b}{\sin^2\theta} = -\frac{A}{8}\sin^2\theta \tag{4.55}$$

$$\frac{\partial^2 h_c}{\partial\theta^2} + \left(\cot\theta - \frac{1}{kT}\frac{\partial U}{\partial\theta}\right)\frac{\partial h_c}{\partial\theta} - \frac{h_c}{\sin^2\theta} = -\frac{A}{8}\sin\theta\cos\theta \tag{4.56}$$

and $h_d = h_c$, $h_{p4} = h_{p2}$, $h_{p8} = h_{p7}$.

We wish to obtain some qualitative features of the modified stress tensor due to the non-linear gyroscopic term, therefore we make the following approximations:

(1.) We use the one-constant approximation, that is, we assume that all the h for a given combination of velocity gradients are equal, $h_{p1} = h_{p2} = h_{p3} = h_{p4}$, $h_{p5} = h_{p6} = h_{p7} = h_{p8}$, and $h_a = h_b = h_c = h_d = h_3'$, so that

$$\begin{aligned} h^{(2)} &= h_{p1} g^s_{\alpha\beta} g^a_{ij} \epsilon_{\gamma ij} (\boldsymbol{n}\times\boldsymbol{e})_\alpha \left[n_\gamma n_\beta + n_\gamma e_\beta + e_\gamma e_\beta + e_\gamma n_\beta\right] \\ &+ h_{p5} g^s_{m\alpha} g^a_{m\beta} \left[n_\alpha n_\beta + e_\alpha e_\beta + n_\alpha e_\beta + e_\alpha n_\beta\right] \\ &+ h_{a1} g^a_{\alpha\beta} g^a_{ij} \epsilon_{\gamma ij} (\boldsymbol{n}\times\boldsymbol{e})_\alpha \left[n_\gamma n_\beta + n_\gamma e_\beta + e_\gamma e_\beta + e_\gamma n_\beta\right] \\ &+ h_{a5} g^a_{m\alpha} g^a_{m\beta} \left[n_\alpha n_\beta + e_\alpha e_\beta + n_\alpha e_\beta + e_\alpha n_\beta\right] \\ &+ h_3' \epsilon_{\alpha jk} \epsilon_{\beta mn} g^a_{jk} g^a_{mn} \left[n_\alpha n_\beta + e_\alpha e_\beta + n_\alpha e_\beta + e_\alpha n_\beta\right] \end{aligned} \tag{4.57}$$

where h_{a1} and h_{a5} are terms due to h_2.

(2.) We assume that in all cases, the term $h_p/\sin^2\theta$ will be negligible if the relaxation time of the molecular rotation about the director $\boldsymbol{n}$ is much smaller than the time of reorientation with respect to the angle θ. In this way the microscopic antisymmetric stress tensor due to gyroscopic effects can be calculated using the formula

$$\sigma^a_{\alpha\beta} = \int w\sigma^{micro}_{\alpha\beta} du = \int w_0 h\sigma^{micro}_{\alpha\beta} du. \tag{4.58}$$

Substituting Equation (4.57) and carry out the calculation explicitly, we eventually arrive at (after restoring dimensional variables):

$$\begin{aligned}\sigma^a_{\alpha\beta} = \int \Big(& -\rho\lambda_\perp (\boldsymbol{n}\cdot\nabla\times\boldsymbol{v}) \left[n_\alpha (\boldsymbol{n}\times g^s\cdot\boldsymbol{n})_\beta - n_\beta (\boldsymbol{n}\times g^s\cdot\boldsymbol{n})_\alpha \right] \cdot h_{p1} \\ & - \frac{\rho\lambda_\perp}{2} \Big\{ n_\alpha \left[g^a\cdot g^s\cdot\boldsymbol{n}\right]_\beta + n_\alpha \left[g^s\cdot g^s\cdot\boldsymbol{n}\right]_\beta \\ & \quad + n_\beta \left[g^a\cdot g^s\cdot\boldsymbol{n}\right]_\alpha + n_\beta \left[g^s\cdot g^s\cdot\boldsymbol{n}\right]_\alpha \Big\} \cdot h_{p5} \\ & + 4\rho\lambda_\perp (\boldsymbol{n}\cdot\nabla\times\boldsymbol{v}) \left[n_\alpha (\nabla\times\boldsymbol{v})_\beta - n_\beta (\nabla\times\boldsymbol{v})_\alpha \right] \cdot h'_3 \Big) \frac{\partial U}{\partial\theta} w_0 \sin\theta d\theta.\end{aligned} \tag{4.59}$$

The coefficients h_{p1}, h_{p5} and h'_3 are derived to take the values:

$$h_{p1} \sim \frac{A}{8}\,\tilde{p}\,\frac{\pi}{q}\,e^q\ ; \quad h_{p5} \sim -\frac{A}{4}\tilde{p}q^{-3/2}e^q\ ; \quad h_2 \sim -\frac{A}{8}q^{-3/2}e^q. \tag{4.60}$$

It can be seen that all terms without the potential derivatives on the right hand side, i.e., Equations (4.51–4.56), retain the same functional dependence of q. The exponential dependence on q, however, exists for all coefficients of h. The non-linear rotational viscosity γ'_{p1} due to h_{p1} is found to be:

$$\gamma'_{p1} = \int h_{p1} \frac{\partial U}{\partial\theta} w_0 \sin\theta d\theta \simeq \frac{1}{16}\rho\lambda_\perp e^{2q}\, q^{-\frac{1}{2}} A\tilde{p} \tag{4.61}$$

which only appears in discotic nematics with non-vanishing $A = \sqrt{I_\parallel/I_\perp}$.

4.6. Preliminary discussion points. Despite the crude assumptions made in the previous section, we managed to obtain some qualitative features of the solutions for kinetic coefficients and constitutive relations.

Equation (4.59) suggests that at a higher flow rate, there is strong coupling of the director with the vorticity. Such effects becomes irrelevant for a rod-like nematic when A vanishes. On the other hand due to the non-vanishing $I_\parallel$ and hence the additional gyroscopic effects in disk-like molecules, there is a non-linear correction to the antisymmetric stress tensor and the rotational viscosity γ_1.

The more general approach to find the complete solution to the non-linear viscosity is to write down the full general expression for h which satisfies all symmetries of the problem, and explicit matchings of the coefficients can be made and determined. This process is however very laborious which does not necessarily yield new physical insights to the solution. Instead we had adopted a more pragmatic approach by focusing on only a subgroup of the complete expression (see Equation (4.57)) using the one-constant approximation.

The conventional intuitive picture to explain the non-linear effects in viscosity is to visualize flow alignment of the molecules along the flow. This microscopic rearrangement of the molecules that results in a decrease of viscosity at higher velocity gradient is commonly known as *shear-thinning mechanism.* Most suspensions of non-spherical particles which are dilute enough tend to be shear-thinning at modest rates of shear. No doubt flow alignment is always partly responsible but there is an additional factor due to rotation of suspended particles by planar shear to adopt a layered arrangement which favors easy shear.[5] The removal of misaligned domains therefore results in a drop in the viscosity. Strain rate then speeds up and eventually a steady state is achieved at an alignment angle (monodomain). However above a critical shear rate there is no solution for the steady state angle of alignment and this results in instabilities such as the tumbling phenomena discussed before.

As was briefly mentioned above, we can derive γ_1 in the linear regime (due to $h^{(1)}$ correction) using the method described in this section. This method predicts an exponential dependence on q, where q represents mean-field coupling strength which is proportional to the order parameter S_2. The apparent contradiction with the result of (4.29) can be resolved if one expands the denominator in the limit of small q.

5. Rotational friction constant. *In this section the rotational friction constant $\lambda_\perp$ for a discotic nematic liquid crystal is derived from microscopic interactions. The expression for this constant suggests an Arrhenius exponential dependence on the isotropic part of the intermolecular potential.*

5.1. Generalized fluctuation-dissipation theorem. So far we have studied Brownian motion as a physical realization of a random process. For our model of a molecule in rotational motions, the nature of the medium entered our consideration only through one parameter, the friction constant. We know, however, that that the medium comprises other molecules that are ultimately subjected to deterministic, not statistical evolution. Therefore we ought to be able to derive the friction constant from basic atomic dynamics. The appropriate formalism requires us to work within the framework of generalized fluctuation-dissipation theorem. To begin with, we consider the generalized Langevin equation, also known as the Mori Equation [17] that relates the friction coefficient to a memory

[5] In some cases, at even higher rates of shear, the layers may break up due to *shear-thickening*, when the particles form a less regular structure such that they occupy a larger volume and the bulk structure becomes stiffer.

function $K(t)$ (in the absence of external forces),

$$\frac{d\boldsymbol{A}}{dt} = \int_o^t K(s)\boldsymbol{A}(t-s)ds + \boldsymbol{F}(t) \tag{5.1}$$

where $\boldsymbol{A}(t)$ is the dynamical variable of the problem, and $\boldsymbol{F}(t)$ is the stochastic source. The memory function $K(t)$ is related to the correlation function of the stochastic force:

$$K(t) = \frac{\langle \boldsymbol{F}(t)\boldsymbol{F}(0)\rangle}{MkT} \tag{5.2}$$

where M is the mass of the Brownian particle.

Equation (5.1) is useful when there is a good separation of time scales for the motions of the components of the system. We note that compared to the Langevin equation, the friction constant λ has become a friction kernel, $K(t)$, which if decays to zero sufficiently rapidly, leads to

$$\int_o^t K(s)\boldsymbol{A}(t-s)ds \approx \boldsymbol{A}(t)\int_0^t K(s)ds \approx \boldsymbol{A}(t)\int_0^\infty K(s)ds. \tag{5.3}$$

Thus the friction term in the Mori equation can be approximated by the friction term in the Langevin equation, provided that the correlation time of the random force is short compared to the time in which $\boldsymbol{A}(t)$ changes appreciably. We thus have a molecular expression for the friction constant λ, which is the time integral of the autocorrelation function of the intermolecular force exerted by the bath particles on the Brownian particle:

$$\lambda = \frac{1}{MkT}\int_0^\infty \langle \boldsymbol{F}(t)\boldsymbol{F}(0)\rangle dt. \tag{5.4}$$

Here the random force is interpreted as the intermolecular force on the Brownian particle exerted by the bath particles when they move in the field of the fixed Brownian particle (notice that there is nothing intrinsically random in the random force). Equation (5.2) states that, if we consider the Langevin's limit, when the correlation time of the stochastic force is so short as to approximate it as a δ-function,

$$\langle \boldsymbol{F}(t)\boldsymbol{F}(0)\rangle = \Xi\delta(t) \tag{5.5}$$

where Ξ is the stochastic strength. Substituting it to the Mori equation, we recover the fluctuation-dissipation theorem $\Xi = \lambda MkT$ (the parameter M can be rescaled to 1 depending on the definition friction constant).

In the context of rotational motion, the friction constant arises as a consequence of the Brownian particle experiencing a field of random external torques. These instantaneous torques arise from fluctuations in the random intermolecular forces surrounding the particle. As a result, the particle executes random rotational motion with arbitrary angular velocity at any time-step. We can write down a similar form for the rotational friction constant $\lambda_\perp$:

$$\lambda_\perp \approx \frac{1}{kT}\int_0^\infty \langle \mathbf{\Gamma}(t)\mathbf{\Gamma}(0)\rangle dt \tag{5.6}$$

where $\mathbf{\Gamma}(t)$ is the stochastic torque at time t.

For a dense system like a nematic liquid, we have to consider two distinct types of averaging processes in Equation (5.6). The first type concerns **ensemble-averaging** which is performed with respect to particle distribution and takes into account short-range correlation effects etc. The second type is the **time-averaging**, where temporal correlations of the stochastic torques are considered. In this section we will consider these two processes separately to arrive at a microscopic expression for $\lambda_\perp$.

The above analysis assumes that the decay rate of the torque correlation function is rapid with respect to the rate of change of the distribution function of the Brownian particle (see Equation (5.3)). It can be shown that the correlation time for the stochastic torques is the effective collision time t_c which is of the same order of magnitude as the relaxation time for the angular velocity. This however is true only in the stretched limit of the Brownian particle being truly small (e.g. molecular liquids). For massive molecules one expects the correlation time to be much longer than the collision time (often neglected in this case)and the correlation time is equated to the *Brownian motion time* t_w. Strictly speaking this only holds true when[6], $t_c \ll t \ll t_w$.

An example is that of a dilute gas as first suggested by Kirkwood [1]. Here two widely different time scales are easily identified as the duration of a collision and the mean free flight time. The collision time may be interpreted as the time in which the motion of a molecule is predictable from a knowledge of its initial momentum and the force on it at the initial instant. For times longer than this collision time, a second collision may occur, completely uncorrelated with the first. The mean free time describes this regime well and it may be interpreted as the decay time of the particle's momentum correlation function.

5.2. Time averaging. As mentioned before, the temporal averaging of the stochastic torques $\langle \mathbf{\Gamma}(t)\mathbf{\Gamma}(0)\rangle$ corresponds to finding the autocorrelation function of the angular velocity [1]. Such correlation function typically follows an exponentially decaying function, where the decay time is often

[6]There is an observation of 'long time-tail' in molecular correlation functions [34, 17] i.e. decay of certain molecular correlation function has an asymptotic slow inverse power law; not the rapid exponential decay that had been assumed. Hence the assumption that K has a short lifetime relative to A may not strictly be true. This can be explained via the existence of slow fluid variables, and the general theoretical framework is known as mode-mode coupling theory. We shall, however not deal with this aspect since it is beyond the scope of this article.

called the Brownian motion time since it is the time above which the motion becomes diffusive, and below which the motion is ballistic. By finding this correlation time for the particle's angular momentum, one effectively finds the correlation time for the stochastic torques. However, for a nematic executing rotational motion, the situation is complicated by the presence of multiple-correlation times due to its non-trivial tensorial formalism and the anisotropy of the molecules. This contrasts with a typical isotropic liquid where the molecular relaxation process can usually be described in terms of a single correlation time τ_c.

Some insights on the Brownian motion time can be drawn from the translational motion of a Brownian particle of mass M, such as in the case of a colloid particle. In this case we consider the dynamical variable velocity $\boldsymbol{v}(t)$, whose correlation function obeys:

$$\langle \boldsymbol{v}(t)\boldsymbol{v}(t')\rangle = \frac{kT}{M}e^{-|t-t'|/\tau} \tag{5.7}$$

where $\tau = M/\lambda_t$ is the Brownian motion time, or the velocity correlation time. λ_t denotes the translational friction constant. Analogously, we can define a similar correlation time τ_ω for the angular velocity $\boldsymbol{\omega}$ for rotational motion. For $t < \tau_\omega$, the rotational motion is 'ballistic' in the sense that the angular velocity is maintained without 'collisions', which come in the form of contacts with random external torques acting on the system. For $t > \tau_w$, the rotational motion becomes diffusive and the particle distribution function eventually reaches the equilibrium Maxwell velocity distribution $e^{-I\omega^2/2kT}$, hence the name rotational Brownian motion time [17]. For a simple spherical molecule, the dynamics can be described with a single Brownian motion time $\tau_r = I/\lambda_r$, where λ_r is now the rotational friction constant and I is the moment of inertia (analogous to the mass in translational motion).

The Brownian motion time for non-spherical molecules are however complicated by both the rotations around the long molecular axis and the large-angle rotations around the shorter molecular axis. It is clear that the larger the moment of inertia the longer the particle maintains its correlation in angular velocity before it enters the diffusive regime [7].

The rotational Brownian motion time is not to be confused with another relaxation time τ_u which is the time it takes to relax slowly to the Boltzmann distribution over the angular coordinates, given by $\exp[-U(\mathbf{n}\cdot\mathbf{u})/kT]$ where $U(\mathbf{n}\cdot\mathbf{u})$ is the mean-field potential experienced by the molecule. The characteristic time-scale τ_u is approximately $\sqrt{I_\perp/kT}$ if we assume rotational motions to be dominated by large-angle rotations about the short molecular axis. This is the regime where our kinetic equa-

[7] The cross-over from Brownian to non-Brownian behavior in a flowing suspension is controlled by the *rotational Peclet number* $Pe = \dot{\gamma}/D_r$ where D_r is the rotary diffusivity of the particles and $\dot{\gamma}$ is the strain rate.

tion is based upon and leads to the mode relaxation times evaluated in Section 2.

It might be tempting to think that the solution for a disk-like molecule in rotational Brownian motion will yield only a trivial modification to the correlation time of a long rod (where the relaxation times for rotations around the long molecular axis is very small and can be neglected). In fact, as we shall see, due to both rotations along and perpendicular to the molecular axis in discotic nematics, non-trivial solutions for the Brownian motion time can be found.

For a discotic nematic phase, due to the significant moment of inertia parallel to the director axis $I_{\|}$, the axial angular momentum $I_{\|}\dot{\phi}$ along the molecular axis may be comparable to or larger than that perpendicular to the axis. From our analysis of the kinetic equation (2.11) before, we have the following equation:

$$I_{\perp}\dot{\omega}_{\alpha} = -\lambda_{\perp}\omega_{\alpha} + \zeta_{\alpha} - I_{\|}\dot{\psi}(\boldsymbol{\omega} \times \boldsymbol{u})_{\alpha} \tag{5.8}$$

where for simplicity we have assumed zero external torque and external flow. $I_{\|}\dot{\psi}$ can be assumed to be almost constant since there is virtually no torque acting on the axis and hence no angular acceleration along the director. Clearly the description for rotational motion is more complicated due to its vectorial form and the precessional term $I_{\|}\dot{\psi}(\boldsymbol{\omega} \times \boldsymbol{u})$, giving non-trivial solutions as $I_{\|}/I_{\perp}$ is non-negligible. Expanding the cross product in tensorial form, Equation (5.8) becomes

$$\dot{\omega}_{\alpha} = -\lambda_{\alpha\beta}\omega_{\beta} + \frac{\zeta_{\alpha}}{I_{\perp}} \tag{5.9}$$

where

$$\lambda_{\alpha\beta} = \frac{\lambda_{\perp}}{I_{\perp}}\delta_{\alpha\beta} + B\dot{\psi}\,\epsilon_{\alpha\beta\gamma}u_{\gamma} \quad \text{with} \quad B = I_{\|}/I_{\perp}. \tag{5.10}$$

The problem is similar to solving small oscillation dynamics using normal mode expansion. The general motion is then a superposition of the various normal modes with the mode frequencies and their amplitudes given by the eigenvalues and eigenvectors of the matrix respectively. Setting the equation in homogeneous form ($\zeta_{\alpha}/I_{\perp} = 0$) and assuming that the director coordinate u_{γ} remains approximately stationary on the fast time scale of ω, we have, writing $K_{\gamma} = B\dot{\psi}u_{\gamma}$,

$$\begin{pmatrix} \dot{\omega}_1 \\ \dot{\omega}_2 \\ \dot{\omega}_3 \end{pmatrix} = \begin{pmatrix} -\lambda_{\perp}/I_{\perp} & -K_3 & K_2 \\ K_3 & -\lambda_{\perp}/I_{\perp} & -K_1 \\ -K_2 & K_1 & -\lambda_{\perp}/I_{\perp} \end{pmatrix} \begin{pmatrix} \omega_1 \\ \omega_2 \\ \omega_3 \end{pmatrix}. \tag{5.11}$$

Here we note that the matrix is non-symmetric, and complex eigenvalues are to be expected. Direct diagonalization of the matrix numerically gives the following eigenvalues and eigenvectors:

$$\lambda_1 = -\frac{\lambda_\perp}{I_\perp}\,; \qquad \lambda_{2,3} = -\frac{\lambda_\perp}{I_\perp} \pm iB\dot{\psi} \tag{5.12}$$

and the eigenvectors are

$$\begin{aligned} \boldsymbol{v}_1 &= \begin{pmatrix} 1 \\ K_2/K_1 \\ K_3/K_1 \end{pmatrix}; \\ \boldsymbol{v}_{2,3} &= \begin{pmatrix} (\pm iB\,\dot{\psi}K_3 - K_1K_2)/\,(K_1^2+K_3^2) \\ 1 \\ (\mp iB\,\dot{\psi}K_1 - K_2K_3)/\,(K_1^2+K_3^2) \end{pmatrix}. \end{aligned} \tag{5.13}$$

It is clear that the eigenvalues become degenerate $\lambda_1 = \lambda_2 = \lambda_3 = -\lambda_\perp/I_\perp$ for the case of rod-like molecules, indicating the presence of a single Brownian motion time corresponding to only rotations of the long molecular axis.

The angular velocity components can be written as following:

$$\begin{aligned} \omega_1 &= e^{-t/\tau}\left[1 - 2\frac{K_1K_2}{K_1^2+K_3^2}\cos(B\,\dot{\psi}t) \;-\; 2\frac{K_3B\,\dot{\psi}}{K_1^2+K_3^2}\sin(B\,\dot{\psi}t)\right] \\ \omega_2 &= e^{-t/\tau}\left[\frac{K_2}{K_1} + 2\cos(B\,\dot{\psi}t)\right] \\ \omega_3 &= e^{-t/\tau}\left[\frac{K_3}{K_1} - 2\frac{K_2K_3}{K_1^2+K_3^2}\cos(B\,\dot{\psi}t) \;+\; 2\frac{K_1B\,\dot{\psi}}{K_1^2+K_3^2}\sin(B\,\dot{\psi}t)\right] \end{aligned} \tag{5.14}$$

where we have defined $\tau = I_\perp/\lambda_\perp$ as the rotational Brownian motion time.

The stochastic force ξ_α introduces inhomogeneity into the equation and the full solution of the inhomogeneous equation can be obtained by integrating over the stochastic term in Equation (5.9). For simplicity we consider just one of the angular velocity components ω_2 and its correlation function $\langle\omega_2(t)\omega_2(0)\rangle$. Careful analysis leads to the following expression for the angular velocity correlation function:

$$\begin{aligned} \langle\omega_2(t)\omega_2(0)\rangle = {} & \frac{kT}{I_\perp}\left(\frac{K_2}{K_1}\right)^2 e^{-t/\tau} + \frac{kT}{I_\perp}e^{-t/\tau}\cos(t/\tau_\psi) \\ & + \frac{kT}{I_\perp}\left(\frac{K_2}{K_1}\right)e^{-t/\tau}\left[\frac{\tau\tau_\psi}{\tau_\psi^2+\tau^2}\right]\left[2\frac{\tau_\psi}{\tau}\cos(t/\tau_\psi) - \sin(t/\tau_\psi)\right] \\ & + \frac{kT}{I_\perp}e^{-t/\tau}\left[\frac{\tau\tau_\psi}{\tau_\psi^2+\tau^2}\right]\left[\frac{\tau_\psi}{\tau}\cos(t/\tau_\psi) - \sin(t/\tau_\psi)\right]. \end{aligned} \tag{5.15}$$

The first term gives the natural decay of the correlation function for the angular velocity which depends on the geometrical projection ratio K_2/K_1.

The imaginary part of the eigenvalues gives rise to the precessional term $\cos(t/\tau_\psi)$ where the precessional period $\tau_\psi = 1/B\dot{\psi}$ is introduced. Explicit comparison between the two time-scales can be made:

$$\frac{\tau}{\tau_\psi} = \frac{kTI_\perp}{\lambda_\perp^2}\frac{I_\parallel}{I_\perp}. \tag{5.16}$$

The first term on the right hand side corresponds to the small parameter α we had defined in Equation (2.16), therefore we conclude that τ is significantly smaller than τ_ψ for disk-like molecules. Rearranging the terms we have for $t < \tau \ll \tau_\psi$,

$$\langle\omega_2(t)\omega_2(0)\rangle = \frac{kT}{I_\perp}e^{-t/\tau}\left[\left(\frac{K_2}{K_1}\right)^2 + 2\frac{K_2}{K_1}\left(1+\frac{K_2}{K_1}\right)\right]. \tag{5.17}$$

Note that except for the prefactor in terms of K's, Equation (5.17) resembles (5.7) in translational motion. We conclude that the system remains heavily damped[8], and the effective correlation or relaxation time is the shorter time-scale τ. Note that the rotational Brownian motion time has a linear dependence on the friction constant $\lambda_\perp$ which is to be contrasted with the mode relaxation time found in Section 2 which has an inverse dependence on $\lambda_\perp$.

Equation (5.6) therefore becomes:

$$\lambda_\perp \approx \frac{1}{kT}\int_0^\infty \langle\Gamma^2(0)\rangle_{\mathrm{ens}} e^{-t/\tau_w} dt \tag{5.18}$$

giving the final form of friction constant after time-averaging:

$$\lambda_\perp \approx \sqrt{\frac{I_\perp}{kT}}\sqrt{\langle\Gamma^2(0)\rangle_{\mathrm{ens}}} \tag{5.19}$$

where $\langle\Gamma^2\rangle_{\mathrm{ens}}$ denotes ensemble averaging of the stochastic torques acting on the molecules.

5.3. Ensemble averaging. As mentioned before, the idea of ensemble averaging is essential when considering macroscopic properties of any dense liquid. The ensemble averaging of the torques in Equation (5.18) describes microscopically the interactions of the molecules with various random potentials exerted by the surrounding molecules. A reasonable expression for $\langle\Gamma^2(0)\rangle_{ens}$ for molecule 1 can be written as:

$$\langle\Gamma^2\rangle_{\mathrm{ens}} \simeq N^2\int \partial_1 U(1,2)\cdot\partial_1 U(1,3)\; W_3(1,2,3)\; d(1)\; d(2)\; d(3) \tag{5.20}$$

[8]This corresponds to the assumption that thermalization occurs on a time scale much shorter with respect to the time for appreciable changes in positional distances. This occurs when the rotational friction constant $\lambda_\perp$ is large and is often called the high friction limit.

where $\partial_{1k} = \epsilon_{kij} u_{1i} \partial / \partial u_{1j}$, $d(1) = d\boldsymbol{r}_1 d\boldsymbol{u}_1$, and $W_3(1,2,3)$ is the three-particle angular distribution function. $\partial_1 U(1,2)$ describes the torque exerted by molecule 2 on molecule 1 due to their pair interaction potential $U(1,2)$, and the same holds for $\partial_1 U(1,3)$.

To evaluate this ensemble we need to have a microscopic model of the molecular pair potential that acts on a particular pair of nematic molecules. We begin this by giving a brief review of the mean-field description of the intermolecular forces, followed by an attempt to build a phenomenological model that goes beyond mean-field regime and consider more realistic effects such as short-range orientational correlations and the excluded volume effects, paying specific attention to the case of a discotic nematic.

5.3.1. Nematic intermolecular forces. The molecular theory of the nematics has been an intense field of research in the past decades, using advanced statistical theories such as the density functional theory. There two main approaches pursued to derive the various thermodynamic quantities that agree with the experimental results:

1) Treating the intermolecular forces as anisotropic and the intermolecular repulsions as isotropic to first approximation, which serves as a positive pressure. The result is temperature- dependent of course.

2) In suspensions of anisotropic particles the nematic order arises purely from short-range anisotropic repulsive forces (exclusion volume effects in the Onsager approach). The high density of the liquid is established by the intermolecular attractions, which are assumed to be isotropic.

In this work, we take into account both anisotropic intermolecular attractive forces depending on the orientation of the interacting molecules, but at the same time consider an anisotropic repulsive potential which arises from the exclusion volume effects. We consider the following expression for the effective uniaxial potential [35]

$$\begin{aligned} V_{\text{eff}}(\boldsymbol{u}_i, \boldsymbol{r}_{ij}, \boldsymbol{u}_j) = V_{\text{iso}} &- J_1(\boldsymbol{r})(\boldsymbol{u}_i \cdot \boldsymbol{u}_j)^2 \\ &- J_2(\boldsymbol{r}) \left[(\boldsymbol{u}_i \cdot \boldsymbol{r}_{ij})^2 + (\boldsymbol{u}_j \cdot \boldsymbol{r}_{ij})^2 \right] \\ &- J_3(\boldsymbol{r}) \left[(\boldsymbol{u}_i \cdot \boldsymbol{u}_j)(\boldsymbol{u}_i \cdot \boldsymbol{r}_{ij})(\boldsymbol{u}_j \cdot \boldsymbol{r}_{ij}) \right] \end{aligned} \tag{5.21}$$

where V_{iso} represents an isotropic dispersion potential independent of $\boldsymbol{u}$'s and the various J's represent the orientation-dependent coupling strengths which can be expressed in terms of the electric dipole and quadrupole matrix elements. $\boldsymbol{u}$ denotes the molecular director while $\boldsymbol{r}_{ij} = \boldsymbol{r}_i - \boldsymbol{r}_j$ denotes the molecular distance from particle i to j. See Fig. 4 below for the geometric illustration. Chandrasekhar *et. al.* [36] had argued that the potential arises mainly from the dispersion forces which have r^{-6} or $r^{-8/3}$ dependence on the intermolecular separation for dipole-dipole and dipole quadrupole interactions respectively. The permanent dipoles are found to play a minor role in providing the stability of the nematic phase (although dipole-dipole forces are much stronger than the van der Waals, in practice, dipolar molecules in liquid always form very strong dimers).

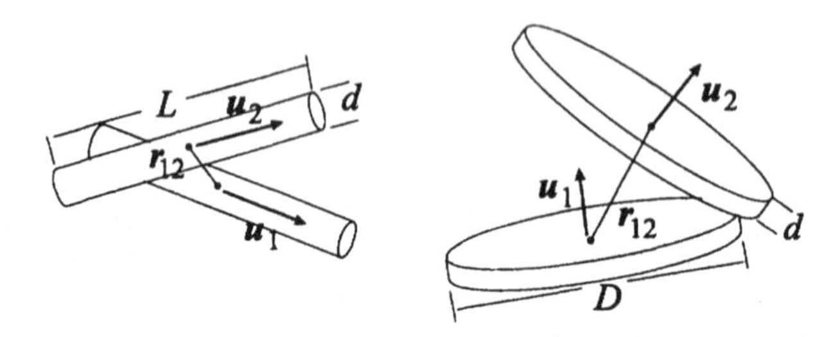

FIG. 4. *The scheme of excluded volume modelling for rod-like and disk-like particles, leading to the orientation dependent expression for the minimal distance separating the two centers of mass, $\xi_{12}(u_1, u_2, u_{12})$ of Equation (5.30). The analogous expression for two rods would read: $\xi_{12} = d + \frac{1}{2}(L-d)\left[(u_1 \cdot u_{12})^2 + (u_2 \cdot u_{12})^2\right]$ and has a minimum ($\xi_{12} = d$) when both u_1 and $u_2 \perp u_{12}$.*

As shown by Gelbart and Gelbart [37], the predominant orientational interaction in nematics must be the isotropic dispersion attraction modulated by the anisotropic molecular hard core. The isotropic part of the dispersion interaction is generally greater than the anisotropic part because it is proportional to the average molecular polarizability. The anisotropy of this overall potential comes mainly from the asymmetric molecular shape. Thus this effective potential is a combination of intermolecular attraction and repulsion,

$$V_{\text{eff}}(1,2) = J_{\text{att}}(r_{12})\Theta(r_{12} - \xi_{12}) \tag{5.22}$$

where the step function $\Theta(r_{12} - \xi_{12})$ determines the steric cut-off.

5.3.2. Mean-field theory: the Maier-Saupe potential. The simplest molecular theory of the nematics can be developed in the context of a mean-field approximation. By mean-field approximation we mean that all correlations between different molecules, such as the fluctuations in the short-range order (mutual alignment of two neighboring molecules), are ignored. This is obviously a crude and unrealistic approximation but it does enable one to obtain very simple and useful expressions for the free energy.

In this section we demonstrate how the mean-field approximation can be established starting with a completely general pairwise intermolecular interaction potential. One appropriate model potential is to write it as an expansion in terms of Legendre polynomials (spherical invariants) P_{lm} [35]:

$$U(u_1, u_{12}, u_2) = \sum_{l,m} J_{lm}(u_{12}) P_{lm}(u_1, u_{12}, u_2). \tag{5.23}$$

To obtain the single molecule potential U in the mean field approximation it is necessary to take successive averages of the intermolecular potential U_{12}. Firstly we note that in the nematic phase there is no positional order and the molecular centres are distributed randomly. If one neglects the positional correlations, the interaction potential can be further simplified by averaging over all values of the intermolecular unit vector $\boldsymbol{u}_{12}$:

$$\tilde{U}(\boldsymbol{u}_1, \boldsymbol{u}_2) = \int U(\boldsymbol{u}_1, \boldsymbol{u}_{12}, \boldsymbol{u}_2) d\boldsymbol{u}_{12}. \tag{5.24}$$

The final mean-field potential $U_{MF}(\boldsymbol{u}_1)$ is obtained by averaging over all orientations of the second molecule $\boldsymbol{u}_2$:

$$U_{MF}(\boldsymbol{u}_1) = \int \tilde{U}(\boldsymbol{u}_1, \boldsymbol{u}_2)\, w_1(\boldsymbol{u}_2) d\boldsymbol{u}_2 \tag{5.25}$$

where $\boldsymbol{u}_i$ specifies the orientation of the molecule i, and

$$w_1(\boldsymbol{u}_i) = \frac{1}{Z} e^{-\beta U_{MF}(\boldsymbol{u}_i)} \tag{5.26}$$

denotes the single-particle distribution function that depends only on the molecular orientation. Equation (5.26) says that in the mean-field approximation, that is, neglecting paiur correlations between $\boldsymbol{u}_1$ and $\boldsymbol{u}_2$, each molecule feels some average angular potential produced by all other molecules in the system. The usual Maier-Saupe potential $JP_2(\boldsymbol{n} \cdot \boldsymbol{u}_i)$ is obtained via this averaging process with respect to the first non-polar term in the Legendre polynomial expansion of the intermolecular potential.

5.3.3. A model potential for discotic nematics. Realistic intermolecular interaction potentials for mesogenic molecules can be very complex and are generally unknown. At the same time molecular theories based on simple model potentials usually offer good qualitative solutions when describing some general properties of liquid crystals that are not sensitive to the details on the interaction. In this section we propose a simple nematic potential to model molecular interactions in a discotic nematic liquid crystal within the mean-field approximation. This leads to an explicit expression of the torque autocorrelation function in Equation (5.20).

1. **A model pair potential**: Previous investigations on the intermolecular interaction potential of a discotic nematic had focused mainly on the regime close to nematic-isotropic (N-I) transition [38]. A reasonable assumption is that the nematic order arises primarily from the short-range and highly anisotropic repulsive forces between the molecules. We consider a modification of the nematic potential in Equation (5.21) which captures the essential physics of the molecular interactions in a discotic nematic phase [35, 37]:

$$\begin{aligned} U(1,2) \simeq & -\frac{G}{r_{12}^6} - \frac{1}{r_{12}^6} J_1 (\boldsymbol{u}_1 \cdot \boldsymbol{u}_2)^2 \\ & - \frac{1}{r_{12}^6} \Big\{ J_2 \left[(\boldsymbol{u}_1 \cdot \boldsymbol{u}_{12})^2 + (\boldsymbol{u}_2 \cdot \boldsymbol{u}_{12})^2 \right] \\ & + J_3 \left[(\boldsymbol{u}_1 \cdot \boldsymbol{u}_2)(\boldsymbol{u}_1 \cdot \boldsymbol{u}_{12})(\boldsymbol{u}_2 \cdot \boldsymbol{r}_{12}) \right] \Big\} \end{aligned} \tag{5.27}$$

where G describes the isotropic attraction and the constants J_1, J_2 and J_3 describe the anisotropic contribution to the pair interaction potential and depend on the anisotropy of the molecular shape. Following previous discussion, we know that a more precise description of the intermolecular interaction has to include higher order Legendre's polynomials such as the P_4 terms [36], but such terms are usually sufficiently small to be ignored in our model. J_2 accounts for the different interaction energies corresponding to different orientation configurations of the two molecules. For instance for long rods, J_2 has to be negative since the orientation configuration is not energetically favorable, and likewise positive for disk-like molecules. The most important weakness of the model (5.27) is its uniform r_{12}^{-6} dependence on molecular separation. We shall see that the specific form of this power law is truly irrelevant, since the dominant contribution to the final integrals is arising from the potential cutoff at the molecular excluded volume cutoff. However, particular dependence on the molecular thickness (the closest approach distance) may not be captured accurately in such a model.

2. **Excluded volume effects**: These effects are determined by hard-core repulsion that does not allow molecules to penetrate each other. It is interesting to note that by doing so we already go beyond the formal mean-field approximation, since at low densities it is possible to express the free energy of the system in the form of the virial expansion [39]:

$$\begin{aligned} \beta F &= \rho \ln \rho + \rho \int w_1(\boldsymbol{u}_1)\,[\,\ln w_1(\boldsymbol{u}_1) - 1\,]\, d\boldsymbol{u}_1 \\ &+ \frac{1}{2}\rho^2 \int w_1(\boldsymbol{u}_1) w_1(\boldsymbol{u}_2) \cdot B(\boldsymbol{u}_1, \boldsymbol{u}_2) d\boldsymbol{u}_1 d\boldsymbol{u}_2 + \ldots \end{aligned} \tag{5.28}$$

where $B(\boldsymbol{u}_1, \boldsymbol{u}_2)$ is the excluded volume for the two disks:

$$B(\boldsymbol{u}_1, \boldsymbol{u}_2) = \int d\boldsymbol{r}_{12} (e^{-\beta U_{\text{steric}}(1,2)} - 1) \tag{5.29}$$

and $U_{\text{steric}}(1,2)$ is the steric repulsion potential.

In Equation (5.28) all terms are purely entropic in origin since the system is athermal by definition. The second term is the additional orientational entropy which is a consequence of the anisotropic shape of the rigid bodies, and are thus absent in an isotropic liquid. The third term is the packing entropy that can be thought of intuitively as a result of the excluded volumes effects that restrict the molecular motion and therefore reducing the total entropy of the liquid. At low volume fraction of the disks, the higher order terms in the expansion can be neglected. The steric repulsion potential is equivalent to introducing a steric cut-off length. For two disk-like molecules this can be expressed phenomenologically as:

$$\xi_{12} = D + \frac{(d-D)}{2}\,[\,(\boldsymbol{u}_1 \cdot \boldsymbol{u}_{12})^2 + (\boldsymbol{u}_2 \cdot \boldsymbol{u}_{12})^2\,] \tag{5.30}$$

where d and D are the thickness and diameter of the disks respectively, and $\boldsymbol{u}_{12}$ is the unit vector along molecular separation line. The expression can be checked by considering the extreme limits of the molecular directors being parallel or perpendicular to the intermolecular unit vector $\boldsymbol{u}_{12}$. For instance, the shortest separation, $\xi_{12} = d$ is achieved when $\boldsymbol{u}_1 \| \boldsymbol{u}_2 \| \boldsymbol{u}_{12}$.

3. **Three-particle correlation functions**: The simplest form of the three-particle angular distribution function in Equation (5.20) can be expressed in the Kirkwood approximation [1], which neglects three-body collisions:

$$W_3(1,2,3) \simeq W_2(1,2)\, W_2(2,3)\, W_2(1,3). \tag{5.31}$$

Such approximation are known to work well at short and long ranges, but is less accurate at medium range of separation. For a long rod system, its has been shown by Onsager [39] that virial coefficients higher than second order vanish in the asymptotic limit as the length of the rod goes to infinity. This means that higher order correlations such as the three-body correlations are negligible, and the Kirkwood's approximation is a good approximation for infinitely long rods system. Such may not be the case for a discotic phase, when three-body correlations has to be taken into account [38]. One might envisage the use of a better approximation scheme using integral equations such as the Percus-Yevick or Hyper-Netted Chain approximations.

For simplicity we neglect the three-particle collisions, which gives:

$$W_3(1,2,3) \simeq e^{-\beta U(1,2)} e^{-\beta U(1,3)} e^{-\beta U(2,3)} w(\boldsymbol{u}_1) w(\boldsymbol{u}_2) w(\boldsymbol{u}_3) \tag{5.32}$$

where $w(\boldsymbol{u}_i)$ is the single-particle equilibrium orientation distribution function for molecule i, and $\beta = 1/kT$ as usual.

We return to the evaluation of the integral in Equation (5.20). We first change the variables from $d\boldsymbol{r}_1$, $d\boldsymbol{r}_2$ and $d\boldsymbol{r}_3$ to $d\boldsymbol{r}_{12}$, $d\boldsymbol{r}_{13}$ and $d\boldsymbol{r}_1$. The integrand can be expressed in terms of $\boldsymbol{r}_{12}$ and $\boldsymbol{r}_{13}$ only, which can be integrated over $\boldsymbol{r}_{12}$ and $\boldsymbol{r}_{13}$ (only the relevant terms are shown):

$$\int \frac{1}{r_{12}^4 r_{13}^4} \exp\left\{ \frac{k(1,2)}{r_{12}^6} + \frac{k(1,3)}{r_{13}^6} + \frac{k(2,3)}{|\boldsymbol{r}_{13} - \boldsymbol{r}_{12}|^6} \right\} d\boldsymbol{r}_{12} d\boldsymbol{r}_{13} \tag{5.33}$$

where

$$\begin{aligned} k(1,2) = \beta\{G + J_1(\boldsymbol{u}_1 \cdot \boldsymbol{u}_2)^2 + J_2\left[(\boldsymbol{u}_1 \cdot \boldsymbol{u}_{12})^2 + (\boldsymbol{u}_2 \cdot \boldsymbol{u}_{12})^2\right] \\ + J_3\left[(\boldsymbol{u}_1 \cdot \boldsymbol{u}_2)(\boldsymbol{u}_1 \cdot \boldsymbol{u}_{12})(\boldsymbol{u}_2 \cdot \boldsymbol{r}_{12})\right]\}\ , \quad \text{etc.} \end{aligned} \tag{5.34}$$

The integrand clearly approaches a maximum towards the cutoff length $r_{12} = \xi_{12} = d$, $r_{13} = \xi_{13} = d$. From Equation (5.30) this requires all the molecular axes to be parallel to intermolecular vectors $\boldsymbol{u}_{12}$, $\boldsymbol{u}_{13}$ and $\boldsymbol{u}_{23}$. If we take small deviations from this conformation only, and $\boldsymbol{u}_{12}$ being in the middle between $\boldsymbol{u}_1$ and $\boldsymbol{u}_2$ etc., then Equation (5.34) simplifies to terms containing only two constants, $\tilde{G} = G + \frac{1}{2}J_2 + \frac{1}{4}J_3$ and $\tilde{J} = J_1 + \frac{1}{2}J_2 + \frac{1}{4}J_3$. Equation (5.20) can be evaluated approximately by observing the sharp rise of the integrand at the end of the integration interval. We therefore obtain:

$$(5.35) \qquad \langle \Gamma^2(0) \rangle \approx \frac{(kT)^2}{6d^3} \tilde{J}^2 e^{3\beta\tilde{G}} \int d\boldsymbol{u}_1 d\boldsymbol{u}_2 d\boldsymbol{u}_3$$

$$\cdot \frac{\partial_1(\boldsymbol{u}_1\cdot\boldsymbol{u}_2)^2 \partial_1(\boldsymbol{u}_1\cdot\boldsymbol{u}_3)^2 \, e^{\beta\tilde{J}[(\boldsymbol{u}_1\cdot\boldsymbol{u}_2)^2+(\boldsymbol{u}_1\cdot\boldsymbol{u}_3)^2+(\boldsymbol{u}_2\cdot\boldsymbol{u}_3)^2]} w(\boldsymbol{u}_1)w(\boldsymbol{u}_2)w(\boldsymbol{u}_3)}{\left[2\tilde{G}' + \tilde{J}'(\boldsymbol{u}_1\cdot\boldsymbol{u}_2)^2 + \tilde{J}'(\boldsymbol{u}_2\cdot\boldsymbol{u}_3)^2\right]\left[3\tilde{G}' + 2\tilde{J}'(\boldsymbol{u}_1\cdot\boldsymbol{u}_3)^2 + \tilde{J}'(\boldsymbol{u}_2\cdot\boldsymbol{u}_3)^2\right]}$$

where $\tilde{G}' = \tilde{G}/d^6$ and $\tilde{J}' = \tilde{J}/d^6$ have the dimensionality of energy.

The equilibrium single-particle orientation distribution function $w(1)$ is proportional to the mean-field nematic potential $U(\boldsymbol{n}\cdot\boldsymbol{u}_1)$, where

$$(5.36) \qquad U(\boldsymbol{u}_1\cdot\boldsymbol{n}) = \int U(\boldsymbol{u}_1, \boldsymbol{u}_2, \boldsymbol{r}_{12}) w(\boldsymbol{u}_2\cdot\boldsymbol{n}) d\boldsymbol{r}_{12} d\boldsymbol{u}_2.$$

That is, the mean-field potential experienced by the first molecule is just the pair interaction energy averaged over the position and orientation of the second molecule. For the discotic nematic phase with interaction energy defined in Equation (5.27) we obtain the mean-field potential with the Maier-Saupe form:

$$(5.37) \qquad U(\boldsymbol{u}_1\cdot\boldsymbol{n}) \approx \text{const.} - \frac{4\pi}{9d^3}(2J_1 + J_3) S_2 (\boldsymbol{u}_1\cdot\boldsymbol{n})^2.$$

Equation (5.35) can be evaluated using again the saddle-point approximation. The integrand possesses a clear maximum point when all molecular axes $\boldsymbol{u}_1$, $\boldsymbol{u}_2$ and $\boldsymbol{u}_3$ are: 1) parallel to each other and 2) aligned parallel to the average macroscopic director $\mathbf{n}$. Another simplification derives from the fact that the anisotropic contribution to the pair potential is usually much smaller than the isotropic contribution $I_1 \ll G$ [35]. With these in mind we obtain a final estimate for the microscopic rotational friction constant $\lambda_\perp$:

$$(5.38) \qquad \lambda_\perp \approx C\sqrt{\frac{I_\perp}{kT}} \exp\left\{\frac{3(\tilde{G}' + \tilde{J}')}{2kT}\right\}$$

where the constant C contains a few microscopic parameters that are not of interest to us. The crucial result from the above analysis is the Arrhenius dependence with the activation energy which corresponds to overcoming the nematic barrier given by the isotropic potential $\tilde{G}'$ and the much weaker anisotropic correction $\tilde{J}'$ during ensemble averaging. The factor $\sqrt{I_\perp/kT}$ takes into account the time averaging process.

To summarize the main results of this section:

1) Due to the gyroscopic effects, the disks exhibit more complicated form of velocity correlation function. The dynamical evolution of the particle rotation exhibit multiple time-scales, but its rotational Brownian motion time is described predominantly by the ratio of the moment of inertia to the rotational frictional constant.
2) The microscopic friction coefficient shows an exponential temperature dependence with a large activation energy determined mainly by the isotropic part of the interaction potential. This seems to account for the observed temperature variation of the Leslie coefficients [40]. Incorporating orientational correlation effects generates a more precise mean-field potential which can be determined self-consistently via numerical methods. One can foresee higher order correlations such as three-body or four-body correlation effects to render even more accurate results and an improved approximation for the friction constant and the Leslie coefficients.

6. Spatial inhomogeneities and domain structure. *In this section we consider the effects of spatial inhomogeneities by incorporating distortional elasticity, using a non-local nematic potential. In the limit of weak flow and mild spatial distortion, this reveals the microscopic origin of the Ericksen stress in the complete Leslie-Ericksen (LE) theory.*

The original LE theory assumes that the molecules have a short relaxation time so that the molecular orientation distribution always retains its uniaxial equilibrium 'shape', while the local axis of symmetry gets rotated by the flow. The rotational dynamics of the nematics are characterized by the local director $\boldsymbol{n}$ and the constant order parameter. However, it seems plausible that at a higher shear rate, the flow may induce significant gradients in the continuous director field and creates spatial inhomogeneities or textures in the sample. In this case the orientation distribution may be distorted by flow into a non-uniaxial configuration, and the formulation of stress tensor in director variable may not be feasible. Instead, a more adequate formalism will be to consider the dynamics of nematics in terms of the evolution of order parameter tensor as in Equation (4.15) [41, 33].

Another circumstance where distortional effects might become important arise in nematics ridden with defects such as disclination lines and point defects which may be generated due to shear flow. Another more common situation arises due to anchoring condition, when the the director field near the surface is forced to align with the walls. This disrupts the molecular packing and incur a free energy penalty, the minimization of which determines the equilibrium or static dependence of $\boldsymbol{n}(\boldsymbol{r})$. Indeed, the neglect of such distortional stress leads to failures to account for rheological properties of liquid crystalline polymers with domain structures. It was shown that the microscopic theory described so far predicts the formation of disclinations due to inhomogeneous director tumbling which are however constantly annihilated and reformed [33]. Clearly, without distortional elasticity, one can not describe the disclinations and eventually explain a steady-state network of disclinations, which seems to arise in reality.

Previous attempts to account for the distortional elasticity are based mainly on phenomenological models. After the original formulation of the Ericksen stress, Edwards and Beris [42] constructed an ad hoc general expression of Frank elasticity in tensorial form. More recently Tsuji and Rey [41] added distortional energy using the Landau-de Gennes free energy to the kinetic equation but their work does not discuss the stress tensor. Furthermore the use of Landau-de Gennes energy expansion proves to be doubtful for systems with moderately high order parameters typical of a nematic liquid crystal. This highlights the necessity of a molecular theory. In this section we demonstrate that by using a non-local nematic potential to model the effects of distortional elasticity, we can derive a new stress tensor and kinetic equation governing the time evolution of the order parameter tensor. The final results are consistent with the complete LE theory in the limit of weak flow and small distortions.

6.1. Ericksen stress. In contrast to a globally uniform director field, a positional variation in the director $\boldsymbol{n}(\boldsymbol{r})$ introduces new distortion free energy in the system which tends to minimize the spatial gradient of the director. The result is an additional contribution to the stress known as the Ericksen stress [21]. In the usual small-motion approximation, this distortional stress gives rise to second order spatial deviations of $\boldsymbol{n}$ which can be discarded in the formulation of stress tensor. This picture however breaks down at a sufficiently strong shear flow when the local variation in **n** becomes non-vanishing. Before we embark on a microscopic description of this new effects, we shall first give a brief outline of the definitions of Frank's elastic energy and the Ericksen stress.

We consider the distortion free energy in the following form [21],

$$F_d = \frac{1}{2}K_1(\nabla\cdot\boldsymbol{n})^2 + \frac{1}{2}K_2(\boldsymbol{n}\cdot\nabla\times\boldsymbol{n})^2 + \frac{1}{2}K_3(\boldsymbol{n}\times\nabla\times\boldsymbol{n})^2 \tag{6.1}$$

where the constants K_i $(i = 1, 2, 3)$ are associated with respect to the three basic types of deformation: splay, bend and twist. For simplicity, we make a useful one-constant approximation: $K_1 = K_2 = K_3 = K$. The free energy then takes the form

$$F_d = \frac{1}{2}K\left\{(\nabla\boldsymbol{n})^2 + (\nabla\times\boldsymbol{n})^2\right\} = \frac{1}{2}K(\nabla_\alpha n_\beta)(\nabla_\alpha n_\beta) \tag{6.2}$$

after integration by parts in which we assumed the surface terms are unimportant in our analysis. We can consider a small change in the total free energy δF_{tot} due to a local change in the director, and a material distortion of the fluid which leaves the director orientation invariant [21]. Any changes in the system may be decomposed into these independent changes.

1. *Variation in embedded order*: Consider the variation $\boldsymbol{n}(\boldsymbol{r}) \rightarrow \delta\boldsymbol{n}(\boldsymbol{r})$, at a fixed point in space, which produces a change in the free energy

$$\delta F_d = \int \left\{ \frac{\partial F}{\partial n_\beta} \delta n_\beta + \frac{\partial F_d}{\delta(\partial_\alpha n_\beta)} \partial_\alpha \left(\delta n_\beta\right) \right\} d\boldsymbol{r}. \tag{6.3}$$

Integrate the second term by parts and neglect the surface term

$$\delta F_d = \int \left\{ \frac{\partial F}{\partial n_\beta} - \nabla_\alpha \left(\frac{\partial F_d}{\delta\left(\nabla_\alpha n_\beta\right)} \right) \right\} \delta n_\beta \, d\boldsymbol{r}. \tag{6.4}$$

We can define the terms in the bracket as a molecular field

$$h_\beta = -\frac{\partial F_d}{\partial n_\beta} + \partial_\alpha \pi_{\alpha\beta}, \quad \text{where } \pi_{\alpha\beta} = \frac{\delta F_d}{\delta(\partial_\alpha n_\beta)} = \frac{\delta F_d}{\delta g_{\alpha\beta}}. \tag{6.5}$$

Equation (6.5) implies that in equilibrium (in the absence of external fields), δF_d must vanish and the director must be at each point parallel to the molecular field.

2. *Material Distortion*: We now consider a distortion of the material which preserves the value of the director $\boldsymbol{r} \longrightarrow \boldsymbol{r}' = \boldsymbol{r} + \boldsymbol{\varepsilon}(\boldsymbol{r})$ with $\boldsymbol{n}'(\boldsymbol{r}') = \boldsymbol{n}(\boldsymbol{r})$. The change in the free energy then becomes

$$\delta F_d = \int \sigma^d_{\alpha\beta} \partial_\beta \varepsilon_\alpha \, d\boldsymbol{r}, \quad \text{where } \sigma^d_{\alpha\beta} = -\pi_{\alpha\gamma} \partial_\beta n_\gamma \tag{6.6}$$

is the distortion stress tensor. If we impose the incompressibility condition for the fluid, we have to introduce a Lagrange multiplier, the pressure p, then the Ericksen stress tensor arises as a result:

$$\sigma^e_{\alpha\beta} = \sigma^d_{\alpha\beta} - p\delta_{\alpha\beta}. \tag{6.7}$$

Using the one-constant approximation and substituting Equation (6.2) into (6.5), we have the full stress tensor acting on the element of nematic fluid,

$$\sigma_{\alpha\beta} = \sigma^e_{\alpha\beta} + \sigma^v_{\alpha\beta} = -K \nabla_\alpha n_i \nabla_\beta n_i + \sigma^v_{\alpha\beta} \tag{6.8}$$

where $\sigma^v_{\alpha\beta}$ is the viscous stress given by Equation (3.5), and the symmetric Ericksen stress is written in the limit of elastic isotropy. One of the aims of this section is to demonstrate that this term can be accounted for by a suitable microscopic theory, and an approximate microscopic expression for the Frank constant K can be obtained.

6.2. Kinetic equation with distortions. We can extend the original theory to include distortional energy. We expect modifications to two major components: the kinetic equation and the microscopic stress tensor. For a nematic with distorted director configurations, we consider an additional non-local nematic potential, proposed by Marrucci and Greco [43]. This potential accounts for spatial variations of the molecular orientation distribution, and represents the molecular interaction energy in a gradually varying orientational mean-field. The effective nematic potential in the presence of spatial inhomogeneities therefore consists of the Maier-Saupe mean-field potential *and* the Marrucci-Greco nematic potential U_{MG}:

$$\tilde{U}(\boldsymbol{u}) = U_{MS} + U_{MG} = -\frac{3}{2} UkTS_{ij}u_iu_j - \frac{1}{16} UkTL^2\nabla^2 S_{ij}\, u_iu_j \tag{6.9}$$

where U is a non-dimensional constant representing the nematic strength and kT is the Boltzmann temperature. L denotes the characteristic length-scale for molecular interaction. S_{ij} is the order parameter tensor defined in Equation (4.15). This particular expression of modified Maier-Saupe potential is the generalized version of Equation (4.23) which applies to non-equilibrium and spatially homogeneous case [9, 43]. On the other hand U_{MG} takes care of distortion over the neighbourhood of the nematic molecules, and it can be derived in the limit of small distortion expansion.

Note that we have chosen to write the potential in second order tensorial form since its relation to the stress tensor can be established more easily. This approach is completely equivalent to our microscopic theory using the distribution function discussed in previous sections. From our kinetic equation in Equation (4.3), we see that the Marrucci-Greco potential generates an additional term:

$$\begin{aligned} &\alpha^2 \int \partial_k \left(w \frac{\partial_k U_{MG}}{kT} \right) \left(u_iu_j - \frac{1}{3}\delta_{ij} \right)\, d\boldsymbol{u} \\ &= \alpha^2 \frac{UL^2}{8} \left(\nabla^2 S_{i\alpha} \langle u_\alpha u_j \rangle + \langle u_i u_\alpha \rangle \nabla^2 S_{\alpha j} - 2\nabla^2 S_{\alpha\beta} \langle u_\alpha u_\beta u_i u_j \rangle \right) \end{aligned} \tag{6.10}$$

6.3. Nonlocal stress tensor. A natural approach to find the modified stress tensor is to return to Equation (3.22). We see that the U_{MG} term generates an additional stress due to distortions:

$$\begin{aligned} &\rho \left\langle \frac{p^2}{p^2+1} u_\alpha \frac{\partial U_{MG}}{\partial u_\beta} - \frac{1}{p^2+1} u_\beta \frac{\partial U_{MG}}{\partial u_\alpha} - \tilde{p} u_\alpha u_\beta u_m \frac{\partial U_{MG}}{\partial u_m} \right\rangle = \\ &-\frac{1}{8} \rho UkTL^2 \left(\frac{p^2}{p^2+1} S_{i\alpha} \nabla^2 S_{i\beta} - \frac{1}{p^2+1} S_{i\beta} \nabla^2 S_{i\alpha} - \tilde{p} \langle u_\alpha u_\beta u_i u_j \rangle \nabla^2 S_{ij} \right). \end{aligned} \tag{6.11}$$

Following our earlier approach, we take the symmetric part of this contribution and discover that it can be related to Equation (6.11). This can be written explicitly in terms of the velocity gradient and gives the symmetric part of the viscous stress tensor. This approach however generates no additional terms which can account for the symmetric Ericksen stress $K\nabla_\alpha n_i \nabla_\beta n_i$ in Equation (6.8). We conclude that this straightforward approach does not give a self-consistent microscopic theory that can account for the Ericksen stress, and a more elaborate formalism is required to evaluate the microscopic stress tensor.

Considering that the Ericksen stress can be regarded as an elastic stress due to distortions, we can invoke the principle of virtual work [11], and calculate the elastic stress $\sigma^E_{\alpha\beta}$ from the reaction of the nematic to a rapid virtual deformation $\delta\varepsilon_{\alpha\beta}(\boldsymbol{r})$.

$$\delta F = \int_v \sigma^E_{\alpha\beta}\delta\varepsilon_{\alpha\beta}\, dV \tag{6.12}$$

where V is the volume of the bulk sample. By calculating the change in the free energy, we can extract the elastic stress.

The free energy of the nematic liquid crystal can be written in terms of the molecules orientation distribution function $w(\boldsymbol{u})$

$$F = \rho \int_v dV \int d\boldsymbol{u}\ (kTw\ln w + w\tilde{U}). \tag{6.13}$$

This gives

$$\delta F = \rho \int_v dV \int d\boldsymbol{u}\ \left[kT\delta w \ln w + kT\delta w + \delta(wU_{MS}) + \delta(wU_{MG})\right]. \tag{6.14}$$

We only have to concern ourselves with the U_{MG} term, since the other terms produce exactly the same microscopic stress tensor as given in Equation (3.22). We therefore have,

$$\begin{aligned}
\delta F_{MG} &= \rho \int_V dV \int \delta\,(wU_{MG})d\boldsymbol{u} \\
&= -\frac{\rho U kTL^2}{16}\int_V dV \int \delta\left(w\nabla^2 S_{ij}u_i u_j\right)d\boldsymbol{u} \\
&= -\frac{\rho U kTL^2}{16}\int_V dV\delta\left(\nabla^2 S_{ij}S_{ij}\right) \\
&= -\frac{\rho U kTL^2}{16}\int_V dV\left(\nabla^2 S_{ij}\delta S_{ij} + \delta\nabla^2 S_{ij}\ S_{ij}\right).
\end{aligned} \tag{6.15}$$

The term $\delta\nabla^2 S_{ij}$ represents the energy for additional spatial distortions due to the strain field. This can be calculated explicitly to give:

$$\begin{aligned}
\delta\nabla^2 S_{ij} &= \left(\frac{\partial\nabla^2 S_{ij}}{\partial t} + \boldsymbol{v}\cdot\nabla\nabla^2 S_{ij}\right)\delta t = \nabla^2\delta S_{ij} + \boldsymbol{v}\cdot\nabla\nabla^2 S_{ij}\delta t \\
&= \nabla^2\delta S_{ij} - \delta\varepsilon_{\alpha\beta}\nabla_\alpha\nabla_\beta S_{ij} - \nabla_\alpha\delta\varepsilon_{\alpha\beta}\nabla_\beta S_{ij}
\end{aligned} \tag{6.16}$$

where we have used integration by parts and neglecting the surface terms. Integrating by parts again we have:

$$\begin{aligned}
\int\left(\delta\nabla^2 S_{ij}\right) S_{ij}\, dV &= \int dV(\nabla^2 S_{ij}\delta S_{ij} - \delta\varepsilon_{\alpha\beta}\nabla_\alpha\nabla_\beta S_{ij}\ S_{ij} + \delta\varepsilon_{\alpha\beta}A_{\alpha\beta}) \\
&\quad + \int_\Sigma dS_\alpha(\nabla_\alpha\delta S_{ij}S_{ij} - \nabla_\alpha S_{ij}\delta S_{ij} - \delta\varepsilon_{\alpha\beta}\nabla_\beta S_{ij}\ S_{ij})
\end{aligned}$$

where $A_{\alpha\beta} = \nabla_\alpha S_{ij} \nabla_\beta S_{ij}$, and the contribution to the virtual work

$$\delta F_{MG} = -\frac{\rho U k T L^2}{16} \left\{ \int dV (2\nabla^2 S_{ij} \delta S_{ij} - \delta\varepsilon_{\alpha\beta} \nabla_\alpha \nabla_\beta S_{ij} S_{ij} + \delta\varepsilon_{\alpha\beta} A_{\alpha\beta}) \right.$$

$$\left. + \int_\Sigma dS_\alpha (\nabla_\alpha \delta S_{ij} S_{ij} - \nabla_\alpha S_{ij} \delta S_{ij} - \delta\varepsilon_{\alpha\beta} \nabla_\beta S_{ij} S_{ij}) \right\}. \tag{6.17}$$

The surface integral can be put to zero since $\delta\varepsilon$, δS_{ij} and $\nabla \delta S_{ij}$ vanish on the surface boundary.

The variation in the order parameter tensor S_{ij} can be calculated from the kinetic equation (4.3) by neglecting the diffusion and potential terms for a rapid virtual deformation [11]. In this case,

$$\delta w = \frac{\partial w}{\partial t} \delta t = -\alpha \partial_k (\Omega_k w) \delta t \tag{6.18}$$

where $\boldsymbol{\Omega} = \frac{\tilde{p}}{2} \boldsymbol{u} \times g^s \cdot \boldsymbol{u} + \frac{1}{2} \boldsymbol{u} \times g^a \cdot \boldsymbol{u}$ is the residue flow field. We then have

$$\delta S_{ij} = \int u_i u_j \delta w d\boldsymbol{u} \quad = \alpha \int \partial_k (u_i u_j) \Omega_k \delta t \; w \; d\boldsymbol{u}$$

$$= \alpha \left\{ \frac{\tilde{p}}{2} \delta\varepsilon^s_{io} S_{jo} + \frac{\tilde{p}}{2} \delta\varepsilon^s_{jo} S_{io} - \tilde{p} \delta\varepsilon^s_{no} \langle u_n u_o u_i u_j \rangle + \frac{1}{2} \delta\varepsilon^a_{io} S_{jo} + \frac{1}{2} \delta\varepsilon^a_{jo} S_{io} \right\} \tag{6.19}$$

where $\delta\varepsilon^s$ and $\delta\varepsilon^a$ are the symmetric and asymmetric strain tensor respectively. Using these in Equation (6.12), we finally obtain the part of the elastic stress due to distortion only:

$$\sigma^E_{\alpha\beta} = -\frac{\rho U k T L^2}{8} \left\{ \frac{p^2}{p^2+1} S_{\alpha j} \nabla^2 S_{\beta j} - \frac{1}{p^2+1} S_{\beta j} \nabla^2 S_{\alpha j} \right.$$

$$\left. -\tilde{p} \nabla^2 S_{ij} \langle u_i u_j u_\alpha u_\beta \rangle \right\} - \frac{\rho U k T L^2}{32} (A_{\alpha\beta} - \nabla_\alpha \nabla_\beta S_{ij} S_{ij}). \tag{6.20}$$

The free energy approach therefore produces nearly identical stress tensor as in Equation (6.11), except with an additional term $-\frac{\rho U k T L^2}{32}(A_{\alpha\beta} - \nabla_\alpha \nabla_\beta S_{ij} S_{ij})$ which is symmetrical. We shall see that this term possesses the correct symmetry, as the Ericksen stress, and justifies our use of the virtual deformation principle. This approach also shows explicitly that the addition of distortional elasticity introduces non-local effects into the stress tensor, which now depends on position due to the non-homogeneous Marrucci-Greco potential. This situation differs from a uniform nematic liquid crystal when the principle of locality is assumed which means that both the flow and the nematic configurations are homogeneous [11].

We can gain some physical insights by considering the symmetric part of the stress tensor due to Marrucci-Greco potential U_{MG}. From Equation

(6.20) this may be written as

$$\sigma^s_{\alpha\beta} = -\frac{\rho UkTL^2\tilde{p}\alpha}{16}\{S_{\alpha j}\nabla^2 S_{\beta j} + S_{\beta j}\nabla^2 S_{\alpha j} - 2\nabla^2 S_{ij}\langle u_i u_j u_\alpha u_\beta\rangle\}$$
$$(6.21)\qquad -\frac{\rho UkTL^2}{32}(A_{\alpha\beta} - \nabla_\alpha\nabla_\beta S_{ij}S_{ij}).$$

The first term on the right hand side, together with the original Maier-Saupe term in Equation (4.7), can be replaced by the flow term in the kinetic equation (6.10). The full symmetric stress tensor with distortions then becomes

$$(6.22)\quad \sigma^s_{\alpha\beta} = -\frac{\rho kT\tilde{p}}{2\alpha^2}\left(\frac{\partial S_{\alpha\beta}}{\partial t} - G_{\alpha\beta}\right) - \frac{\rho UkTL^2}{32}(A_{\alpha\beta} - \nabla_\alpha\nabla_\beta S_{ij}S_{ij})$$

where G_{ij} follows from Equation (4.11). Comparing with the full Leslie-Ericksen stress tensor in Equation (6.8) we see that the first term on the right hand side produces the viscous stress, while the second term must be equivalent to the Ericksen stress. Equation (6.22) therefore expresses the reaction of the nematic liquid crystal in terms of velocity gradient (as in the original LE theory) *and* the local variation of nematic configurations, which gives rise to non-local nature of the Ericksen stress.

Using the expression for the uniaxial order parameter tensor S_{ij} and assuming the magnitude of scalar order parameter S_2 is constant, we have

$$\nabla_\alpha S_{ij}\nabla_\beta S_{ij} - \nabla_\alpha\nabla_\beta S_{ij}S_{ij} = \frac{8}{3}S_2^2\ \nabla_\alpha n_i\nabla_\beta n_i - \frac{2}{3}S_2\nabla_\alpha n_i\nabla_\beta n_i$$
$$(6.23)\qquad - \frac{4}{3}S_2^2\ n_i\nabla_\alpha\nabla_\beta n_i - \frac{2}{3}S_2 n_i\nabla_\alpha\nabla_\beta n_i$$
$$= 4S_2^2\ \nabla_\alpha n_i\nabla_\beta n_i$$

where $n_i\nabla_j n_i = 0$ is frequently used and the last line is obtained from integration by parts.[9] We therefore obtain a microscopic expression for the average Frank constant:

$$(6.24)\qquad K = \frac{1}{8}\rho UkTL^2S_2^2$$

in the limit of elastic isotropy (one constant approximation) and assuming a type of Marrucci-Greco distortional energy.

This expression has several nice features. It depends on the molecular interaction length and the order parameter. However we would expect by intuition that the Frank constants depends on the molecular aspect ratio

[9] It is important to note at this point that in recent years a number of theories have appeared, which examine the additional effects of variation ∇S_2, or leaving the nematic variables in the tensor form, as $S_{ij}(\boldsymbol{r}, t)$ [44]. Clearly, our approach is adaptable for these continuum theories although here we rigidly follow the path towards the LE model.

p and differ in general for discotic or rod-like nematic phase. However this only applies in the limit of elastic anisotropy when $K_1 \neq K_3$. Marrucci and Greco [43] demonstrated that this is a result of assuming that the length of the rods is long compared to the molecular interaction length L. On the other hand if the interaction length is much larger than the molecular length, the 'interaction neighborhood' becomes essentially spherical and the one-constant approximation becomes fairly accurate. We did not attempt here to pursue the more accurate derivation of Frank elasticity from the kinetic theory, firstly because an excellent equilibrium microscopic models already exist [45, 46] and secondly because our limited aim has been the LE theory of viscosity. No doubt this is an interesting possible avenue of new research.

We note that with the incorporation of Ericksen stress, the Leslie stress tensor becomes non-symmetric in general, and hence angular momentum is not conserved in the usual sense. This gives rise to a mean-field torque which is to be expected since the mean-field potential exerts a torque on the molecules when they are forced away by flow. We therefore expect the usual balance of torque equation to be modified [21]:

$$\int \epsilon_{\alpha\mu\rho}\{r_\rho \sigma^e_{\beta\alpha} + n_\rho \pi^e_{\beta\alpha}\} dS_\beta - \int (\boldsymbol{n} \times \boldsymbol{h})_\mu d\boldsymbol{r} = 0. \tag{6.25}$$

This balance is required for the conservation of the total angular momentum in static equilibrium. The first term on the left hand side denotes the surface torque due to elastic distortions in the director field $\boldsymbol{h}$ given by Equation (6.5). There are two contributions to the surface torques, one from the Ericksen stress and one deriving from the tensor π. The second term denotes the torque due to viscous processes, where $\boldsymbol{h}'$ is given by Equation (3.6). In static equilibrium this term vanishes when the director is aligned parallel to the molecular field, but the total torque becomes non-vanishing due to the surface torque.

REFERENCES

[1] J.G. Kirkwood, *Selected Topics in Statistical Mechanics.* Gordon and Breach Science Publication, New York, 1967.

[2] P. Resibois and M. De Leener, *Classical Kinetic Theory.* Wiley, New York, 1977.

[3] S.A. Rice and P. Gray, *The Statistical Mechanics of Simple Liquids.* Wiley, New York, 1965.

[4] R.G. Larson, *The structure and rheology of complex fluids.* Oxford University Press, New York, 1999.

[5] F.M. Leslie, Some constitutive equations for anisotropic fluids. *Quart. J. Mech. Appl. Math.*, **19**: 357, 1966.

[6] E.J. Hinch and L.G. Leal, The effect of Brownian motion on the rheological properties of a suspension of non-spherical particles. *J. Fluid Mech.*, **52**: 683, 1972.

[7] J.L. Ericksen, Anisotropic fluids. *Arch. Ration. Mech. Analysis*, **4**: 231, 1960.

[8] A.C. DIOGO AND A.F. MARTINS, Order parameter and temperature-dependence of the hydrodynamic viscosities of nematic liquid-crystals. *J. Physique*, **43**: 779, 1982.
[9] N. KUZUU AND M. DOI, Constitutive equation for nematic liquid-crystals under weak velocity-gradient derived from a molecular kinetic-equation. *J. Phys. Soc. Japan*, **52**: 3486, 1983.
[10] M.A. OSIPOV AND E.M. TERENTJEV, Rotational diffusion and rheological properties of liquid-crystals. *Z. Naturforsch. (a)*, **44**: 785, 1989.
[11] M. DOI AND S.F. EDWARDS, *Theory of Polymer dynamics*. Oxford Publisher, Oxford, 1986.
[12] B. HAMMOUDA, J. MANG, AND S.KUMAR, Shear-induced orientational effects in discotic-liquid-crystal micelles. *Phys. Rev. E*, **51**: 6282, 1995.
[13] Y. FARHOUDI AND A.D. REY, Ordering effects in shear flows of discotic polymers. *Rheol. Acta*, **32**: 207, 1993.
[14] S.M. JOGUN AND C.F. ZUKOSKI, Rheology and microstructure of dense suspensions of plate-shaped colloidal particles. *J. Rheol.*, **43**: 847, 1999.
[15] G.B. JEFFREY, The motion of ellipsoidal particles immersed in a viscous fluid. *Proc. Roy. Soc. London*, **102A**: 161, 1922.
[16] R. GRAHAM, Covariant formulation of non-equilibrium statistical thermodynamics. *Z. Physik. B*, **26**: 397, 1977.
[17] R. MAZO, *Brownian Motion, Fluctuations, Dynamics and applications*. Oxford University Press, Oxford, 2000.
[18] A.V. ZAKHAROV AND A. MALINIAK, Structure and elastic properties of a nematic liquid crystal: A theoretical treatment and molecular dynamics simulation. *Eur. Phys. J.E.*, **4**: 435, 2001.
[19] N.G. VAN KAMPEN, *Stochastic processes in physics and chemistry*. North Holland, Armsterdam, 1992.
[20] H. RISKEN, *The Fokker-Planck equation: Methods of solution and application*. Springer Verlag, New York, 1989.
[21] P.G. DE GENNES AND J. PROST, *The Physics of Liquid Crystals*. Oxford University Press, Oxford, 1993.
[22] H.R. ZELLER, Dielectric-relaxation and the glass-transition in nematic liquid-crystals. *Phys. Rev. Lett.*, **48**: 334, 1982.
[23] P.C. MARTIN, P.J PERSHAN, AND J. SWIFT, New elastic-hydrodynamic theory of liquid crystals. *Phys. Rev. Lett.*, **25**: 844, 1970.
[24] I. HALLER AND J.D. LITSTER, Temperature dependence of normal modes in a nematic liquid crystal. *Phys. Rev. Lett.*, **25**: 1550, 1970.
[25] P. MAZUR AND S.R. DE GROOT, *Non-equilibrium thermodyanmics*. Dover, New York, 1984.
[26] O. PARODI, Stress tensor for a nematic liquid crystal. *J. Physique*, **31**: 581, 1970.
[27] T. CARLSSON, Remarks on the flow-alignment of disk like nematics. *J. Physique*, **44**: 909, 1983.
[28] G.E. VOLOVIK, Relationship between molecular shape and hydrodynamics in a nematic substance. *Pis'ma Zh. Eksp. Teor. Fiz. (JETP Lett.)*, **31**: 297, 1980.
[29] A. CHRZANOWSKA AND K. SOKALSKI, Microscopic description of nematic liquid crystal viscosity. *Phys. Rev. E*, **52**: 5228, 1995.
[30] R.G. LARSON, Arrested tumbling in shearing flows of liquid-crystal polymers. *Macromolecules*, **23**: 3983, 1990.
[31] CH. GÄHWILLER, Temperature dependence of flow alignment in nematic liquid crystals. *Phys. Rev. Lett.*, **28**: 1554, 1972.
[32] P. PIERANSKI AND E. GUYON, Two shear-flow regimes in nematic $p-n-$hexyloxybenzilidene$-p'-$aminobenzonitrile. *Phys. Rev. Lett.*, **32**: 924, 1974.
[33] J. FENG AND L.G. LEAL, Simulating complex flows of liquid-crystalline polymers using the Doi theory. *J. Rheol.*, **41**: 1317, 1997.
[34] I.R. MCDONALD AND J.-P. HANSEN, *Theory of simple liquids*. Academic Press,

New York, 1986.

[35] B.W. VAN DER MEER AND G. VERTOGEN, *Molecular physics of Liquid Crystals, edited by G.R. Luckhurst, G.W. Gray.* Academic Press, New York, 1979.

[36] S. CHANDRASEKHAR AND N.V. MADHUSUDANA, Molecular statistical theory of nematic liquid crystals. *Acta Crystallogr.*, **27**: 303, 1971.

[37] B.A. BARON AND W.M. GELBART, Molecular shape and volume effects on the orientational ordering of simple liquid crystals. *J. Chem. Phys.*, **67**: 5795, 1977.

[38] R. EPPENGA AND D. FRENKEL, Monte carlo study of the isotropic-nematic transition in a fluid of thin hard disks. *Phys. Rev. Lett.*, **40**: 1089, 1982.

[39] L. ONSAGER, The effects of shape on the interaction of colloidal particles. *Ann. N. Y. Acad. Sci.*, **51**: 627, 1949.

[40] S.T. WU AND C.S. WU, Experimental confirmation of the osipov-terentjev theory on the viscosity of nematic liquid-crystals. *Phys. Rev. A*, **42**: 2219, 1990.

[41] T. TSUJI AND A.D. REY, Effect of long range order on sheared liquid crystalline materials. 1. compatibility between tumbling behavior and fixed anchoring. *J. Non-Newtonian Fluid Mech.*, **73**: 127, 1997.

[42] B.J. EDWARDS AND A.N. BERIS, Order parameter representation of spatial inhomogeneities in polymeric liquid-crystals. *J. Rheol.*, **33**: 1189, 1989.

[43] G. MARRUCCI AND F. GRECO, The elastic-constants of maier-saupe rodlike molecule nematics. *Mol. Cryst. Liq. Cryst.*, **26**: 17, 1991.

[44] A.M. SONNET, P.L. MAFFETTONE, AND E.G. VIRGA, Continuum theory for nematic liquid crystals with tensorial order. *J. Non-Newtonian Fluid Mech.*, **119**: 51, 2004.

[45] J.P. STRALEY, Frank elastic constants of the hard-rod liquid crystal. *Phys. Rev. A*, **8**: 2181, 1973.

[46] W.M. GELBART AND A. BEN-SHAUL, Molecular theory of curvature elasticity in nematic liquids. *J. Chem. Phys.*, **77**: 916, 1982.

ANISOTROPY AND HETEROGENEITY OF NEMATIC POLYMER NANO-COMPOSITE FILM PROPERTIES

M. GREGORY FOREST*, RUHAI ZHOU†, QI WANG‡, XIAOYU ZHENG*, AND ROBERT LIPTON§

Abstract. Nematic polymer nanocomposites (NPNCs) are comprised of large aspect ratio rod-like or platelet macromolecules in a polymeric matrix. Anisotropy and heterogeneity in the effective properties of NPNC films are predicted in this article. To do so, we combine results on the flow-processing of thin films of nematic suspensions in a planar Couette cell, together with homogenization results for the effective conductivity tensor of spheroidal inclusions in the low volume fraction limit. The orientational probability distribution function (PDF) of the inclusions is the central object of Doi-Hess-Marrucci-Greco theory for flowing nematic polymers. From recent simulations, the PDF for a variety of anisotropic, heterogeneous thin films is applied to the homogenization formula for effective conductivity. The principal values and principal axes of the effective conductivity tensor are thereby generated for various film processing conditions. Dynamic fluctuations in film properties are predicted for the significant parameter regime where the nematic polymer spatial structure is unsteady, even though the processing conditions are steady.

1. Introduction. Nematic (liquid crystalline) polymers, because of their extreme aspect ratio, impart anisotropy in properties through the orientational probability distribution of the molecular ensemble. This is well-known in fibers, where the rod-like molecules strongly align with the centerline of the fiber during flow processing. In these cases, one can anticipate that the assumption of perfectly aligned spheroidal inclusions is a reasonable approximation, and apply the numerical tools of Gusev and collaborators [1] to estimate fiber effective properties. At very dilute concentrations, the assumption of isotropic orientation is accurate for bulk, quiescent mesophases of nematic polymers, which is the other extreme typically assumed.

In shear-dominated flows typical of film processes, however, the orientational distribution of nematic polymers has been the object of intense theory, modeling and simulations for at least two decades. The monographs of deGennes & Prost [2] and Larson [7, 8] provide excellent treatments, as well as the recent review by Rey & Denn [9]. In such confined flows, the orientational distribution of the inclusions is neither random nor perfectly aligned, and furthermore there are lengthscales of distortion in the distribution, which are evident but poorly understood.

*Applied & Interdisciplinary Mathematical Sciences Center and Institute for Advanced Materials, University of North Carolina at Chapel Hill, Chapel Hill, NC 27599.

†Department of Mathematics and Statistics, Old Dominion University, Norfolk, VA 23529.

‡Department of Mathematics, Florida State University, Tallahassee, FL 32306; and Nankai University, Tianjin, P.R. China.

§Department of Mathematics, Louisiana State University, Baton Rouge, LA 70803.

Our goal in this article is to combine two recent advances to predict effective property tensors of nematic polymer nano-composites. We use results from [3] on the orientational probability distribution function (PDF) of nematic polymer films, which have been processed in a plane Couette cell. The simulations only allow spatial variation between the parallel plates, and restrict the molecular orientation function to so-called in-plane symmetry, in which the principal axes of the orientational distribution lie in the flow-flow gradient plane. The results of these simulations are then sampled across a range of plate speeds and a range of distortional elasticity strength of the nematic polymer liquid. We then use recent results from [11] which determine the effective conductivity tensor for spheroidal inclusions in an isotropic matrix in the low volume fraction limit. The key result in this paper is that only the second moment of the PDF is required to predict the leading order property tensor. The application to predict anisotropy of bulk monodomains is extended here to heterogeneous films.

2. Plane Couette film flow of nematic polymers. We recall results for film flows in a plane Couette shear cell. This device is mathematically convenient in that one can self-consistently assume one-dimensional variations in the gap between moving parallel plates. We summarize the model only to the extent necessary to explain the fundamental parameters, Deborah and Ericksen numbers, used to represent the phase diagram of spatial film structures [12, 3].

The nematic polymer liquid is trapped between plates located at $y = \pm h$, in Cartesian coordinates $\mathbf{x} = (x, y, z)$, and moving with corresponding velocity $\mathbf{v} = (\pm v_0, 0, 0)$, respectively. Even though kinetic theory now exists to include a viscoelastic polymeric solvent [4], the numerical codes have yet to be written. Following all other model simulations to date, cf. the review by Rey and Denn [9] and references in [12, 3], we model nematic polymers in a viscous solvent. There are two apparent length scales in this problem: the gap width $2h$, an external length scale, and the finite range l of molecular interaction, an internal length scale, set by the distortional elasticity in the Doi-Marrucci-Greco (DMG) model. The plate motion sets a bulk flow time scale ($t_0 = h/v_0$); the nematic average rotary diffusivity (D_r^0) sets another (internal) time scale ($t_n = 1/D_r^0$) and the ratio t_n/t_0 defines the *Deborah number De*. The nematic liquid is also elastic, with a short-range excluded volume potential of dimensionless strength N, along with a distortional elasticity potential, which has a persistence length l, and a degree of anisotropy θ. The *Ericksen number* is then defined by:

$$Er = \frac{8h^2}{Nl^2}, \tag{1}$$

which measures short-range nematic potential strength relative to distortional elasticity strength, and θ is a fraction between -1 and ∞ that corresponds to equal ($\theta = 0$) or distinct ($\theta \neq 0$) elasticity constants.

The dimensionless Smoluchowski equation for the probability distribution function (PDF) $f(\mathbf{m}, \mathbf{x}, t)$ is

$$\begin{aligned} \frac{Df}{Dt} &= \mathcal{R} \cdot [(\mathcal{R}f + f\mathcal{R}V)] - \mathcal{R} \cdot [\mathbf{m} \times \dot{\mathbf{m}} f], \\ \dot{\mathbf{m}} &= \Omega \cdot \mathbf{m} + a[\mathbf{D} \cdot \mathbf{m} - \mathbf{D} : \mathbf{mmm}], \end{aligned} \tag{2}$$

where D/Dt is the material derivative $\partial/\partial t + \mathbf{v} \cdot \nabla$, $\mathcal{R}$ is the rotational gradient operator:

$$\mathcal{R} = \mathbf{m} \times \frac{\partial}{\partial \mathbf{m}}, \tag{3}$$

$\mathbf{D}$ and Ω are the symmetric and anti-symmetric parts of $\nabla \mathbf{v}$, $a = (r^2 - 1)/(r^2 + 1)$ is the molecular shape parameter for spheroidal macromolecules of aspect ratio r. The extended Doi-Marrucci-Greco potential is

$$V = -\frac{3N}{2}\left[\left(\mathbf{I} + \frac{1}{3Er}\Delta\right)\mathbf{M} : \mathbf{mm} + \frac{\theta}{3Er}(\mathbf{mm} : (\nabla\nabla \cdot \mathbf{M}))\right], \tag{4}$$

where the second moment projection tensor $\mathbf{M}$ of f is

$$\mathbf{M} = \mathbf{M}(f) = \int_{||\mathbf{m}||=1} \mathbf{mm} f(\mathbf{m}, \mathbf{x}, t) d\mathbf{m}. \tag{5}$$

The dimensionless forms of the balance of linear momentum, stress constitutive equation, and continuity equation are

$$\begin{aligned} \frac{d\mathbf{v}}{dt} &= \nabla \cdot (-p\mathbf{I} + \tau) \\ \tau &= \left(\frac{2}{Re} + \mu_3(a)\mathbf{D} + a\alpha\left(\mathbf{M} - \frac{\mathbf{I}}{3} - n\mathbf{M} \cdot \mathbf{M} \cdot \mathbf{M} + N\mathbf{M} : \mathbf{M}_4\right)\right. \\ &\quad - a\frac{\alpha}{6Er}(\Delta\mathbf{M} \cdot \mathbf{M} + \mathbf{M} \cdot \Delta\mathbf{M} - 2\Delta\mathbf{M} : \mathbf{M}_4) \\ &\quad - \frac{\alpha}{12Er}\Big(2(\Delta\mathbf{M} \cdot \mathbf{M} - \mathbf{M} \cdot \Delta\mathbf{M}) + (\Delta\mathbf{M} : \Delta\mathbf{M} - (\Delta\Delta\mathbf{M}) : \mathbf{M})\Big) \\ &\quad - a\frac{\alpha\theta}{12Er}[\mathbf{M} \cdot \mathbf{M}_d + \mathbf{M}_d \cdot \mathbf{M} - 4(\nabla\nabla \cdot \mathbf{M}) : \mathbf{M}_4] \\ &\quad - a\frac{\alpha\theta}{12Er}[\mathbf{M}_d \cdot \mathbf{M} - \mathbf{M} \cdot \mathbf{M}_d + (\nabla\nabla \cdot \mathbf{M}) \cdot \mathbf{M}_{\beta j,\alpha}\mathbf{M}_{ij,i}] \\ &\quad + [\mu_1(a)(\mathbf{D} \cdot \mathbf{M} + \mathbf{M} \cdot \mathbf{D}) + \mu_2(a)\mathbf{D} : \mathbf{M}_4], \\ \nabla \cdot \mathbf{v} &= 0 \end{aligned} \tag{6}$$

where

$$\begin{aligned} \mathbf{M}_d &= \nabla\nabla \cdot \mathbf{M} + (\nabla\nabla \cdot \mathbf{M})^T, \\ \mathbf{M}_4 &= \int_{||\mathbf{m}||=1} \mathbf{mm} f(\mathbf{m}, \mathbf{x}, t) d\mathbf{m}. \end{aligned} \tag{7}$$

Rheological properties of nematic polymers arise from the total stress τ, which is important for mechanical properties of soft matter materials that are locked in during processing. These effects have not yet been explored; rather we will focus on volume-averaged conductivity, for which only the orientational distribution of the nematic polymer molecules is needed. The other parameters above are solvent and nematic viscosities and a molecular entropy parameter, defined in [10, 12, 3], and not relevant for this paper. The extra stress involves the moments of the PDF f and their gradients, the fourth moment $\mathbf{M}_4$ and the second moment $\mathbf{M}$. It is traditional to define $\mathbf{Q}$ (a second order, symmetric, traceless tensor) known as the orientation tensor:

$$\mathbf{Q} = \mathbf{M} - \frac{1}{3}\mathbf{I}. \tag{8}$$

The boundary conditions of the velocity $\mathbf{v} = (v_x, 0, 0)$ are given by the Deborah number

$$v_x(y = \pm 1, t) = \pm De. \tag{9}$$

We assume *homogeneous anchoring at the plates*, given by the quiescent nematic equilibrium,

$$f(\mathbf{m}, y = \pm 1, t) = f_e(\mathbf{m}), \tag{10}$$

where $f_e(\mathbf{m})$ is a nematic equilibrium without flow, $\mathbf{v} = 0$. At nematic concentrations, equilibria $f_e(\mathbf{m})$ are invariant under orthogonal rotations; the peak axis of orientation is experimentally set by mechanical rubbing, chemical properties, or applied fields. We only consider tangential and normal anchoring, where the peak orientation (so called major director) on the boundary is aligned with the plate motion axis, or normal to it. The initial condition for the PDF is given by

$$f(\mathbf{m}, y, t = 0) = f_e(\mathbf{m}), \tag{11}$$

modeling experiments that start from a homogeneous liquid in a statistically uniform, thermal equilibrium.

The PDF is expanded in a spherical harmonic representation

$$f(\mathbf{m}, \mathbf{x}, t) \approx \sum_{l=0}^{L} \sum_{m=-l}^{l} a_l^m(\mathbf{x}, t) Y_l^m(\mathbf{m}). \tag{12}$$

We then apply a standard Galerkin scheme to arrive at a system of 65 coupled, nonlinear partial differential equations for a_l^m, corresponding to the truncation order $L = 10$. Spatial derivatives are discretized using 4th order finite difference methods, and an adaptive moving mesh algorithm is important for efficiency and to capture localized internal and boundary layers with strong defocusing of the PDF. Spectral deferred corrections are used for time integration to achieve 4th order convergence, and thereby remove dynamic sensitivity especially near transition phenomena.

3. Conductivity properties across the phase diagram of flow-induced film structures. In [3], the nematic concentration is fixed at $N = 6$, which corresponds to a volume fraction of about 1% for rod-like spheroids with aspect ratio $r = 200$. One then numerically determines the structure attractors, that is, the convergent space-time solutions of the above system of model equations for imposed shear flow between the two plates. This assumes the experiment is running at steady state, and the structure attractors are compiled versus Ericksen number Er and Deborah number De for two different anchoring conditions. Four distinct spatio-temporal attractors arise, listed in Table 1, repeated from [3]. Two are steady state structures, whereas the other two are periodic responses to steady plate motion. This dynamic response to steady driving conditions has been recognized since the experiments of Kiss and Porter [5, 6]. For material properties, addressed next, there are significant implications since the fluctuations in anisotropy and heterogeneity of the PDF translate to property fluctuations. The timescale and timing of the quench process then becomes an intriguing issue in film processing of these materials.

TABLE 1

(From [3]). In-plane structure attractors and phase transitions for 3 decades of Deborah number (De) and Ericksen number (Er). ES and VS stand for elastic (E) and viscous (V) dominated steady (S) states. T or W indicates a transient structure in which the peak orientation axis at each height between the plates either oscillates with finite amplitude (wagging) or rotates continuously (tumbling).

$De \backslash Er$	5	10	15	50	180	500	2000	5000	∞
0.01	ES	ES	Es	ES	ES	ES	TW	TW	T
0.05	ES	ES	ES	ES	ES	TW	TW	TW	T
0.10	ES	ES	ES	ES	**W**	TW	TW	TW	T
0.50	ES	ES	ES	**W**	TW	TW	TW	TW	T
1.00	ES	ES	**W**	TW	TW	TW	TW	TW	T
3.00	ES	ES	**W**	TW	TW	TW	TW	TW	T
5.50	ES	ES	**W**	TW	TW	TW	TW	TW	T
6.00	ES	ES	**W**	**W**	**W**	**W**	**W**	**W**	**W**
8.00	ES	ES	**W**	**W**	**W**	**W**	**W**	**W**	**W**
8.50	ES	ES	ES	VS	VS	VS	VS	VS	FA
10.00	ES	ES	ES	VS	VS	VS	VS	VS	FA
12.00	ES	ES	VS	VS	VS	VS	VS	VS	FA

One of the purposes of the present study is to quantitatively predict the steady and dynamic property fluctuations of these one-dimensional film structures.

We now recall the homogenization theory result from [11], based on volume averaging of spheroidal inclusions in the low volume fraction limit.

In that paper, we illustrated the anisotropy of steady, homogeneous monodomains of nematic polymer composites. Here, we generalize to spatially heterogeneous PDFs of the spheroidal inclusions, for both steady and unsteady attractors. The basic assumption is that the lengthscales of distortions in the PDF are much larger than the volume averaging scale. Since there are on the order of a million macromolecules in a cubic micron, this assumption seems quite reasonable.

The effective conductivity tensor in closed form is

$$
\begin{aligned}
\Sigma^e_{\theta_2} = \Sigma_0 + \sigma_1\theta_2(\sigma_2-\sigma_1)\Bigg(& \frac{2}{(\sigma_2+\sigma_1)-(\sigma_2-\sigma_1)L_a}\mathbf{I} \\
& + \frac{(\sigma_2-\sigma_1)(1-3L_a)}{((\sigma_2+\sigma_1)-(\sigma_2-\sigma_1)L_a)(\sigma_1+(\sigma_2-\sigma_1)L_a)}\mathbf{M}(f)\Bigg) \\
& + O(\theta_2^2).
\end{aligned} \tag{13}
$$

where σ_1, σ_2 are the conductivity of the matrix and the inclusions, respectively. The conductivity contrast between the nano phase and the matrix solvent is specified as $\sigma_2/\sigma_1 = 10^6$. θ_2 is the volume fraction of nematic polymers, $\mathbf{I}$ is the 3 by 3 identity matrix, L_a is the spheroidal depolarization factor depending on the aspect ratio r of the molecular spheroids through the relation

$$
L_a = \frac{1-\varepsilon^2}{\varepsilon^2}\left[\frac{1}{2\varepsilon}\ln\left(\frac{1+\varepsilon}{1-\varepsilon}\right)-1\right], \qquad \varepsilon = \sqrt{1-r^{-2}}, \tag{14}
$$

f is the orientational PDF of the inclusions, and $\mathbf{M}(f)$ is the second-moment of the PDF, defined earlier.

The three principal axes of the effective conductivity are identical with the principal axes of the second moment tensor $\mathbf{M}(f)$. The three principal conductivity values (eigenvalues of $\Sigma^e_{\theta_2}$) are denoted by $\sigma^e_1 \geq \sigma^e_2 \geq \sigma^e_3$. We define the *relative principal value enhancements* by

$$
\varepsilon_i = \frac{\Sigma^e_{\theta_2} - \sigma_1\mathbf{I}}{\sigma_1} : \mathbf{n}_i\mathbf{n}_i = \frac{\sigma^e_i - \sigma_1}{\sigma_1}, \qquad i = 1, 2, 3. \tag{15}
$$

We now apply this formula directly to the PDF attractors of Table 1. As in [11], we highlight the key property features for each attractor, focusing on the maximum relative principal value enhancement $\varepsilon_{\max}$ of the effective conductivity tensor.

3.1. Elasticity-dominated steady states (ES structure attractors). Spatial elasticity dominates the viscous driving force from the plates when the Ericksen number and Deborah number are both sufficiently small. In this parameter regime, the experiment saturates in a steady structure in which stored elastic stresses balance the viscous stress. For fixed Ericksen number, Figures 1 and 2 show heterogeneity in the maximum relative principal value enhancement of effective conductivity at each gap height, as a

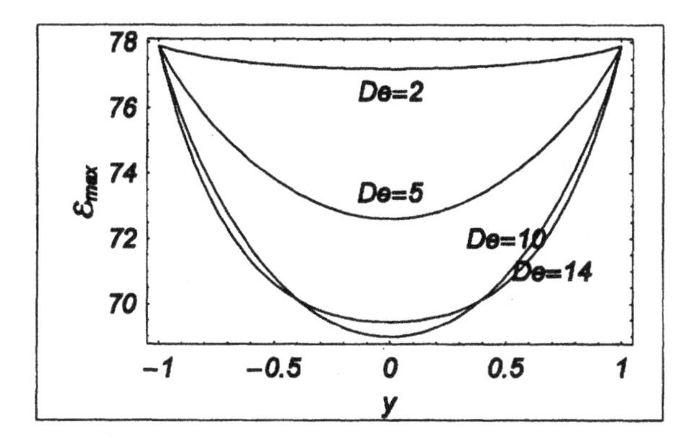

FIG. 1. *Maximum relative principal value enhancement of the effective conductivity tensor of steady states for fixed small Ericksen number $Er = 5$, and increasing Deborah number, $De = 2, 5, 10, 14$, with parallel anchoring.*

function of variable plate speeds (Deborah number). The comparison of Figure 1 and 2 shows the effect of boundary anchoring on properties. We underscore that the principal axis associated with the maximum conductivity is also varying across the film thickness, and that this principal axis follows the eigenvector associated with the maximum eigenvalue of $\mathbf{M}(f)$, called the major director or peak axis of orientation of the molecular distribution. Figure 2 illustrates the non-monotone Deborah number dependence of $\varepsilon_{\text{max}}(y)$, with increasing gradient morphology as De increases from 2 to 10, followed by a transition to more uniform spatial variation for $De = 15$.

3.2. Viscous-dominated steady states (VS structure attractors). For sufficiently high Deborah number, the viscous driving forces induced by the moving plates overwhelm short-range elasticity (which governs bulk monodomain dynamics), and the molecular distribution at each gap height aligns at some preferred direction. The anchoring conditions do not promote transient responses, so the material settles again into a steady structure between the plates. The maximum relative principal value enhancement of effective conductivity, analogous to Figures 1 and 2, are illustrated in Figures 3 and 4.

3.3. Composite tumbling-wagging periodic states (TW structure attractors). Dynamic structures occupy a large fraction of the parameter domain in Table 1. The so-called tumbling-wagging attractors have the distinguished feature of being periodic in time, with the PDF oscillating with finite amplitude of peak axis variation in layers near the plates, while rotating continuously in a mid-gap layer. The finite oscillation mode is called wagging, whereas the continuous rotation of the peak orientation axis is called tumbling. The distinguished optical and rheological feature, which has impact on mechanical as well as conductive properties, is that a small boundary layer between the tumbling and wagging layers emerges

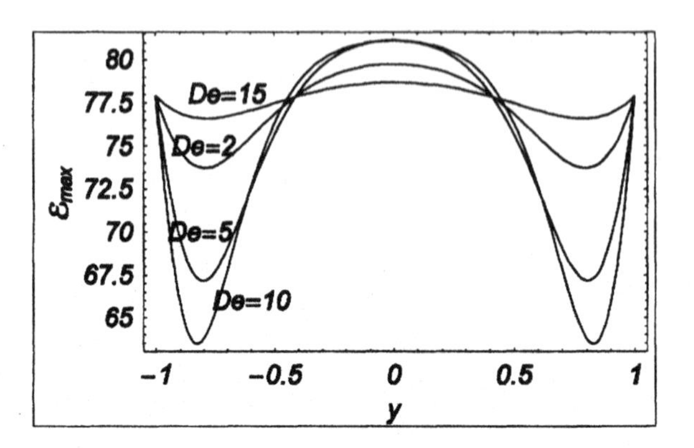

FIG. 2. *Maximum relative principal value enhancement of the effective conductivity tensor of steady states for fixed small Ericksen number* $Er = 5$*, and increasing Deborah number,* $De = 2, 5, 10, 15$*, with normal anchoring.*

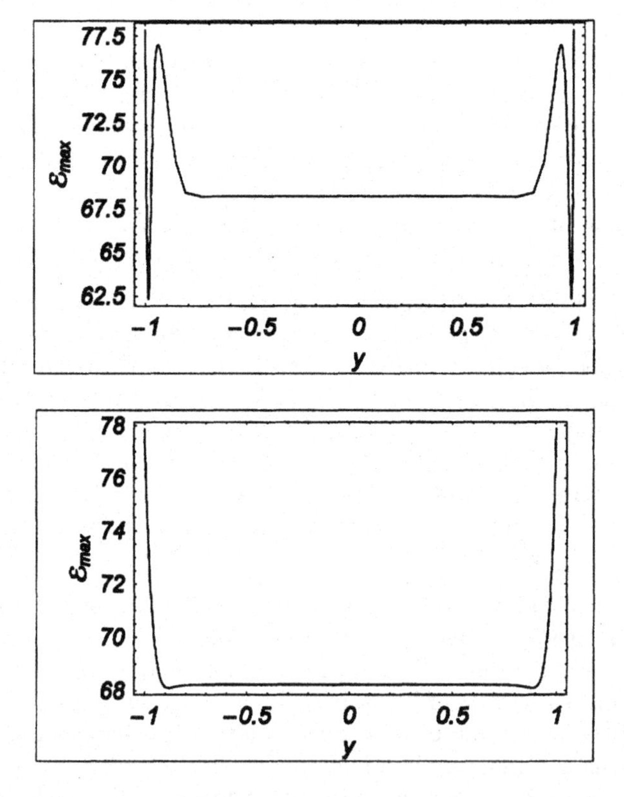

FIG. 3. *Maximum relative principal value enhancement of effective conductivity across the gap for* $Er = 1000$*,* $De = 10$*. Top: normal anchoring. Bottom: parallel anchoring.*

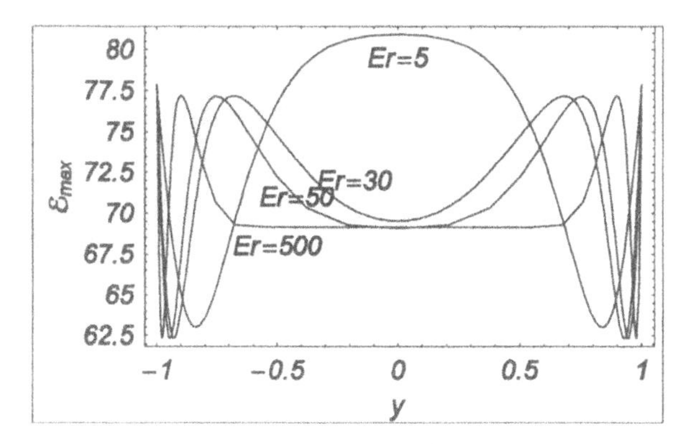

FIG. 4. *Maximum relative principal value enhancement of effective conductivity across the gap with fixed $De = 12$ and varying Er, for normal anchoring.*

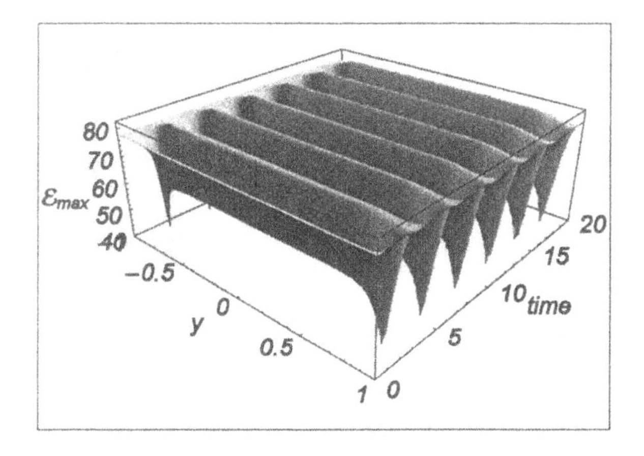

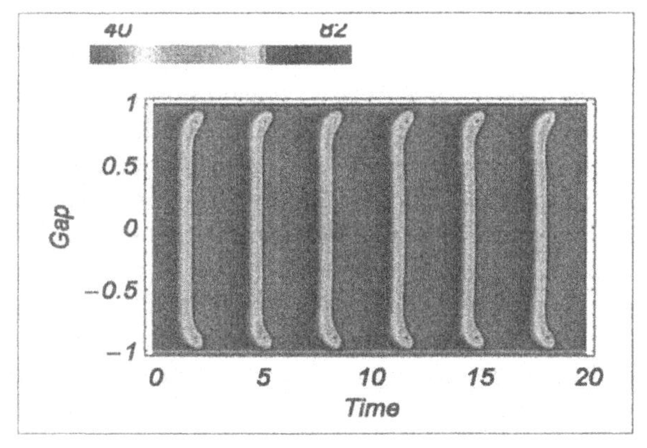

FIG. 5. *Spatio-temporal variations of maximum relative principal value enhancement of effective conductivity with $Er = 500$, $De = 4$, for parallel anchoring.*

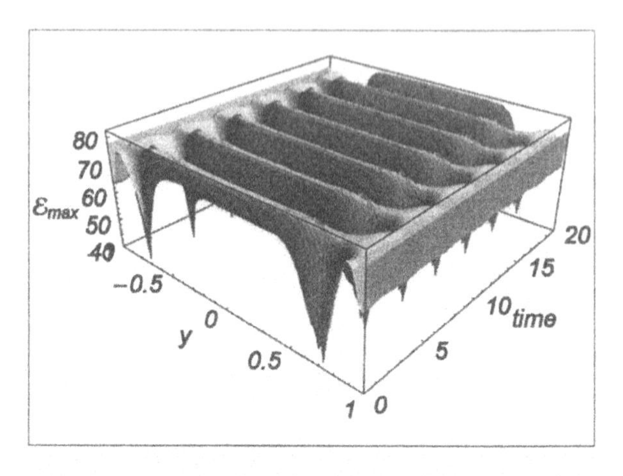

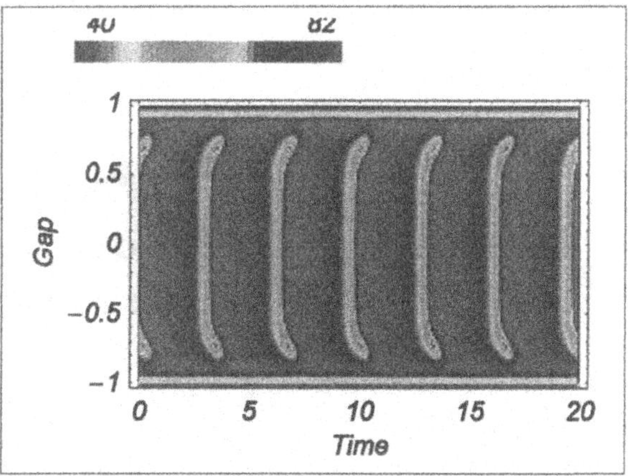

FIG. 6. *Tempo-spatial structure of maximum relative principal value enhancement of effective conductivity with $Er = 500$, $De = 4$, for normal anchoring.*

periodically. This layer experiences a precipitous drop in the degree of orientation, indeed the PDF goes isotropic in this layer, which is called a defect. Our purpose here is to amplify the consequences of these defect fluctuations on properties. Figures 5 and 6 show representative features. Since the orientation order parameter enters strongly into the principal conductivity values, one finds a dramatic drop of the maximum relative principal value enhancement of effective conductivity. Not shown is the related effect in which the principal axis of maximum conductivity becomes degenerate, and an entire plane of directions is associated with this drop in degree of orientation.

3.4. Wagging periodic states (W structure attractors). The other periodic attractor, in which the entire gap experiences finite amplitude oscillation of the PDF at each gap height, is called a wagging structure.

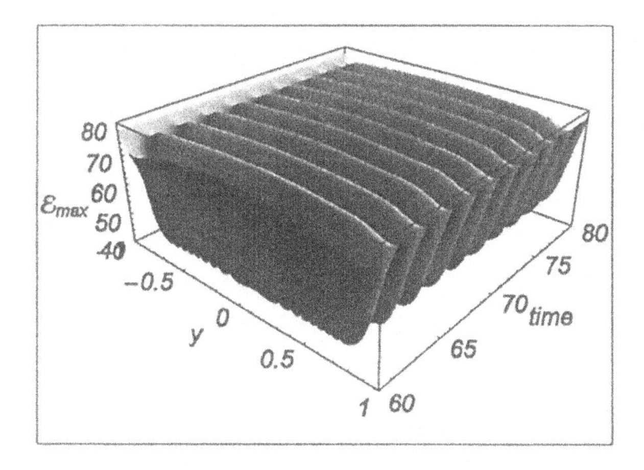

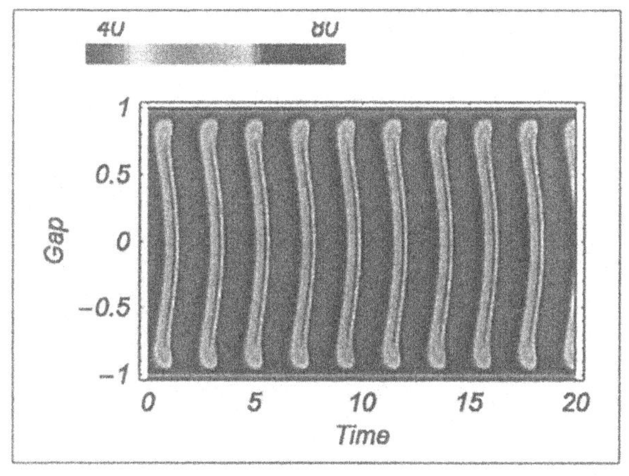

FIG. 7. *Spatial-temporal structure of maximum relative principal value enhancement of effective conductivity with $Er = 500$, $De = 6$, for parallel anchoring.*

The interesting feature of the PDF of wagging oscillations versus tumbling is that energy shifts into focusing and defocusing of the PDF rather than rotation. This means that there is apt to be more oscillation and variability in the principal values of the effective conductivity tensor, rather than tortuous paths of the principal axis. Figures 7 and 8 show representative features of these attractors.

4. Conclusions. We have connected two central features of nematic polymer nano-composite films: processing-induced orientational anisotropy and heterogeneity of the spheroidal inclusions, and the corresponding volume-averaged effective conductivity. Recent numerical simulations of film structures in Couette cells versus plate driving conditions and nematic elasticity have been translated into film properties. These are proof-of-principle results, in that idealized assumptions have been made which need

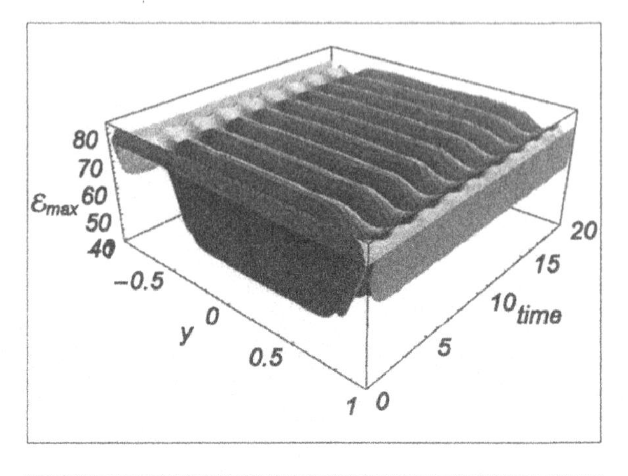

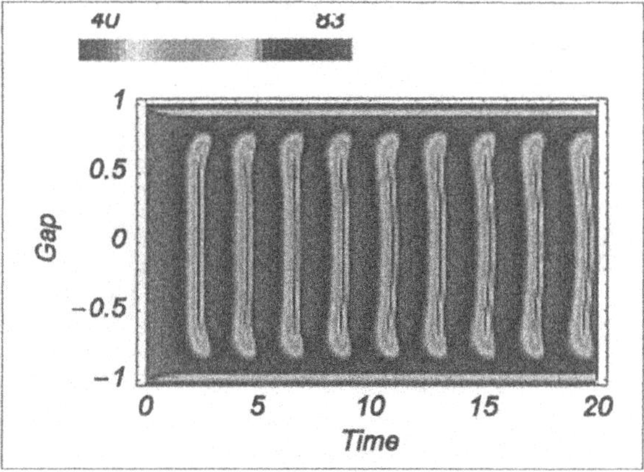

FIG. 8. *Spatial-temporal structure of maximum relative principal value enhancement of effective conductivity with $Er = 500$, $De = 6$, for normal anchoring.*

to be generalized to actual high performance materials. Examples include polymeric solvents, flexibility and concentration variability of the macromolecular ensembles, higher dimensional orientational configurations and spatial structures. Nonetheless, these are the first results to our knowledge, which give a sense of the property anisotropy and heterogeneity in shear-dominated processing of nematic polymer materials.

Acknowledgements. Effort sponsored by the Air Force Office of Scientific Research, Air Force Materials Command, USAF, under grant numbers F49620-02, 1-0086 and 6-0088, and the National Science Foundation through grants DMS-0204243, DMS-0308019. This work is supported in part by the NASA University Research, Engineering and Technology Institute on Bio Inspired Materials (BIMat) under award No. NCC-1-02037, and the Army Research Office Materials Division.

REFERENCES

[1] R.H. LUSTI, P.J. HINE, AND A.A. GUSEV, Direct numerical predictions for the elastic and thermoelastic properties of short fibre composites, *Composites Science and Technology*, 2002, **62**: 1927–1934.

[2] P.G. DE GENNES AND J. PROST, The Physics of Liquid Crystals, 1993, *Oxford University Press.*

[3] M.G. FOREST, R. ZHOU, AND Q. WANG, Kinetic structure simulations of nematic polymers in plane Couette cells, II: In-plane structure transitions, *SIAM Multiscale Modeling and Simulation*, 2005, accepted.

[4] M.G. FOREST AND Q. WANG, Hydrodynamic theories for blends of flexible and nematic polymers, Phys. Rev. E, 2005, accepted.

[5] G. KISS AND R.S. PORTER, Rheology of concentrated solutions of poly (γ-Bensyl-Glutamate), *J. Polymer Sci., Polym. Symp.*, 1978, **65**: 193–211.

[6] G. KISS AND R.S. PORTER, Rheology of concentrated solutions of helical polypetides, *J. Polymer Sci., Polym. Phys. Ed.*, 1980, **18**: 361–388.

[7] R.G. LARSON, The Structure and Rheology of Complex Fluids, 1999, *Oxford University Press.*

[8] R.G. LARSON AND D.W. MEAD, The Ericksen number and Deborah number cascades in sheared polymeric nematics, *Liquid Cryst.*, 1993, **15**: 151–169.

[9] A.D. REY AND M.M. DENN, Dynamical phenomena in liquid crystalline materials, *Annual Rev. Fluid Mech.*, 2002, **34**: 233–266.

[10] Q. WANG, A hydrodynamic theory of nematic liquid crystalline polymers of different configurations, *J. Chem. Phys.*, 2002, **20**: 9120–9136.

[11] X. ZHENG, M.G. FOREST, R. LIPTON, R. ZHOU, AND Q. WANG, Exact scaling laws for electrical conductivity properties of nematic polymer nano-composite monodomains, *Adv. Func. Mat.*, 2005, **15**: 627–638.

[12] R. ZHOU, M.G. FOREST, AND Q. WANG, Kinetic structure simulations of nematic polymers in plane Couette cells, I: The algorithm and benchmarks, *SIAM Multiscale Modeling and Simulation*, 2005, **3**: 853–870.

NON-NEWTONIAN CONSTITUTIVE EQUATIONS USING THE ORIENTATIONAL ORDER PARAMETER

HARALD PLEINER*, MARIO LIU†, AND HELMUT R. BRAND‡

Abstract. Nonlinear hydrodynamic equations for non-Newtonian fluids are discussed. We start from the recently derived hydrodynamic-like nonlinear description of a slowly relaxing orientational order parameter tensor. The reversible quadratic nonlinearities in this tensor's dynamics are material dependent due to the generalized nonlinear flow alignment effect that comes in addition to the material independent corotational convected derivative. In the entropy production these terms are balanced by linear and nonlinear orientational-elastic contributions to the stress tensor. These can be used to get a nonlinear dynamic equation for the stress tensor (sometimes called constitutive equation) in terms of a power series in the variables. A comparison with existing phenomenological models is given. In particular we discuss how these ad-hoc models fit into the hydrodynamic description and where the various non-Newtonian contributions are coming from. We also discuss the connection to the hydrodynamic-like description of non-Newtonian effects that employs a relaxing strain tensor.

Key words. Constitutive equations, orientational order parameter, non-Newtonian effects, hydrodynamics, flow alignment, relaxing strain tensor.

AMS(MOS) subject classifications. Primary 76A05, 74D10, 80A17, 76A15.

1. Introduction.

Hydrodynamics is a well established field to describe macroscopically simple fluids by means of the Navier-Stokes, continuity, and heat conduction equations. However, it applies also to more complex fluids that are fully characterized by conservation laws and broken symmetries. It is based on (the Gibbsian formulation of) thermodynamics [1, 2], symmetries and well-founded physical principles [3]. A detailed description of this method can be found in [4, 5]. This method can be generalized to include slowly relaxing variables that are relevant on experimental macroscopic time scales albeit being non-hydrodynamic. Examples are the soft mode near phase transitions [6, 7], the magnetic degree of freedom in ferrofluids [8, 9] and the relative velocity in 2-fluid descriptions [10]. The derivation of such macroscopic nonlinear dynamic equations is still based on first principles, making use of thermostatics, linear irreversible thermodynamics, symmetries and broken symmetries, and invariance principles. Only the choice of the slowly varying variable is heuristic and material dependent. In that sense non-Newtonian fluids are non-universal.

On the other hand, a host of different empirical models have been proposed [11–17] to cope with the rheology of such substances. Typically these models are formulated as generalizations of the linear, Newtonian relation between stress and deformational flow allowing for additional time

*Max Planck Institute for Polymer Research, 55021 Mainz, Germany.
†Theoretical Physics, University Tübingen, 72076 Tübingen, Germany.
‡Theoretische Physik III, Universität Bayreuth, 95440 Bayreuth, Germany.

derivatives and nonlinearities. They are tailored to accommodate empirical findings or are based on principles [15] that are ad-hoc and generally insufficient.

Quite recently we have derived a nonlinear hydrodynamic description of elastic media [18, 19] that has been confirmed within the GENERIC formalism [20]. Allowing in this hydrodynamic description the strains to relax (and not only to diffuse) a generalized hydrodynamic description of nonlinear viscoelasticity is obtained in terms of a dynamic equation for the (Eulerian) strain tensor [18, 19]. This strain tensor description can be transformed approximately into one that uses a dynamic equation for the stress tensor [21] and can thus be directly compared with many of the empirical models proposed to describe non-Newtonian rheology. The comparison reveals possible inconsistencies and connects the various ad-hoc additions of those models with physical relevant processes, like strain relaxation, elasticity and viscosity [21].

In this communication we use a different approach that relates non-Newtonian behavior to fluctuating, transient, and slowly relaxing orientational order. This has been used e.g. for describing the dynamics of semiflexible polymers, where long-lived polymer alignments and entanglements lead to viscoelastic effects [22]. The relaxational dynamics of the orientational order parameter tensor has been used in the isotropic phase of low molecular weight nematogens [23] describing orientational fluctuations that become important as pre-transitional effects near the phase transition. The relaxational (and non relaxational) dynamics of the orientational tensor has been derived and rederived pretty often [24–29]. Here, we will rely on the hydrodynamic description [30] that e.g. makes the clear distinction between reversible and irreversible processes and avoids any detours via additional auxiliary and unphysical dynamic variables. In Sec. 2 orientational elasticity and the phenomenological material tensors describing reversible and irreversible transport (flow alignment, viscosity, and relaxation) that are part of the hydrodynamic description are given as an expansion in powers of the orientational tensor. The back-flow effect in the stress tensor (Sec. 3), which is required for thermodynamic reasons, as well as the part of the viscosity that depends on the orientational tensor provide a coupling between the stress and the orientational tensor. This can be used to generate a dynamic equation for the stress tensor from that of the orientational tensor (Sec. 4). This translation is achieved by a power series expansion in the variables and can be done only approximately, since, generally, nonlinear equations cannot be inverted analytically. The power series is truncated after the quadratic order, since most of the phenomenological constitutive models, which we compare with in Sec. 5, are of that form. A summary (Sec. 6) of the main results concludes the paper.

2. Dynamics of the orientational order parameter tensor. The transient orientational order is described by a symmetric, traceles second rank tensor Q_{ij} ($Q_{ij} = Q_{ji}$ and $Q_{ii} = 0$). In contrast to the case of a nematic phase with spontaneous and permanent orientational order, there is no nematic order in equilibrium and a director does not exist. The relaxational dynamics of Q_{ij} can be written as [30]

$$\dot{Q}_{ij} + v_k \nabla_k Q_{ij} + Q_{jk}\Omega_{ki} + Q_{ik}\Omega_{kj} - \lambda_{ijkl}A_{kl} = -\alpha_{ijkl}\psi_{kl} \tag{2.1}$$

with $2A_{ij} = \nabla_j v_i + \nabla_i v_j$ and $2\Omega_{ij} = \nabla_j v_i - \nabla_i v_j$ the symmetric and antisymmetric velocity gradients characterizing deformational and rotational flow, respectively. The orientational elastic stress tensor ψ_{kl} is defined by the Gibbs relation [4]

$$d\epsilon - T d\sigma = v_i dg_i + \psi_{ij} dQ_{ij} + \mu d\rho. \tag{2.2}$$

as the conjugate to Q_{ij}. It has to be taken as symmetric and traceless, since only that part enters the Gibbs relation and has a physical meaning. The Gibbs relation contains all the other variables (density ρ, momentum density g_i, energy density ϵ or entropy density σ) and defines their conjugates (temperature T, velocity v_i, and chemical potential μ), where the latter are related to the more familiar (thermodynamic) pressure p by the Gibbs-Duhem equation

$$dp = \sigma dT + g_i dv_i - \psi_{ij} dQ_{ij} + \rho d\mu \tag{2.3}$$

In Eq. (2.1) the nonlinear reversible coupling terms to flow are a priori of the corotational or Jaumann derivative type (containing only Ω_{ij} the rotational flow – suitably for the orientational order involved), but there is in addition a phenomenological reversible coupling to symmetric velocity gradients that makes the effective convective derivative material dependent [30]. The phenomenological material tensor λ_{ijkl} (a kind of generalized flow alignment tensor) is given as a power series expansion in Q_{ij}

$$\begin{aligned}\lambda_{ijkl} &= \lambda_1 \left(\delta_{ik}\delta_{jl} + \delta_{jk}\delta_{il} - \frac{2}{3}\delta_{ij}\delta_{kl}\right) + \lambda_3 \delta_{kl} Q_{ij} \\ &+ \lambda_2 \left(\delta_{ik}Q_{jl} + \delta_{jk}Q_{il} + \delta_{jl}Q_{ik} + \delta_{il}Q_{jk} - \frac{4}{3}\delta_{ij}Q_{kl}\right) + O(2)\end{aligned} \tag{2.4}$$

where higher order terms $O(2)$ have been discussed in [30], but are not needed here. It contains one phenomenological, material dependent, reversible reactive coefficient in linear, and two additional ones in quadratic order. If in Eq. (2.1) the Jaumann terms are combined with the quadratic contribution (2.4) for the special value $\lambda_2 = \frac{1}{2}$ $(= -\frac{1}{2})$ one gets something that looks like an upper (lower) convected derivative – with some additional correction terms that ensure $\dot{Q}_{ii} = 0$. However, there is no general reason why such a relation should hold for all different materials nor can it hold for all temperatures and pressures, since $\lambda_{1,2,3}$ generally depend on all scalar state variables, like ρ, σ (or p, T) and on the invariants $Q_{ij}Q_{ij}$

and $Q_{ij}Q_{jk}Q_{ki}$. Within the quadratic approximation used here, the latter dependencies do not show up.

In [30] the relaxation of Q_{ij} has been given in linear approximation. More generally, the dissipative material tensor α_{ijkl} reads in a power series expansion in Q_{ij} (with $\alpha_{iikl} = 0 = \alpha_{ijkk}$)

$$\begin{aligned}\alpha_{ijkl} &= \alpha_1\left(\delta_{ik}\delta_{jl} + \delta_{jk}\delta_{il} - \frac{2}{3}\delta_{ij}\delta_{kl}\right) \\ &+ \alpha_2\left(\delta_{ik}Q_{jl} + \delta_{jk}Q_{il} + \delta_{jl}Q_{ik} + \delta_{il}Q_{jk} - \frac{4}{3}[\delta_{ij}Q_{kl} + \delta_{kl}Q_{ij}]\right) \\ &+ O(2)\end{aligned} \tag{2.5}$$

with the relaxation parameters $\alpha_{1,2}$ being functions of the scalar state variables. It should be noted that we stay very well inside the framework of "linear irreversible thermodynamics" that has a solid foundation in statistical mechanics, although the expressions (2.4, 2.5) and (3.3) below are genuinely nonlinear due to the dependence on state variables.

The orientational elastic stress is derived from an energy functional by the variational derivative $\psi_{ij} = \delta \int \epsilon \, dV/\delta Q_{ij}$, where only the trace free part enters Eqs. (2.1–2.3), which is given in quadratic order by

$$\psi_{ij} = c_1 Q_{ij} + c_2\left(Q_{ik}Q_{jk} - \frac{1}{3}\delta_{ij}Q_{kl}Q_{kl}\right) + O(2) \tag{2.6}$$

neglecting gradient terms. Near a phase transition the rotational elastic moduli c_1, c_2 can be interpreted as Landau parameters. Generally they are still functions of all scalar state variables.

Putting together Eqs. (2.1–2.6) the final dynamic orientational order parameter equations, quadratic in the variables, is obtained as

$$\begin{aligned}\dot{Q}_{ij} + v_k\nabla_k Q_{ij} + Q_{jk}\Omega_{ki} + Q_{ik}\Omega_{kj} - 2\lambda_1\left(A_{ij} - \frac{1}{3}\delta_{ij}A_{kk}\right) \\ -2\lambda_2\left(A_{il}Q_{jl} + A_{jl}Q_{il} - \frac{2}{3}\delta_{ij}A_{kl}Q_{kl}\right) - \lambda_3 Q_{ij}A_{kk} \\ = -\frac{1}{\tau_1}Q_{ij} - \frac{1}{\tau_2}\left(Q_{il}Q_{jl} - \frac{1}{3}\delta_{ij}Q_{kl}Q_{kl}\right)\end{aligned} \tag{2.7}$$

where the relaxation times are related to the elastic moduli and the relaxation parameters by $1/\tau_1 = 2c_1\alpha_1$ and $1/\tau_2 = 2c_2\alpha_1 + 4c_1\alpha_2$.

3. Stress tensor. In the preceding sections we discussed nonlinear reversible terms in the dynamic equation for the orientational order (2.1) that describe couplings to flow. In the Navier-Stokes or momentum conservation equation

$$\dot{g}_i + \nabla_j(v_j g_i + \delta_{ij}p + \sigma_{ij}) = 0, \tag{3.1}$$

on the other hand, there must be appropriate counter terms describing couplings to orientational order, due to the requirement of zero or positive entropy production, in the case of reversible and irreversible terms, respectively [4, 5, 31]. Their form can equivalently be derived from Onsager relations [32]. For the stress tensor σ_{ij} this leads to the expression

$$\sigma_{ij} = -\lambda_{klij}\psi_{kl} - \nu_{ijkl}A_{kl} \tag{3.2}$$

The counter term to the linear deformational flow term in (2.1), $\sim \lambda_{klij}$, leads to a symmetric part of the stress tensor, while there are no counter terms to the nonlinear Jaumann terms, since the latter do not at all contribute to the entropy production [30]. The viscosity tensor is again expanded in Q_{ij} as

$$\begin{aligned}\nu_{ijkl} = &\frac{\nu_1}{2}(\delta_{ik}\delta_{jl} + \delta_{il}\delta_{jk}) + \frac{\nu_2}{2}(Q_{ik}\delta_{jl} + Q_{jk}\delta_{il} + Q_{il}\delta_{jk} + Q_{jl}\delta_{ik}) \\ &+ \nu_3\delta_{ij}\delta_{kl} + \nu_4(\delta_{ij}Q_{kl} + \delta_{kl}Q_{ij})\end{aligned} \tag{3.3}$$

with the viscosities generally being functions of the scalar state variables.

Taking together Eqs. (2.5, 3.2, 3.3) the stress is given by

$$\begin{aligned}\sigma_{ij} = &-\nu_1 A_{ij} - \nu_2(Q_{ik}A_{jk} + Q_{jk}A_{ik}) - \nu_3\delta_{ij}A_{kk} \\ &- \nu_4(\delta_{ij}Q_{kl}A_{kl} + Q_{ij}A_{kk}) - \bar{\lambda}_1 Q_{ij} - \bar{\lambda}_2 Q_{ik}Q_{jk} - \bar{\lambda}_3\delta_{ij}Q_{kl}Q_{kl}\end{aligned} \tag{3.4}$$

where we have used the abbreviations $\bar{\lambda}_1 = 2c_1\lambda_1$, $\bar{\lambda}_2 = 2c_2\lambda_1 + 4c_1\lambda_2$, and $\bar{\lambda}_3 = c_1\lambda_3 - (2/3)c_2\lambda_1$. In the incompressible limit, which we will use below, $A_{kk} = 0$, and the viscosity ν_3 does not appear in the stress tensor, while λ_3 drops out of Eq. (2.7). If we allow for a "redefinition" of the pressure, $p \to p - \nu_4 Q_{kl}A_{kl} - \bar{\lambda}_3 Q_{kl}Q_{kl}$, also ν_4 and $\bar{\lambda}_3$ do not show up explicitly in the final equations. However, in that case p looses its simple physical meaning. For a general discussion of the incompressible limit and its connection to redefining the pressure cf. [33].

4. Dynamic stress tensor equation. Eqs. (2.7, 3.1, 3.4) constitute an (isothermal) description of viscoelasticty based on a relaxing orientational order parameter tensor. This hydrodynamic-like description contains as special cases [30] some of the well-known model-based descriptions of viscoelasticity that also employ the orientational order parameter tensor, like e.g. the Doi-Edwards model for isotropic semiflexible polymers [22]. However, most of the heuristic constitutive models are written in terms of a dynamic equation for the stress tensor, very often quadratic in the variables and under the assumption of incompressibility. In order to compare with those models we have to translate our $\dot{Q}_{ij}/\dot{g}_i$ into a $\dot{\sigma}_{ij}/\dot{g}_i$ description by replacing the orientational order parameter tensor (and its derivatives) by the stress tensor (and its derivatives). This can only be done in an approximate way, since the equations are nonlinear. We will set up a power series expansion up to second order in the (old and new) variables. Of course, the resulting equations are less general than the starting ones and only applicable, if quadratic nonlinearities are sufficient for the problem at hand. This procedure is similar in spirit to that in [21], where we used the hydrodynamic-like description of viscoelasticity in terms of a relaxing Eulerian strain tensor [18, 19] and translated it into a dynamic stress tensor description, again to facilitate comparison.

Taking the derivative $d/dt = \partial/\partial t + v_i \nabla_i$ of σ_{ij} in (3.4) and replacing dQ_{ij}/dt according to Eq. (2.7) we get

$$\frac{d}{dt}\sigma_{ij} = -f\left(Q_{ij}, A_{ij}, \frac{d}{dt}A_{ij}, \Omega_{ij}\right) \tag{4.1}$$

in terms of the orientational order parameter tensor and flow. To convert this into the desired dynamic equation for the stress tensor, we have to invert $\sigma_{ij} = \sigma_{ij}(Q_{ij}, A_{ij})$, Eq. (3.4), into $Q_{ij} = Q_{ij}(\sigma_{ij}, A_{ij})$. This is done approximately by the power expansion $Q_{ij} = Q_{ij}^{(lin)} + Q_{ij}^{(quad)} + \ldots$, where $Q_{ij}^{(lin)}$ and $Q_{ij}^{(quad)}$ contain expressions linear and quadratic in the variables, respectively. In particular we find

$$\bar{\lambda}_1 Q_{ij}^{(lin)} = -\sigma_{ij}^0 - \nu_1 A_{ij} \tag{4.2}$$

$$\begin{aligned} \bar{\lambda}_1^3 Q_{ij}^{(quad)} &= -\bar{\lambda}_2(\sigma_{ik}\sigma_{jk})^0 + (\bar{\lambda}_1\nu_2 - \bar{\lambda}_2\nu_1)(\sigma_{ik}A_{jk} + \sigma_{jk}A_{ik})^0 \\ &\quad + \nu_1(2\bar{\lambda}_1\nu_2 - \bar{\lambda}_2\nu_1)(A_{ik}A_{jk})^0 \end{aligned} \tag{4.3}$$

where the superscript ^{0} denotes the traceless part of the associated tensor. Since we assume incompressibility, A_{ij} is traceless by itself.

Using these expressions the dynamic equation for the stress tensor takes the final form

$$\begin{aligned} \tau_1 \frac{D_s}{Dt}\sigma_{ij} + \sigma_{ij} = &-\nu_\infty A_{ij} - \nu_1\tau_1 \frac{D_q}{Dt}A_{ij} + \frac{r}{2c_1\lambda_1}\sigma_{ik}\sigma_{jk} + \frac{1}{3}\delta_{ij}\Sigma \\ &+ \frac{\tau_1\nu_2}{2c_1\lambda_1}\left([\sigma_{jk}+\nu_1 A_{jk}]\frac{\partial}{\partial t}A_{ik} + [\sigma_{ik}+\nu_1 A_{ik}]\frac{\partial}{\partial t}A_{jk}\right) \\ &+ O(3) \end{aligned} \tag{4.4}$$

where

$$\nu_\infty = \nu_1 + 4c_1\tau_1\lambda_1^2 \qquad \text{and} \qquad r = \frac{\tau_1}{\tau_2} + \frac{c_2}{c_1} + 2\frac{\lambda_2}{\lambda_1} \tag{4.5}$$

and

$$\frac{D_s}{Dt}T_{ij} \equiv \frac{d}{dt}T_{ij} - s(T_{ik}A_{jk} + T_{jk}A_{ik}) - (T_{ik}\Omega_{jk} + T_{jk}\Omega_{ik}) \tag{4.6}$$

for any tensor T_{ij} and number s. For $s = -1$ ($s = +1$) D_s/Dt is the lower (upper) convected derivative, for $s = 0$ the Jaumann or corotational derivative, while for a general s a linear combination of those is invoked. In our case the numbers s and q are

$$s = -2\lambda_1\left(\frac{c_2}{c_1} + 3\frac{\lambda_2}{\lambda_1}\right) - \frac{\nu_1 r}{2c_1\lambda_1\tau_1} \tag{4.7}$$

$$q = -2\lambda_1\left(\frac{c_2}{c_1} + 3\frac{\lambda_2}{\lambda_1} - \frac{\nu_2}{\nu_1}\right) - \frac{\nu_1 r}{4c_1\lambda_1\tau_1} \tag{4.8}$$

where r is given in Eq. (4.5). The part $\sim \delta_{ij}$ in Eq. (4.4) is due to the fact that Q_{ij} is traceless, while σ_{ij} is not. It can in principle be incorporated into the pressure term by a redefinition $p \to p + (1/3)\Sigma$, where

$$\Sigma = \sigma_{kk} + x\,A_{kl}A_{kl} + y\,\sigma_{kl}\sigma_{kl} + z\,\sigma_{kl}A_{kl} + \frac{3\nu_4\tau_1}{2c_1\lambda_1}(\nu_1 A_{kl} + \sigma_{kl})\frac{\partial}{\partial t}A_{kl} \tag{4.9}$$

with

$$\begin{aligned} x = {} & \frac{\nu_1^2}{2c_1\lambda_1}\left(\frac{\lambda_3}{\lambda_1} - \frac{\tau_1}{\tau_2} + \frac{c_2}{3c_1} + \frac{2\lambda_2}{\lambda_1}\right) - \frac{\nu_1(2\nu_2 + 3\nu_4)}{2c_1\lambda_1} \\ & - 2\nu_1\lambda_1\tau_1\left(\frac{2\lambda_2}{\lambda_1} - \frac{\lambda_3}{\lambda_1} + \frac{2c_2}{3c_1}\right) - 6\lambda_1\tau_1\nu_4 \end{aligned} \tag{4.10}$$

$$y = \frac{1}{2c_1\lambda_1}\left(\frac{\lambda_3}{\lambda_1} - \frac{\tau_1}{\tau_2} + \frac{c_2}{3c_1} + \frac{2\lambda_2}{\lambda_1}\right) \tag{4.11}$$

$$\begin{aligned} z = {} & \frac{\nu_1}{c_1\lambda_1}\left(\frac{2\lambda_3}{\lambda_1} - \frac{2\tau_1}{\tau_2} - \frac{c_2}{3c_1} + \frac{2\lambda_2}{\lambda_1}\right) - \frac{\nu_2}{c_1\lambda_1} \\ & - 2\lambda_1\tau_1\left(\frac{2\lambda_2}{\lambda_1} - \frac{\lambda_3}{\lambda_1} + \frac{2c_2}{3c_1}\right) - \frac{3\nu_4}{2c_1\lambda_1} \end{aligned} \tag{4.12}$$

This "redefinition" of the pressure, however, is rather dubious, since it renders a "pressure" that depends nonlinearly on flow and its time derivative, and even more disturbing on the stress tensor itself. It is completely different from the appropriate "redefinition" in the $\dot{Q}_{ij}/\dot{g}_i$ description of Sec. 3. In a more reasonable description one notices that σ_{kk} and its derivative are at least of quadratic order (for $A_{kk} = 0$) and do not influence the constitutive equation for σ_{ij}^0 in that order. For the latter one then gets finally

$$\begin{aligned} \tau_1\left(\frac{D_s}{Dt}\sigma_{ij}\right)^0 + \sigma_{ij}^0 = {} & -\nu_\infty A_{ij} - \nu_1\tau_1\left(\frac{D_q}{Dt}A_{ij}\right)^0 + \frac{r}{2c_1\lambda_1}(\sigma_{ik}\sigma_{jk})^0 \\ & + \frac{\tau_1\nu_2}{2c_1\lambda_1}\left([\sigma_{jk} + \nu_1 A_{jk}]\frac{\partial}{\partial t}A_{ik} + [\sigma_{ik} + \nu_1 A_{ik}]\frac{\partial}{\partial t}A_{jk}\right)^0 \\ & + O(3) \end{aligned} \tag{4.13}$$

with the coefficients defined above. The time evolution of the trace

$$\begin{aligned}\tau_1 \frac{\partial}{\partial t}\sigma_{kk} &= \left(\frac{r}{2c_1\lambda_1} + y\right)\sigma^0_{kl}\sigma^0_{kl} + x\,A_{kl}A_{kl} + z\,\sigma^0_{kl}A_{kl} \\ &+ \frac{\tau_1}{2c_1\lambda_1}(2\nu_2 + 3\nu_4)(\nu_1 A_{kl} + \sigma^0_{kl})\frac{\partial}{\partial t}A_{kl} + O(3)\end{aligned} \tag{4.14}$$

is completely determined by σ^0_{ij} and A_{ij} in lowest order.

5. Comparison with constitutive models. Eq. (4.4) constitutes the most general form for a constitutive equation (up to quadratic order in the variables) that can be derived from a transient orientational order parameter as source of non-Newtonian behavior. It contains eight material coefficients (four linear and four quadratic ones with subscript 1 and 2, respectively), and two more in the trace part (λ_3, ν_4), characterizing orientational elasticity, relaxation of orientational order, viscosity and flow alignment. These coefficients are still functions of density and temperature. Most of the traditional constitutive models are much simpler than Eq. (4.4). We will now discuss, whether and how these models fit into the frame derived above.

The general case (4.4) contains the relaxation of stresses as well as of flow with relaxation times τ_1 and $\tau_1\nu_1/\nu_\infty$, respectively. Here the effective viscosity ν_∞ is different from the bare one (ν_1) due to the relaxation of orientational order and its coupling to flow via the flow alignment effect. Thus, the Maxwell [15], Johnson-Segalman [16], and Giesekus [14] models, which neglect flow relaxation, implicitly assume $\nu_1 = 0$ and ν_∞ is completely due to flow alignment. The quadratic stress contribution $\sim r$ in (4.4) is nonzero (as in the Giesekus model) only, if at least one of the second order material parameters, $c_2, \lambda_2, 1/\tau_2$, is nonzero. Vice versa, all the other models (including the Oldroyd [11] and Jeffreys [15] models) that have $r = 0$ also implicitly assume $c_2 = 0 = \lambda_2 = 0 = 1/\tau_2$. In principle, it would be possible to have $r = 0$ for a special set of nonzero values of the second order parameters, but this would be highly incidentally and would work only for one special point in phase space (for one combination of density and temperature), but not in general. As a consequence the nature of the convected derivatives of stress and flow, characterized in (4.4) by s, q, is fixed to be of the corotational or Jaumann type ($s = 0 = q$), since in all the models mentioned above there is either $\nu_1 = 0$ or $c_2 = 0 = \lambda_2 = 0 = 1/\tau_2$ or both. Thus, only the Jeffreys and Johnson-Segalman model (the latter in the version with the corotational convective derivative of the stress tensor) are compatible with viscoelasticity due to transient orientational order. These models also consistently lack the complicated nonlinear term in the second line of (4.4), since they have $\nu_2 = 0$; in addition, they miss the trace part $\sim \Sigma$. That means in these models the pressure has to be interpreted as the redefined pressure discussed above, rather than the thermodynamic hydrostatic pressure.

6. Summary. We have explored the hydrodynamic form of non-Newtonian fluid dynamics, if viscoelasticity is due to transient orientational order. The dynamic equation for the orientational order parameter tensor has been converted approximately into a dynamic equation for the stress tensor, which is then compared with traditional constitutive models. Due to the intricate relations among the coefficients of the nonlinearities in this effective constitutive equation some of the models are incompatible with this type of visoelasticity, since they lack one type of nonlinearity, but inconsistently not some other one, or they assume a special type of convective derivative incompatible with other choices of the nonlinear terms. Compatible are a generalized Giesekus model (with the convective derivative of the stress tensor being material dependent, in general), the Jeffreys model and the Johnson-Segalman model with the corotational convective derivative for the stress tensor. This is quite complementary to our recent findings [21] that the latter two models are incompatible with viscoelasticity due to transient elasticity characterized by a relaxing strain tensor, while Maxwell and Oldroyd models (incompatible in the present case) have been found to be compatible. The deeper reason for this difference lies in the type of viscoelasticity used, either a transient elasticity leading to a relaxing strain tensor that contains the lower (upper) convected time derivative in the Eulerian (Lagrangian) case [18, 19], or a transient orientational order leading to a relaxing orientational order parameter tensor that contains the corotational convected time derivative modified by second order flow alignment material parameters [30]. Of course, in nature both (and even other) sources of viscoelasticity can be present allowing all these models to exist, but one should bear in mind that the general effective constitutive equation obtained in that way is by far richer and more complicated than any of the traditional models.

REFERENCES

[1] H.B. CALLEN, *Thermodynamics*, John Wiley, New York, 2nd ed., 1985.

[2] L.E. REICHL, *A Modern Course in Statistical Physics*, Texas University Press, Austin, 1980.

[3] D. FORSTER, *Hydrodynamic Fluctuations, Broken Symmetry and Correlation Functions*, Benjamin, Reading, Mass., 1975.

[4] P.C. MARTIN, O. PARODI, AND P.S. PERSHAN, *Unified hydrodynamic theory for crystals, liquid crystals, and normal fluids*, Phys. Rev. A, **6** (1972), pp. 2401–2420.

[5] H. PLEINER AND H.R. BRAND, *Hydrodynamics and Electrohydrodynamics of Nematic Liquid Crystals*, in Pattern Formation in Liquid Crystals, A. Buka and L. Kramer (eds.), Springer, New York, (1996), pp. 15–67.

[6] I.M. KHALATNIKOV, *Introduction to the Theory of Superfluidity*, Benjamin, New York, 1965.

[7] M. LIU, *Hydrodynamic theory near the nematic - smectic A transition*, Phys. Rev. A **19** (1979), pp. 2090–2094.

[8] H.-W. MÜLLER AND M. LIU, *Shear Excited Sound in Magnetic Fluid*, Phys. Rev. Lett. **89** (2002), no. 67201.

[9] E. Jarkova, H. Pleiner, H.-W. Müller, and H.R. Brand, *Macroscopic Dynamics of Ferronematics*, J. Chem. Phys. **118** (2003), pp. 2422–2430.

[10] H. Pleiner and J.L. Harden, *General Nonlinear 2-Fluid Hydrodynamics of Complex Fluids and Soft Matter*, Nonlinear Problems of Continuum Mechanics, Special issue of Notices of Universities. South of Russia. Natural sciences (2003), pp. 46–61 and AIP Conference Proceedings **708** (2004), pp. 46–51.

[11] J.G. Oldroyd, *On the formulation of equations of state*, Proc. Roy. Soc. A **200** (1950), pp. 523–541 and *The hydrodynamics of materials whose rheological properties are complicated*, Rheol. Acta **1** (1961), pp. 337–344.

[12] B.D. Coleman and W. Noll, *Foundations of linear viscoelasticity*, Rev. Mod. Phys. **33** (1961), pp. 239–249.

[13] C. Truesdell and W. Noll, *The non-linear field theories of mechanics*, Springer, Berlin/New York, 1965.

[14] H. Giesekus, *Die Elastizität von Flüssigkeiten*, Rheol. Acta **5** (1966), pp. 29–35 and *A simple constitutive equation for polymer fluids based on the concept of deformation-dependent tensorial mobility*, J. Non-Newt. Fluid Mech. **11** (1982), pp. 69–109.

[15] R.B. Bird, R.C. Armstrong, and O. Hassager, *Dynamics of Polymeric Liquids*, Vol. 1, John Wiley & Sons, New York, 1977.

[16] M.W. Johnson and D. Segalman, *Model for viscoelastic flow behavior which allows non-affine deformation*, J. Non-Newt. Fluid Mech. **2** (1977), pp. 255–270 and J. Rheol. **22** (1978), pp. 445–446.

[17] R.G. Larson, *Constitutive equations for polymer melts and solutions*, Butterworths, Boston, 1988.

[18] H. Temmen, H. Pleiner, M. Liu, and H.R. Brand, *Convective Nonlinearity in Non-Newtonian Fluids*, Phys. Rev. Lett. **84** (2000), pp. 3228–3231 and **86** (2001), p. 745.

[19] H. Pleiner, M. Liu, and H.R. Brand, *"The Structure of Convective Nonlinearities in Polymer Rheology*, Rheol. Acta **39** (2000), pp. 560–565.

[20] M. Grmela, *Lagrange hydrodynamics as extended Euler hydrodynamics: Hamiltonian and GENERIC structures*, Phys. Lett. A **296** (2002), pp. 97–104.

[21] H. Pleiner, M. Liu, and H.R. Brand, *Nonlinear Fluid Dynamics Description of non-Newtonian Fluids*, Rheol. Acta **43** (2004), pp. 502–508 and *A physicists' view on constitutive equations*, Proc. XIVth Intern. Congress on Rheology, Seoul 2004, pp. 168–170.

[22] M. Doi and S.F. Edwards, *The theory of polymer dynamics*, Clarendon Press Oxford, 1986.

[23] P.G. de Gennes and J. Prost, *The physics of liquid crystals*, Clarendon Press Oxford, 1993.

[24] M. Grmela, *Bracket formulation of dissipative fluid-mechanics equations*, Phys. Lett. A **102** (1984), pp. 355–358.

[25] S. Hess, *Irreversible thermodynamics of nonequilibrium alignment phenomena in molecular liquids and in liquid crystals. I. Derivation of nonlinear constitutive laws, relaxation of the alignment, phase transition*, Z. Naturforsch. **30a** (1975), pp. 728–738.

[26] S. Hess, *Irreversible thermodynamics of nonequilibrium alignment phenomena in molecular liquids and in liquid crystals. II. Viscous flow and flow alignment in the isotropic (stable and metastable) and nematic phases*, Z. Naturforsch. **30a** (1975), pp. 1224–1232.

[27] P.D. Olmsted and P.M. Goldbart, *Isotropic-nematic transition in shear flow: State selection, coexistence, phase transitions, and critical behavior*, Phys. Rev. A **46** (1992), pp. 4966–4993.

[28] A.N. Beris and B.J. Edwards, *Thermodynamics of flowing systems with internal microstructure*, University Press, Oxford (1994).

[29] A.M. Sonnet, P.L. Maffetone, and E.G. Virga, *Continuum theory for nematic liquid crystals with tensorial order*, J. Non-Newt. Fluid Mech. **119** (2004),

pp. 51–59.

[30] H. PLEINER, M. LIU, AND H.R. BRAND, *Convective Nonlinearities for the Orientational Tensor Order Parameter in Polymeric Systems*, Rheol. Acta **41** (2002), pp. 375–382.

[31] M. GRMELA, *Stress tensor in generalized hydrodynamics*, Phys. Lett. A **111** (1985), pp. 41–44.

[32] S.R. DEGROOT AND P. MAZUR, *Nonequilibrium Thermodynamics*, 2nd ed., Dover, New York, 1984.

[33] H. PLEINER AND H.R. BRAND, *Incompressibility Conditions in Liquid Crystals*, Continuum Mech. Thermodyn. **14** (2002), pp. 297–306.

[34] J.G. OLDROYD, *Non-Newtonian effects in steady motion of some idealized elasto-viscous liquids*, Proc. Roy. Soc. A **245** (1958), pp. 278–297.

SURFACE ORDER FORCES IN NEMATIC LIQUID CRYSTALS*

FULVIO BISI† AND EPIFANIO G. VIRGA†‡

Abstract. The notion of surface order force in nematic liquid crystals is presented and contrasted with the notions of similar forces already introduced in the literature. We illustrate how a surface order force could in principle be measured and how it would convey the mechanical signature of an intrinsically nanoscopic phenomenon, often referred to as *order reconstruction.* The relationship between this force and the occurrence of biaxial states of the nematic order tensor is further illuminated.

Key words. Nematic liquid crystals, order reconstruction; biaxial ordering; order forces.

AMS(MOS) subject classifications. 76A15 Liquid crystals; 82B21 Continuum models (systems of particles, etc.)

1. Introduction.

What we shall call here *order* forces have also been given other names in the past, such as *structural* forces, or *solvation* forces, or *hydration* forces—especially in aqueous media [1, 2]. We shall reserve the name of structural forces for the forces so called by Horn, Israelachivili and Perez in their seminal paper [1], which illuminates the basic distinctions between different forces exchanged by solid, smooth surfaces immersed in a nematic liquid crystal, which here serves as a paradigm for ordered fluids in general. As also recalled in [1], when two solid bodies approach one another, they can interact *directly*, for example through electrostatic or Van der Waals forces, or *indirectly*, through forces mediated by a fluid placed between them.

In general, the interactions between a molecularly smooth solid surface and the molecules of a fluid in contact with it affect the molecular order in the fluid; the order can thus be either enhanced or depressed in the vicinity of the surface. This effect on the order propagates in the fluid for some characteristic distance ξ, as a result of the molecular interactions. When two surfaces confine a fluid, the alterations in the free energy resulting from changes in the surface order ultimately contribute to the force between the confining surfaces, as their separation becomes comparable with ξ. Depending on the size of ξ, we can distinguish amongst different types of such indirect forces, which we now proceed to describe in the case where the fluid is a liquid crystal.

Liquid crystals are constituted by elongated molecules. Such a molecular anisotropy would not be sufficient by itself to generate liquid crystal

*This work has been partially funded by the Institute for Mathematics and its Applications (IMA).

†Dipartimento di Matematica, Istituto Nazionale di Fisica della Materia, Università di Pavia, via Ferrata 1, 27100 Pavia, Italy (fulvio.bisi@unipv.it).

‡virga@imati.cnr.it.

phases: it must also be reflected in the intermolecular potential, as, for example, in the mean-field model of Maier and Saupe [3]. On a macroscopic scale, the tendency of liquid crystal molecules to be oriented in a common direction is described by assigning a unit vector field, the nematic director $\boldsymbol{n}$, which represents the local average orientation of molecules.

The classical elastic theory of Oseen [4] and Frank [5] employs the director field $\boldsymbol{n}$ to describe the local state of the fluid and assumes that the free-energy density depends in a quadratic fashion on $\nabla\boldsymbol{n}$. No internal length scale is present in the theory, so that local distortions of $\boldsymbol{n}$, mainly forced upon it by contrasting boundary conditions, decay over a distance comparable with the distance d between the bounding surfaces.

A scalar order parameter S often accompanies $\boldsymbol{n}$ to describe the degree at which molecules are aligned. If $\boldsymbol{\ell}$ is the unit vector along the individual molecular axis, S is defined by

$$\langle(\boldsymbol{\ell}\cdot\boldsymbol{n})^2\rangle =: \frac{1}{3}(2S+1),$$

where $\langle\cdot\rangle$ denotes an ensemble molecular average. The nematic coherence length ξ_n is defined as the length over which disturbances in the equilibrium value of S decay in space. The nematic coherence length depends on the temperature and it is typically nanometric, that is, larger than the molecular size μ: it thus reflects the long-range ordering interactions responsible for the very existence of liquid crystal phases.

In the foregoing discussion, we introduced three separate length scales, namely, d, ξ_n, and μ. We now see how three different order forces can be identified by letting ξ coincide with each of them.

Elastic forces. When the director $\boldsymbol{n}$ is prescribed on smooth, rigid surfaces that bound a nematic liquid crystal, the force exchanged by the surfaces, related to the distortions of $\boldsymbol{n}$ in the liquid crystal, and thus called elastic, can be considered as an order force, as it results from the ability of the surfaces to orient the director. For elastic forces, $\xi \approx d$. All elastic forces are repulsive; their strength increases monotonically as d decreases.

Structural forces. Most surfaces that come in contact with a liquid crystal enhance the order parameter S in a boundary layer [6, 7]. In the absence of other disturbing influences, S decays back to its bulk equilibrium value S_b within a length comparable with ξ_n. Since the difference between the surface value of S and S_b causes a local increase in the free energy, when the region occupied by the liquid crystal is so thin that facing boundary layers come within a distance ξ_n from one another and partially overlap, a repulsive order force, called structural force in [1], is expected to arise between the rigid surfaces nearly brought in contact. For these forces $\xi \approx \xi_n$. The structural forces described in [1] are repulsive and monotonic like the elastic forces.

Positional ordering forces. This class of forces is related to the presence of a certain positional order induced by the surface, which is short-range, and so can only be transmitted to within a few molecular lengths. Seen at the molecular level, the liquid crystal layer adjacent to a solid bounding surface reveals a positional structure that results from the adhesion of the first molecular layer to the surface. This structure clearly depends on the nature of the substrate responsible for the liquid crystal anchoring. The force that manifests itself when the separation between two boundary layers is comparable with the molecular size μ can also fail to be monotonic, as illuminated by the following example. If in a nematic liquid crystal cell the molecules in the first surface layer are aligned orthogonally to the bounding plates, further layers of molecules form close to this and show a local smectic ordering. Upon decreasing the distance d between the plates, an oscillating force may be observed due to the periodical variation of the free energy as a result of the interdigitation of the smectic layers.

Evidence of these phenomena, relating mechanical properties and local ordering, has already been obtained in several experiments [8, 9, 10, 11]. In a force-controlled experiment, the transition from the regime where the order force is increasing for decreasing d to the regime where the order force could also be decreasing is marked by a *snapping* instability in the force-displacement diagram, usually associated with a hysteresis loop. The first occurrence of such an instability is taken as the sign that the force being measured ceases to be structural and thus starts revealing the positional molecular ordering. However, it has recently been shown that this transition from one regime to the other is far more subtle when the biaxial degree of order is also considered, which arises when the local molecular ordering is described within a finer resolution through de Gennes' order tensor $\mathbf{Q}$. The biaxial degree of order relaxes over a characteristic coherence length ξ_b, comparable with ξ_n, at least away from the transition to the isotropic phase. The force associated with the local biaxial ordering can fail to be monotonic as a function of the distance between two approaching surfaces [12]. In a force-controlled experiment, such a lack of monotonicity would cause precisely the same snapping instability so far attributed to the occurrence of positional ordering forces.

Biaxial surface order forces are intimately related to *order reconstruction* in the bulk. This phenomenon was first described in the core of liquid crystal defects by Schopohl and Sluckin [13]. Subsequently, Palffy-Muhoray, Gartland and Kelly [14] recognized the same pattern in a thin cell between two parallel plates with contrasting uniaxial anchorings (one planar and the other homeotropic). It has been shown that in this system the two uniaxial states on the boundaries can be connected both through a director bend and through a transformation which does not involve any director rotation, a transformation where two uniaxial states can be transformed into one another by letting one eigenvalue of $\mathbf{Q}$ grow at the expense of another, until a new uniaxial state, differently oriented, is reached via a wealth of biaxial states, where $\mathbf{Q}$ has different eigenvalues, but the same eigenframe.

In this article we recall the stress fields that describe the distributions of both internal forces and internal torques in a liquid crystal when the order tensor $\mathbf{Q}$ also attains biaxial states. With the aid of these fields we can compute both force and torque transmitted from one plate to the other of a classical nematic twist cell and we thus see how both these mechanical actions are affected by the order reconstruction in the bulk. We shall show that both a force- and a torque-controlled machine would experience a snapping instability, as a distinctive sign of order reconstruction.

2. Energy and stresses. We describe the local nematic state of a liquid crystal by means of the order tensor $\mathbf{Q}$, which is a symmetric, traceless tensor of rank two related to the second moments of the probability distribution of the molecular long axes (see pp. 56–57 of [15]). Since $\mathbf{Q}$ is symmetric, it can be represented in the orthonormal basis of its eigenvectors $\{\boldsymbol{e}_1, \boldsymbol{e}_2, \boldsymbol{e}_3\}$ as

$$\mathbf{Q} = \sum_{i=1}^{3} \lambda_i \boldsymbol{e}_i \otimes \boldsymbol{e}_i,$$

where the eigenvalues λ_i must obey the constraint

$$\lambda_1 + \lambda_2 + \lambda_3 = 0\,.$$

A *uniaxial state* is described by the condition that two eigenvalues coincide, and, in that case, we can write

$$\mathbf{Q} = S\Big(\boldsymbol{n} \otimes \boldsymbol{n} - \frac{1}{3}\mathbf{I}\Big), \tag{2.1}$$

where $S \in [-\frac{1}{2}, 1]$ is the scalar order parameter, $\boldsymbol{n}$ is the nematic director, and $\mathbf{I}$ is the identity tensor. It is worth noticing that the upper bound of S corresponds to the configuration where all molecules are oriented along $\boldsymbol{n}$, whilst the lower bound corresponds to a configuration where the molecules are on average isotropically distributed in the plane orthogonal to $\boldsymbol{n}$. A *biaxial state* is characterized by the condition that all eigenvalues of $\mathbf{Q}$ be distinct; an index of how far a biaxial state is from a uniaxial one is the *degree of biaxiality*, which can be defined as [16]

$$\beta^2 := 1 - 6\frac{(\mathrm{tr}\mathbf{Q}^3)^2}{(\mathrm{tr}\mathbf{Q}^2)^3} \tag{2.2}$$

and ranges in the interval $[0, 1]$. In all uniaxial states, $\beta^2 = 0$, while states with maximal biaxiality correspond to $\beta^2 = 1$; since $\mathrm{tr}\mathbf{Q}^3 = 3\det\mathbf{Q}$, these latter states are precisely those where $\det\mathbf{Q} = 0$, that is, where at least one eigenvalue of $\mathbf{Q}$ vanishes.

The free-energy functional that we consider here is the following

$$\mathcal{F}[\mathbf{Q}] := \int_{\mathcal{B}} W\, \mathrm{d}V\,, \tag{2.3}$$

where $\mathcal{B}$ is the region in space occupied by the liquid crystal, W is defined by

$$W := \frac{L}{2}|\nabla\mathbf{Q}|^2 + f_b(\mathbf{Q}), \tag{2.4}$$

and

$$f_b(\mathbf{Q}) := \frac{A}{2}\mathrm{tr}\mathbf{Q}^2 - \frac{B}{3}\mathrm{tr}\mathbf{Q}^3 + \frac{C}{4}(\mathrm{tr}\mathbf{Q}^2)^2. \tag{2.5}$$

L is the only elastic constant appearing in the gradient term. A, B, and C are the usual coefficients in the Landau-de Gennes bulk potential, which we also refer to as the *ordering* potential, as its role is favoring the uniaxial states. Moreover, we set

$$A = a(T - T^*)$$

where T is the current temperature and T^* is the *supercooling* temperature of the isotropic phase; a, B, and C are positive constants typical of the specific material. The potential f_b is an expansion truncated at the fourth power, and so, in principle, it should be accurate only close to the isotropic-nematic transition, but it has also been used in a wide range of temperatures below T^*, an attitude that we also take here. According to the ordering potential in (2.5), the temperature T_{NI} that marks the transition from the nematic to the isotropic phase is defined by

$$T_{NI} := T^* + \frac{B^2}{27AC}. \tag{2.6}$$

For $T < T_{NI}$, the order tensors that minimize f_b are all uniaxial as in (2.1) with $\boldsymbol{n}$ arbitrary and

$$S = S_b := \frac{B + \sqrt{B^2 - 24AC}}{4C}. \tag{2.7}$$

Our definition of biaxial coherence length is [17, 18]

$$\xi_b := \sqrt{\frac{4LC}{B^2(1+\sqrt{1-\theta})}}, \tag{2.8}$$

where $\theta < 1$ is the reduced temperature defined by

$$\theta := \frac{24AC}{B^2} = \frac{T - T^*}{T^{**} - T^*}, \tag{2.9}$$

with

$$T^{**} := T^* + \frac{B^2}{24aC} \tag{2.10}$$

the *superheating* temperature of the nematic phase. The length ξ_b, which represents the typical distance over which biaxial disturbances die out in space, clearly depends on the temperature. We can easily estimate ξ_b for a typical liquid crystal, that is, 5CB (4–cyano–4′n–pentylbiphenyl), for which the values of the constants appearing in (2.5) are obtained from [19]; they are: $a = 0.20\times10^6\mathrm{J/Km^3}$, $B = 7.2\times10^6\mathrm{J/m^3}$, and $C = 8.8\times10^6\mathrm{J/m^3}$. If for the reduced temperature we choose $\theta = -8$, which is well below the value $\theta_{NI} = \frac{8}{9}$ corresponding to T_{NI}, and for the elastic constant we set $L = 9.075\times10^{-12}\mathrm{N}$, which is the average of the elastic constants given in [20], we obtain $\xi_b = 1.2$ nm. Continuum theories have already been employed successfully at the nanometric scale both to describe pre-transitional effects in thin films [7, 21] and to explore the disclination core [13]: we pursue here this line of thought and we also apply a continuum theory at the ξ_b-scale to describe mechanical properties.

To compute both forces and torques transmitted within liquid crystals, we introduce both stress and couple stress tensors. The former, which we call *Ericksen's* stress tensor $\mathbf{T}^{(E)}$ has been derived within this theory by Gartland and Virga [22]; it parallels the tensor defined originally by Ericksen within the director theory of liquid crystals [23].

$\mathbf{T}^{(E)}$ is defined as

$$\mathbf{T}^{(E)} := W\mathbf{I} - \nabla\mathbf{Q}\odot\frac{\partial W}{\partial\nabla\mathbf{Q}}, \tag{2.11}$$

where W is the energy density, and

$$\left(\nabla\mathbf{Q}\odot\frac{\partial W}{\partial\nabla\mathbf{Q}}\right)_{ij} := Q_{hk,i}\frac{\partial W}{\partial Q_{hk,j}}. \tag{2.12}$$

For clarity, in Equation (2.12) Cartesian components are adopted for tensors; partial derivatives with respect to space variables are therein denoted by commas, and summation on repeated indices is understood. The total force $\boldsymbol{F}(\mathcal{P})$ exerted by the liquid crystal on any submerged body $\mathcal{P}$ is thus expressed as

$$\boldsymbol{F}(\mathcal{P}) = \int_{\partial\mathcal{P}}\mathbf{T}^{(E)}\boldsymbol{\nu}\,\mathrm{d}s, \tag{2.13}$$

where $\boldsymbol{\nu}$ is the outer unit vector normal to the boundary $\partial\mathcal{P}$, and s denotes the area measure.

The couple stress tensor $\mathbf{L}$ appropriate to this theory, which we call *Leslie's* couple stress tensor, as it parallels the one introduced by Leslie in his director theory of liquid crystals, was derived by Sonnet, Maffettone and Virga [24]. The Cartesian components of $\mathbf{L}$ are

$$L_{ij} := 2\varepsilon_{ikl}Q_{km}\frac{\partial W}{\partial Q_{ml,j}}, \tag{2.14}$$

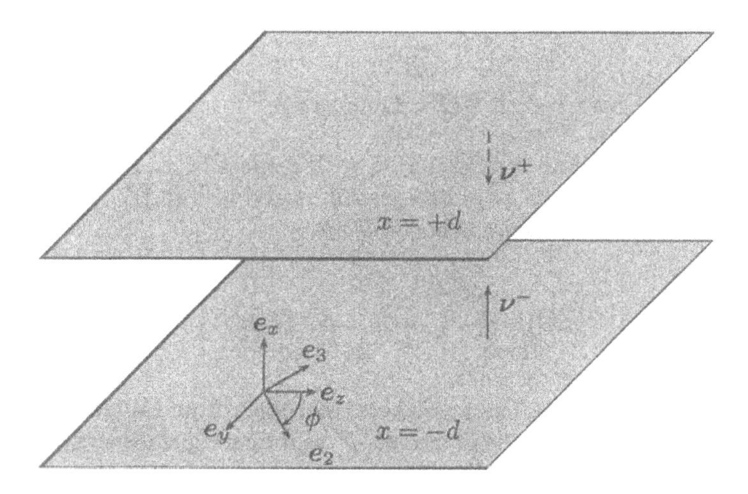

FIGURE 1. *A cell bounded by two parallel plates. The unit vector $\mathbf{e}_x$ is orthogonal to the plates; the origin of the frame $\{\mathbf{e}_x, \mathbf{e}_y, \mathbf{e}_z\}$ is in the middle of the cell. The nematic director $\boldsymbol{n}$ is along $\mathbf{e}_z$ at $x = -d$, whilst it makes the angle ϕ_0 with $\mathbf{e}_z$ on the plate at $x = +d$. The angle ϕ denotes the rotation about $\mathbf{e}_x$ of the eigenframe of $\mathbf{Q}$.*

where ε_{ikl} is the usual Ricci alternator. The total torque $\boldsymbol{M}(\mathcal{P})$ exerted on a submerged body $\mathcal{P}$ by the liquid crystal is thus given by

$$\boldsymbol{M}(\mathcal{P}) = \int_{\partial\mathcal{P}} \mathbf{L}\boldsymbol{\nu}\, ds\,. \tag{2.15}$$

3. Twist cell. We assume that a nematic liquid crystal occupies the region $\mathcal{B}$ bounded by two parallel, infinite plates at a distance $2d$. The symmetry of the system suggests choosing a reference frame $\{\mathbf{e}_x, \mathbf{e}_y, \mathbf{e}_z\}$ with a unit vector $\mathbf{e}_x$ orthogonal to both plates and with the origin of the co-ordinates in the middle of the cell, so that the plates lie at $x = -d$ and $x = +d$. $\mathbf{Q}$ is assumed to be prescribed on both plates as a uniaxial tensor with scalar order parameter S_0 and possibly different nematic directors $\boldsymbol{n}$, each lying parallel to the corresponding plate. In particular, we assume that $\boldsymbol{n}$ is parallel to $\mathbf{e}_z$ on the plate at $x = -d$, so that

$$\mathbf{Q} = \mathbf{Q}^- := S_0\Big(\mathbf{e}_z \otimes \mathbf{e}_z - \frac{1}{3}\mathbf{I}\Big). \tag{3.1}$$

On the plate at $x = d$, the nematic director is rotated by the angle ϕ_0 (normalized so as to be in $[0, \frac{\pi}{2}]$). Therefore, for $x = +d$

$$\mathbf{Q} = \mathbf{Q}^+ := S_0\Big(\boldsymbol{n}_0 \otimes \boldsymbol{n}_0 - \frac{1}{3}\mathbf{I}\Big), \quad \text{with} \quad \boldsymbol{n}_0 := \cos\phi_0\, \mathbf{e}_z + \sin\phi_0\, \mathbf{e}_y. \tag{3.2}$$

Symmetry considerations also suggest assuming that $\mathbf{e}_x$ is everywhere an *eigenvector* of $\mathbf{Q}$ and that all quantities are functions of the variable x. Thus, we write

$$\mathbf{Q} = \mathbf{Q}(x) \tag{3.3}$$

and

$$\nabla \mathbf{Q} = \mathbf{Q}'(x) \otimes \boldsymbol{e}_x \,, \tag{3.4}$$

where a prime denotes differentiation with respect to x. Furthermore, we will use for $\mathbf{Q}$ the same q–representation introduced in [17], according to which, the Cartesian components of $\mathbf{Q}$ are

$$[\mathbf{Q}] = \begin{bmatrix} -2q_1 & 0 & 0 \\ 0 & q_1 - q_2 & q_3 \\ 0 & q_3 & q_1 + q_2 \end{bmatrix}, \tag{3.5}$$

Under the assumptions made here, we obtain from Equation (2.3) the following expression for the functional $\mathcal{F}$ representing the energy stored in the cell per unit area of the bounding plates:

$$\mathcal{F}[\mathbf{Q}] := \int_{-d}^{+d} \left\{ \frac{L}{2} |\mathbf{Q}'|^2 + f_b(\mathbf{Q}) \right\} \mathrm{d}x \,. \tag{3.6}$$

Accordingly, Ericksen's stress tensor in (2.11) becomes

$$\mathbf{T}^{(E)} = \left(\frac{L}{2} |\mathbf{Q}'|^2 + f_b(\mathbf{Q}) \right) \mathbf{I} - L |\mathbf{Q}'|^2 \boldsymbol{e}_x \otimes \boldsymbol{e}_x \,. \tag{3.7}$$

Since the unit outer vectors $\boldsymbol{\nu}^-$ and $\boldsymbol{\nu}^+$ normal to the plates at $x = -d$ and $x = +d$ are such that $\boldsymbol{\nu}^- = -\boldsymbol{\nu}^+ = \boldsymbol{e}_x$, by Equation (3.7) the forces $\boldsymbol{f}^-$ and $\boldsymbol{f}^+$ exerted per unit area on these plates are, correspondingly,

$$\begin{aligned} \boldsymbol{f}^- &= -\left(\frac{L}{2} |\mathbf{Q}'|^2 - f_b(\mathbf{Q}) \right) \boldsymbol{e}_x = -f(-d) \boldsymbol{e}_x \\ \boldsymbol{f}^+ &= \left(\frac{L}{2} |\mathbf{Q}'|^2 - f_b(\mathbf{Q}) \right) \boldsymbol{e}_x = f(+d) \boldsymbol{e}_x \end{aligned} \tag{3.8}$$

where the function f, defined as

$$f(x) := \frac{L}{2} |\mathbf{Q}'|^2 - f_b(\mathbf{Q}) \,, \tag{3.9}$$

is actually constant throughout the cell when computed on a solution of the equilibrium equations [17]. Thus, at equilibrium,

$$\boldsymbol{f}^- = -\boldsymbol{f}^+.$$

In the q–representation introduced in Equation (3.5), f reads as

$$\begin{aligned} f = & L \left[3(q_1')^2 + (q_2')^2 + (q_3')^2 \right] \\ & - \left[A(3q_1^2 + q_2^2 + q_3^2) + 2Bq_1(q_1^2 - q_2^2 - q_3^2) + C(3q_1^2 + q_2^2 + q_3^2)^2 \right] . \end{aligned} \tag{3.10}$$

Similarly, we can compute the torque exerted per unit area of the plate. Leslie's couple stress tensor here takes the form

$$\mathbf{L} = -\boldsymbol{m} \otimes \boldsymbol{e}_x , \tag{3.11}$$

where $\boldsymbol{m}$ is the vector associated with the skew-symmetric tensor

$$\mathbf{M} := 2L(\mathbf{Q}\mathbf{Q}' - \mathbf{Q}'\mathbf{Q}) \tag{3.12}$$

so that

$$\mathbf{M}\boldsymbol{v} = \boldsymbol{m} \times \boldsymbol{v} \qquad \text{for all vectors } \boldsymbol{v} . \tag{3.13}$$

It is easily seen that the vector $\boldsymbol{m}$ is indeed parallel to $\boldsymbol{e}_x$, that is,

$$\boldsymbol{m} = m\boldsymbol{e}_x , \tag{3.14}$$

and that at equilibrium $\boldsymbol{m}$ is constant throughout the cell [12]. Moreover, in the q–representation introduced above

$$m = 4L(q_3 q_2' - q_2 q_3') . \tag{3.15}$$

Finally, the torques $\boldsymbol{m}^-$ and $\boldsymbol{m}^+$ exerted per unit area on the plates at $x = -d$ and at $x = +d$ read as

$$\boldsymbol{m}^- = m\boldsymbol{e}_x , \qquad \boldsymbol{m}^+ = -m\boldsymbol{e}_x . \tag{3.16}$$

As in [17], we scale $\mathbf{Q}$ to

$$S_* := \frac{B}{4C} ,$$

which is the equilibrium value of S at the superheating temperature T^{**}. Moreover, we scale all lengths to d. Thus, we set

$$q_i(d\eta) =: S_* \chi_i(\eta) \qquad \text{for } i = 1, 2, 3 , \tag{3.17}$$

where

$$\eta := \frac{x}{d} . \tag{3.18}$$

In the scaled variables, the equilibrium equations associated with the functional $\mathcal{F}$ in (3.6) read as follows [17]:

$$\begin{aligned} \frac{\xi_b^2}{d^2}(\sqrt{1-\theta}+1)\chi_1'' = \frac{\theta}{6}\chi_1 - \frac{1}{3}\left(\chi_2^2 + \chi_3^2 - 3\chi_1^2\right) \\ + \frac{1}{2}\left(3\chi_1^2 + \chi_2^2 + \chi_3^2\right)\chi_1 , \end{aligned} \tag{3.19a}$$

$$\frac{\xi_b^2}{d^2}(\sqrt{1-\theta}+1)\chi_2'' = \frac{\theta}{6}\chi_2 - 2\chi_1\chi_2 + \frac{1}{2}\left(3\chi_1^2 + \chi_2^2 + \chi_3^2\right)\chi_2 , \tag{3.19b}$$

$$\frac{\xi_b^2}{d^2}(\sqrt{1-\theta}+1)\chi_3'' = \frac{\theta}{6}\chi_3 - 2\chi_1\chi_3 + \frac{1}{2}\left(3\chi_1^2 + \chi_2^2 + \chi_3^2\right)\chi_3 , \tag{3.19c}$$

where, now, a prime denotes differentiation with respect to η, and use has also been made of (2.8). Boundary conditions (3.1) and (3.2) for $\mathbf{Q}$ are translated into the following conditions for χ_1, χ_2, and χ_3:

$$\chi_1(-1) = \frac{1}{6}s_\delta\,, \qquad \chi_1(1) = \frac{1}{6}s_\delta\,, \tag{3.20a}$$

$$\chi_2(-1) = \frac{1}{2}s_\delta\,, \qquad \chi_2(1) = \frac{1}{2}s_\delta\cos 2\phi_0\,, \tag{3.20b}$$

$$\chi_3(-1) = 0\,, \qquad \chi_3(1) = \frac{1}{2}s_\delta\sin 2\phi_0\,, \tag{3.20c}$$

where we have set

$$S_0 = (1+\delta)S_b \qquad \text{and} \qquad s_\delta := (1+\delta)\left(1+\sqrt{1-\theta}\right). \tag{3.21}$$

We will refer to δ as the *incremental surface ordering parameter*, as it measures how the degree of order imposed by the bounding surfaces differs from its bulk value. In particular, when $\delta > 0$, both bounding plates have an enhanced aligning ability on the liquid crystal molecules; they enforce a uniaxial state with a degree of order higher than the equilibrium bulk value S_b. On the contrary, when $\delta < 0$, the aligning ability of the plates is depressed: the enforced state is still uniaxial, but with a degree of order smaller than S_b. When $\delta = 0$, the order parameter enforced on the boundary is precisely S_b. To keep S_0 in the interval $(0,1)$, which corresponds to a pattern in which molecules on the plates tend to align along the director $\boldsymbol{n}$, we shall assume that

$$-1 < \delta < -1 + \frac{1}{S_*}\,. \tag{3.22}$$

Both in this and in the following section we set $\delta = 0$.

In terms of the new variables, f and m in (3.10) and (3.15) read as

$$\begin{aligned} f =f_0\Bigg\{&\frac{\xi_b^2}{d^2}(\sqrt{1-\theta}+1)\left[3(\chi_1')^2+(\chi_2')^2+(\chi_3')^2\right] \\ &-\left[\frac{\theta}{6}(3\chi_1^2+\chi_2^2+\chi_3^2)+2\chi_1(\chi_1^2-\chi_2^2-\chi_3^2)\right. \\ &\left.+\frac{1}{4}(3\chi_1^2+\chi_2^2+\chi_3^2)^2\right]\Bigg\}, \end{aligned} \tag{3.23}$$

and

$$m = m_0\frac{\xi_b}{d}\frac{\chi_2'\chi_3-\chi_2\chi_3'}{4}\,, \tag{3.24}$$

where $f_0 := B^4/64C^4$, and $m_0 := B^3L^{1/2}/C^{5/2}$.

In general, Equations (3.19) subject to (3.20) have more than one solution [17]. First, the same boundary conditions could be obeyed by two twist solutions, corresponding to opposite rotations of the eigenframe of **Q** across the cell. In addition, a third type of solution can be found in which the eigenframe of **Q** stays unchanged throughout the cell, and the uniaxial states at the boundary are reconciled through an order reconstruction in the bulk culminating in a uniaxial state with negative order parameter in the middle of the cell [17]. Our numerical exploration of these equations relied upon MATLAB [25] and AUTO 2000 [26]. MATLAB is a commercial scientific computing and visualization environment, based on a high-level language syntax, and endowed with a wealth of mathematical library software, and integrated graphics. We have used primarily the MATLAB graphics, as well as a code from its ODE Suite, `bvp4c`, for the solution of general nonlinear systems of ordinary differential equations.

The MATLAB BVP solver is not equipped for numerical bifurcation analysis, and for that we relied upon the AUTO package. For ODE boundary value problems, the package has the capabilities to perform parameter continuation, detection of bifurcation points, and stable numerical calculation of limit points. It is also capable of monitoring auxiliary quantities, such as the values of f and m, as well as integral functionals (*e.g.*, the free energy), and of generating a two-parameter locus of fold points. We have used all of these features to obtain the results we discuss below.

Here and in the following we set $\theta = -8$: other smaller values of θ would not affect qualitatively the scenario we shall depict [17]. For $\phi_0 = \frac{\pi}{2}$, a texture bifurcation appears in the twist cell for $d = d_c \approx 2.47\,\xi_b$. For $d < d_c$, there is only one equilibrium texture, which bridges the boundary conditions through order reconstruction: the eigenframe of **Q** remains unchanged throughout the cell, while the transverse eigenvalues in the (y, z)-plane are exchanged. For $d > d_c$, the reconstruction texture becomes unstable and two stable symmetric textures emerge from it, corresponding to opposite twists of the transverse eigenvectors of **Q**. For $\phi_0 = \frac{\pi}{2}$, Fig. 2 illustrates all three types of equilibrium textures by ellipsoids that sketch the **Q** field within the cell.

4. Torque and force. The multiplicity of equilibrium textures and their stability just recalled are mirrored by the diagrams of both torque m and force f as functions of the half-thickness of the cell. In Fig. 3 we present the classical bifurcation pitchfork obtained from the plot of m/m_0 against d/ξ_b; for the typical values of 5CB (4-cyano-4′n-pentylbiphenyl) [19], $m_0 \approx 5\,\mathrm{mJ\,m^{-2}}$ and $\xi_b \approx 1\,\mathrm{nm}$ at $\theta = -8$. It is worth noting that for $d < d_c$ the equilibrium texture bears no torque at all values of d; this is clearly related to the lack of rotation of the eigenframe of **Q**, as also suggested by the pictorial description of the system in Fig. 2. The two symmetric equilibrium twist textures existing for $d \geq d_c$ exhibit a maximum torque m for $d = d_M \approx 3.54\,\xi_b$. Furthermore, as d increases, the order tensor **Q** tends to become uniaxial everywhere in the cell; for any ϕ_0 and θ, the representation of the smaller twist tends asymptotically for $\frac{d}{\xi_b} \gg 1$ to be

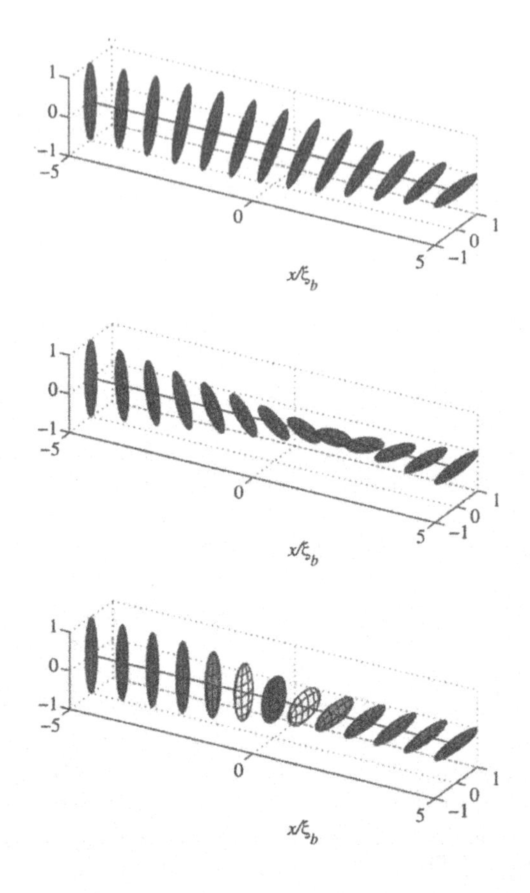

FIGURE 2. *Order-tensor ellipsoids against position across the cell, in units of the biaxial coherence length ξ_b, for three basic solutions: opposite twists (top and middle), order-reconstruction (bottom). Ellipsoids are oriented along the eigenframe of the order tensor* $\mathbf{Q}$ *at each point; their semiaxes are the eigenvalues of* $\mathbf{Q}$ *appropriately augmented and scaled to the largest eigenvalue at the boundary. The gray scale is associated to the degree of biaxiality* β^2 *(the lighter the color, the larger* β^2*). Parameters: twist angle* $\phi_0 = \frac{\pi}{2}$*, reduced temperature* $\theta = -8$*, dimensionless cell half-width* $d/\xi_b = 5$.

$$\chi_1(\eta) = \frac{2}{3} \tag{4.1a}$$

$$\chi_2(\eta) = 2\cos(\phi_0(1+\eta)) \tag{4.1b}$$

$$\chi_3(\eta) = 2\sin(\phi_0(1+\eta)) \; ; \tag{4.1c}$$

it follows from the last two of these equations and (3.24) that m decays to zero like $1/d$.

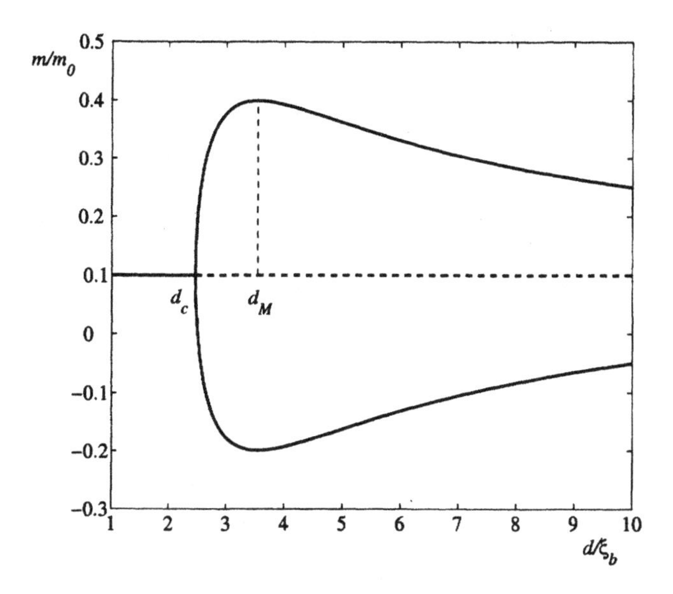

FIGURE 3. *The torque m transmitted per unit area from one plate to the other, scaled to m_0, is plotted against the half-thickness d of the cell, scaled to the biaxial coherence length ξ_b defined in (2.8), for $\theta = -8$ and total twist angle $\phi_0 = \frac{\pi}{2}$. Stable branches are represented by solid lines, whilst the unstable branch is represented by a dashed line. The graphs corresponding to the two symmetric twist textures merge with the reconstruction straight line bearing no torque at $d = d_c \approx 2.47\,\xi_b$; for 5CB, $m_0 = B^3L^{1/2}/C^{5/2} \approx 5\,\mathrm{mJ\ m^{-2}}$ and $\xi_b \approx 1\,\mathrm{nm}$.*

Measuring m as a function of d in a displacement-controlled machine should reproduce the bifurcation diagram in Fig. 3, thus allowing for a direct mechanical measurement of ξ_b. Conversely, if one imagines a machine in which the distance between the plates is progressively decreased by incrementing the applied torque, a snapping instability should be met at the critical value d_M: there, in fact, the equilibrium torque required to further reduce d would decrease, instead of increasing. However, in practice, nanotorque machines have not yet been developed. On the other hand, nanoforce machines (such as, for example, the Surface Force Apparatus [1, 27, 28, 29]), have been established and improved over the years.

This suggests seeing how the texture bifurcation in the twist cell is also reflected onto the force diagram. In accordance with Equations (3.9) and (3.10), for large values of d the force f tends to a constant $-f_m$, where f_m is the minimum of f_b, independent of ϕ_0. We expect f to decay to $-f_m$ as $1/d^2$ decays to zero. The decay of the force f to the residual force $-f_m$ is an artifact of the normalization chosen for f_b: to make the total free energy stored in the infinite cell finite, f_b should be shifted so as to make $f_m = 0$. In the following, all forces are meant to be measured relative to their asymptotic residual value. Figure 4 shows the graph of f

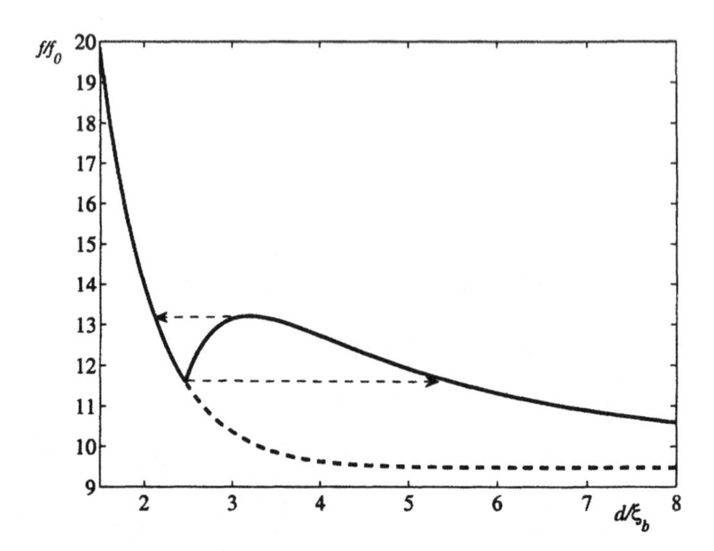

FIGURE 4. *The force f transmitted per unit area from one plate to the other, scaled to f_0, is plotted against the half-thickness d of the cell, scaled to the biaxial coherence length ξ_b defined in (2.8), for $\theta = -8$ and total twist angle $\phi_0 = \frac{\pi}{2}$. Stable branches are represented by solid lines, while the unstable branch is represented by a dashed line. The dashed branch corresponds to the unstable reconstruction texture, which becomes stable at $d = d_c \approx 2.47\,\xi_b$. There, it meets the single stable branch corresponding to the two symmetric twists (unlike the case of torque, these are represented by only one branch, since the force is the same for both). For 5CB, $f_0 = B^4/64C^3 \approx 60\,\text{mPa}$, and $\xi_b \approx 1\,\text{nm}$. The arrows delimit the hysteresis loop described in an ideal force-controlled cycle.*

against $\frac{d}{\xi_b}$ for $\phi = \frac{\pi}{2}$. Upon reducing d, the force exerted by both twist textures increases as long as $d > d_M^* \approx 3.18\,\xi_b$ and then decreases until it meets the force exerted by the reconstruction texture at $d = d_c$. Upon further reducing d below d_c, the reconstruction force keeps increasing and diverges like $1/d^2$. The slope of the graph of f against d is discontinuous at $d = d_c$, where m vanishes. As for the torque diagrams, here the lack of monotonicity would also imply that a snapping instability is expected to happen at d_M^*. The solid line in the diagram of Fig. 4 corresponds to all locally stable configurations. Thus, in an ideal force-controlled experiment, where the force is steadily increased, the equilibrium value of d would be, by continuity, the largest value of d compatible with the applied force, as long as this exceeds d_M^*. When the force is further increased, there is a single value of d compatible with the applied force: the one attained on the reconstruction branch. Upon decreasing the force, this latter branch, equally locally stable, is followed until d reaches d_c; there, as shown by the lower arrow in Fig. 4, the equilibrium value of d jumps again on the branch of the twist textures, which are then the only stable ones.

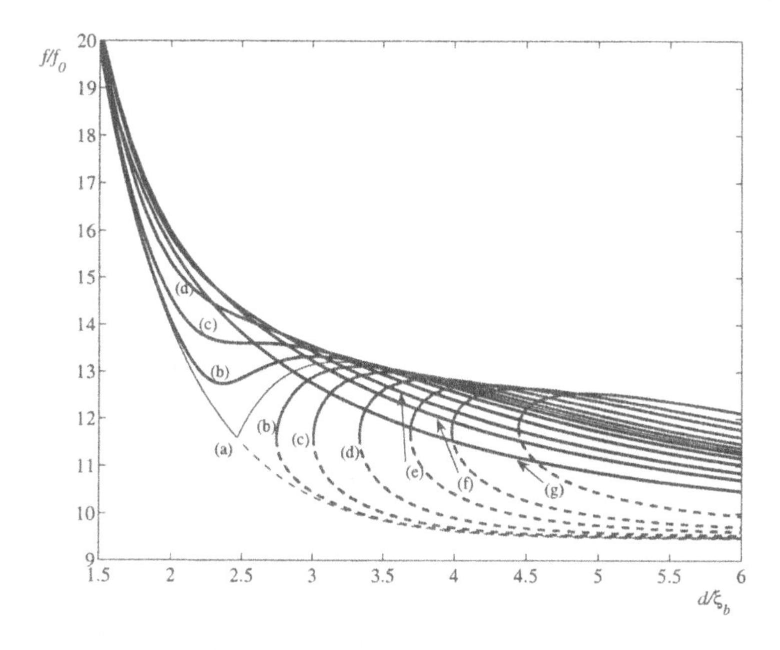

FIGURE 5. *The force f transmitted per unit area from one plate to the other is plotted against the half-thickness d of the cell scaled to the biaxial coherence length ξ_b, for the following values of the total twist angle: (a) $\phi_0 = \frac{\pi}{2}$, (b) $\phi_0 = 0.49\,\pi$, (c) $\phi_0 = \phi_c^* \approx 0.474\,\pi$, (d) $\phi_0 = 0.45\,\pi$, (e) $\phi_0 = 0.42\,\pi$, (f) $\phi_0 = \phi_c \approx 0.394\,\pi$, (g) $\phi_0 = 0.35\,\pi$; all other parameters are as in Fig. 4. The solid lines represent stable textures. The dark gray dashed branches correspond to the unstable reconstruction texture; they meet the stable branches (dark gray solid lines) corresponding to the more twisted texture at $d = d_c(\phi_0)$.*

It is interesting to study the behavior of the bifurcation plots when $\phi_0 \neq \pi/2$: in fact, in a real experiment the value of the total twist angle is known with some uncertainty, and it is unlikely, if not impossible, to have it exactly equal to the desired value. Both the force and the torque diagrams are expected to be unfolded. Figure 5 shows the graphs of f/f_0 against d/ξ_b, for several values of ϕ_0; now two solutions, corresponding to a less twisted texture and to a more twisted one, are stable, with the former existing for all values of d as the absolute minimizer of the free energy, whilst the latter, just metastable, is found only above a critical distance $d_c(\phi_0)$, now depending on ϕ_0 and greater than the limiting value $d_c(\frac{\pi}{2})$. In addition to that, the reconstruction texture, always unstable, can be found for $d > d_c(\phi_0)$; its plot merges into that of the metastable texture. For $\phi_0 < \frac{\pi}{2}$ the angular point at d_c evolves into a regular minimum at d_m^*, which survives until ϕ_0 reaches a critical value $\phi_c^* \approx 0.474\,\pi$. The behavior of the solutions at small and large d mimics that described above in the presence of the bifurcation.

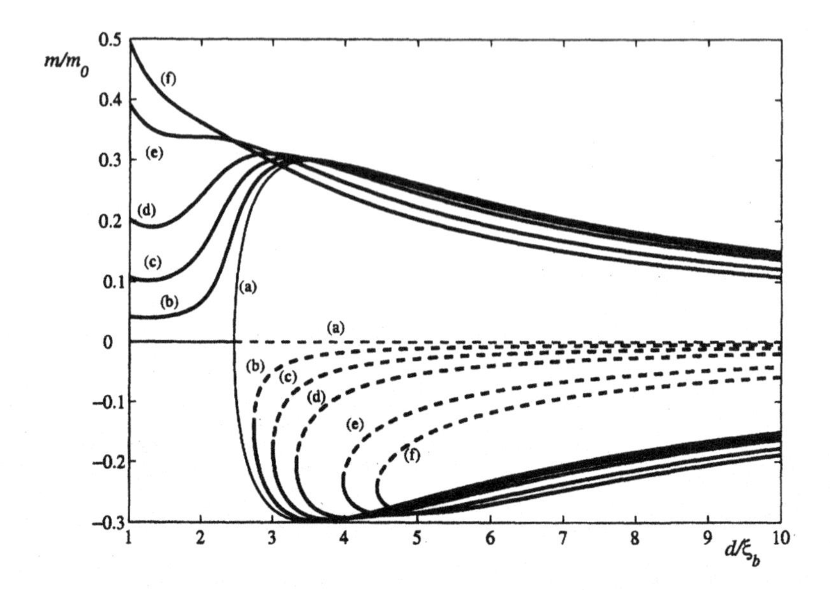

FIGURE 6. *The torque m transmitted per unit area from one plate to the other is plotted against the half-thickness d of the cell scaled to the biaxial coherence length ξ_b, for the following values of the total twist angle: (a) $\phi_0 = \frac{\pi}{2}$, (b) $\phi_0 = 0.49\,\pi$, (c) $\phi_0 = \phi_c^* \approx 0.474\,\pi$, (d) $\phi_0 = 0.45\,\pi$, (e) $\phi_0 = \phi_c \approx 0.394\,\pi$, (f) $\phi_0 = 0.35\,\pi$; all other parameters are as in Fig. 3. Solid lines represent the stable textures. The dark gray dashed branches correspond to the unstable reconstruction texture; they meet the stable branches (dark gray solid lines) corresponding to the more twisted texture at $d = d_c(\phi_0)$.*

Figure 6 illustrates the unfolding of the torque m; this differs somehow from the unfolding of the force. First, for $\phi_0 < \pi/2$ the stable solutions bear nonzero torque, even when d is small; actually, m diverges like $1/d$ as d tends to zero. As a consequence, in addition to the maximum at d_M, a local minimum appears at $d_m < d_M$. As for the force, there exists a critical value ϕ_c for which minimum and maximum merge in an inflection point, and for $\phi > \phi_c$ the torque diagram becomes monotonic; for $\theta = -8$, $\phi_c \approx 0.394\,\pi$.

The unfolded diagrams of both force and torque would also host a hysteresis loop, as long as they fail to be monotonic. We can imagine using a force-controlled apparatus in which the distance between the plates of the cell is decreased by applying an increasing equilibrating force. We would follow the stable branch of the less twisted texture in Fig. 5 until we reach the maximum at d_M^*: a further slight increase of the force would cause the cell to snap and reach the point on the stable branch at $d < d_m^*$ located on the diverging side of the branch, at the same value of f. Now, if we suppose to reduce the applied force in a quasi-static way, we would follow

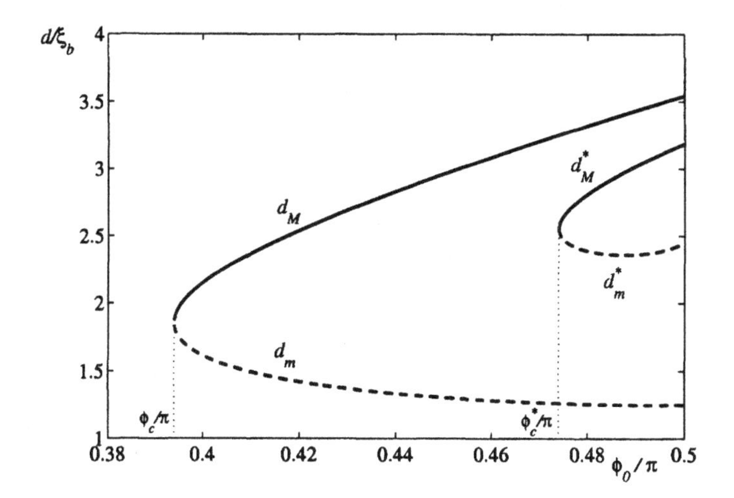

FIGURE 7. *The graphs of the pairs (d_m, d_M) and (d^*_m, d^*_M), scaled to ξ_b, as functions of ϕ_0/π. The graphs of the first pair merge at $\phi_0 = \phi_c$, whilst those of the second pair merge at $\phi_0 = \phi^*_c$.*

the decreasing stable branch until reaching the minimum point at d^*_m; a further attempt to reduce the force, would cause the cell to snap back at the value of $d > d^*_M$ that keeps the same equilibrium force f. The same argument would apply to the torque diagram, where d^*_m and d^*_M would be replaced by the distances d_m and d_M.

It is worth noticing that both d_M and d_m, which also depend on ϕ_0, are generally different from d^*_M and d^*_m. In particular, for our choice of parameters, $d^*_M < d_M$ and $d^*_m > d_m$. In Fig. 7 we plot these four quantities as functions of ϕ_0/π, for $\theta = -8$. The graphs for d_m and d_M meet at $\phi_0 = \phi_c$, whilst the graphs for d^*_m and d^*_M meet at $\phi_0 = \phi^*_c < \phi_c$. The nesting of the graphs in Fig. 7 clearly shows that the lack of monotonicity is more pronounced and more persistent for the torque diagram than for the force one; this is one more reason in favor of employing a torque-controlled experiment to reveal the mechanical signature of order reconstruction.

5. Surface biaxial force. So far we have set $\delta = 0$ in Equation (3.20), that is, we have assumed that the uniaxial order parameter S_0 enforced on both bounding plates of the cell coincides with the equilibrium value S_b. Most substrates, however, enhance the surface degree of alignment, and so the question arises as to whether choosing $\delta > 0$ would alter the properties of the order force f outlined above. In particular, we wonder whether the force diagram drawn in the preceding section could become monotonic, thus reducing the order force described here to the pattern of the classical structural force already shown. For simplicity, henceforth we set $\phi_0 = \frac{\pi}{2}$ and explore the force diagram f for positive values of δ. Fig-

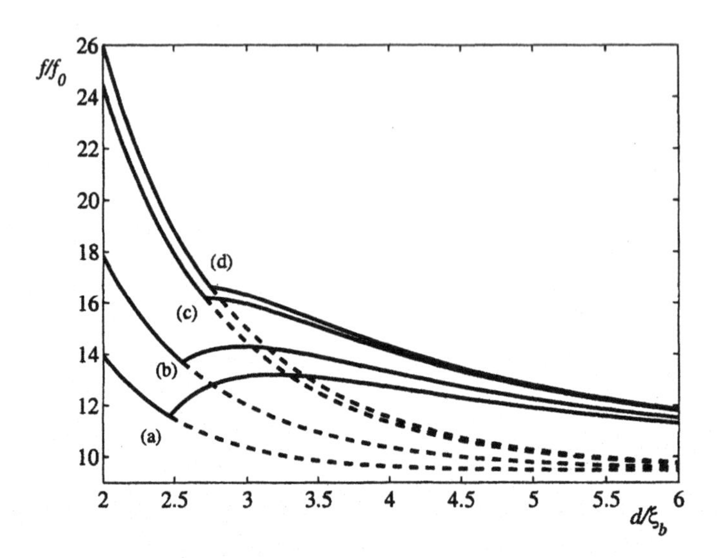

FIGURE 8. *The force f transmitted per unit area from one plate to the other is plotted against the half-thickness d of the cell scaled to the biaxial coherence length ξ_b, for the following values of the surface ordering parameter: (a)$\delta = 0$, (b)$\delta = 0.2$, (c)$\delta = \delta_c = 0.5583$, (d)$\delta = 0.65$. All other parameters are as in Fig. 4. Solid lines represent stable textures. The dashed branches correspond to the unstable reconstruction texture.*

ure 8 shows the graphs of f/f_0 against d/ξ_b for several values of δ, namely, $\delta = 0, 0.2, 0.5583, 0.65$. Upon increasing δ, the force diagram is shifted upwards, while the well and the hill making the graph non-monotonic are drawn closer. To illustrate better how the force diagram eventually becomes monotonic when the surface ordering parameter is sufficiently large, we draw in Fig. 9 the graphs of both the critical point d_c and the maximum point d^*_M, as functions of δ; these points meet at a critical value δ_c, which is $\delta_c \approx 0.5583$ for $\theta = -8$. For this critical value to be actually attained it must fall below the upper bound in Equation (3.22); for example, since for 5CB at $\theta = -8$ $S_b \approx 0.82$, δ_c cannot be reached for that material at that temperature, and so the force would remain non-monotonic. Other materials or other temperatures, however, could make δ_c attainable, thus rendering the force diagram monotonic.

Surface order transitions induced by temperature in a nematic liquid crystal have been theoretically analyzed by adopting a pseudomolecular approach to obtain the surface free energy [30]. In such a model, the surface shows the tendency to impose a scalar order parameter different from the one in the bulk, which depends only on temperature; two second-order phase transitions are possible: one from the homeotropic director alignment to a tilted alignment, and another from this latter to a planar director orientation. However, within the approximation of perfect local order for

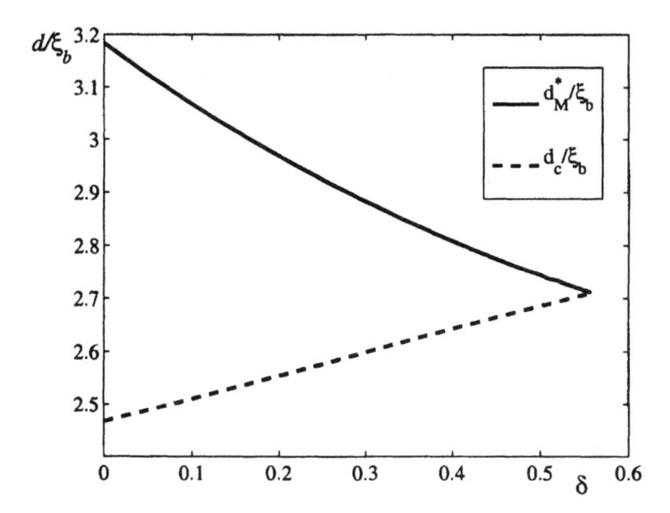

FIGURE 9. *The maximum point* d^*_M *(continuous line) and the critical point* d_c *(dashed line) scaled to the biaxial coherence length* ξ_b *are plotted against the surface ordering parameter* δ*, for* $\theta = -8$ *and total twist angle* $\phi_0 = \frac{\pi}{2}$*. The two plots merge at the critical value* $\delta_c \approx 0.5583$*: above this value, the force profile is monotonic.*

a model nematic liquid crystal composed of ellipsoidal molecules, it has been shown that a tilted equilibrium director orientation is energetically unfavorable [31]. Therefore, only planar or homeotropic orientations would be allowed; moreover, whilst the former alignment is stable under ordinary conditions, the latter could be stable only with the aid of a dominant repulsive contribution to the free energy, which would also imply negative values of the elasticity coefficients. Tilted orientations at the surface might be favored by other contributions to the surface energy, such as that of long-range electrostatic forces. In view of these results, we reckon that negative values of δ, though possible, are not likely to occur in a real system.

An enhanced degree of uniaxiality at the surface has the potential to disrupt the pattern of the force described here, making it more similar to the uniaxial structural force described in [1]. We say that the order force f predicted here is a surface *biaxial* force as it indeed reveals the biaxial structure hidden in the bulk order reconstruction. A quantitative measure of the biaxial character of this force can be gained by recording the maximum degree of biaxiality β^2_M attained within the cell when $d = d^*_M$, that is, at the snapping limit. The graph of β^2_M against δ, shown in Fig. 10, exhibits an increasing function approaching a plateau in the vicinity of δ_c.

6. Conclusion. We computed both the nanotorque m and the nanoforce f transmitted between two parallel plates $2d$ apart, confining a nematic liquid crystal in a ϕ_0-twist cell with infinitely strong anchor-

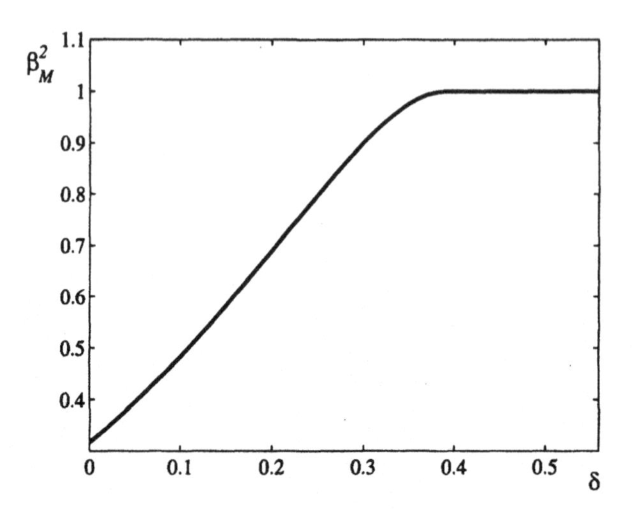

FIGURE 10. *The maximum degree of biaxiality β_M^2 attained within the cell when $d = d_M^*$ is plotted against the surface ordering parameter $\delta \in [0, \delta_c]$. As δ approaches $\delta_c \approx 0.5583$, β_M^2 tends to its maximum value.*

ing. The signature of order reconstruction could be revealed by measuring $m(d)$ for ϕ_0 close to $\frac{\pi}{2}$, which would also provide the most direct evaluation of the biaxial coherence length ξ_b; on the other hand, it may be more practical to employ a nanoforce machine where the order reconstruction appears as an angular point in the diagram of $f(d)$. In fact, for $\phi_0 = \frac{\pi}{2}$, both the torque and the force diagrams are not monotonic. We predicted the existence of two critical twist angles ϕ_c and $\phi_c^* > \phi_c$, below which the force and the torque diagrams, respectively, become strictly monotonic. The ordering effect of the plates also plays a role: the order parameter at boundaries can be increased with respect to that in the bulk without destroying the non-monotonic pattern in the force, as long as a critical value δ_c of the incremental surface ordering parameter is not attained. By this disruptive effect of an enhanced degree of uniaxiality at the surface, the force f studied here may be considered as a surface biaxial force; this is also reflected by the fact that the maximum degree of biaxiality attained within the cell approaches its limiting value 1 when the incremental surface ordering parameter is close to δ_c.

In the light of the model studied here, it is clear that our theory describes only the continuum contribution to nanoforces. A more accurate description should also account for other forces, such as van der Waals's and Casimir's, which are likely to affect the divergence of f predicted here for $d \ll \xi_b$. In principle, these forces should also depend on the underlying liquid crystal texture. Similarly, the oscillating structural forces ascribed to the positional ordering of the molecules on the bounding plates [1] fall outside the scope of this study, but our analysis shows that the onset of a non-monotonic behavior in the force should be ascribed to medium-range surface biaxial forces, and not to short-range positional ordering forces as in [1].

The possibility of using flat surfaces in nanoforce machines could be questioned, but previous experiments have shown that they are indeed possible [32], though they need to be improved to explore nanothicknesses. Although the assumption on infinite anchoring made here could also be questioned, such an assumption seems however to be compatible with nanoscale observations of single nematic molecular layers on cleaved monocrystal surfaces [33, 34, 35, 36].

Recent experiments on forces may be explained in terms of the present model; Zappone *et al.* [37] observed peculiar features in a cell containing two thermotropic nematics (5CB and ME10.5) subject to hybrid anchoring conditions, when the distance between the plates is below 10nm: though in a bend geometry, these experiments provide a clear evidence of a non-monotonic behavior of the force, in agreement with our model.

REFERENCES

[1] R.G. HORN, J.N. ISRAELASCHVILI, AND E. PEREZ, J. Phys. (France), **42**, 39 (1981).
[2] B.W. NINHAM, J. Chem. Phys. **84**, 1423 (1980).
[3] W. MAIER AND A. SAUPE, Z. Naturforsch., **14a**, 882 (1959).
[4] C.W. OSEEN, Trans. Faraday Soc., **29**, 883 (1939).
[5] F.C. FRANK, Discuss. Faraday Soc., **25**, 19 (1958).
[6] H. SCHRÖDER, J. Chem. Phys. **67**, 16 (1977).
[7] P. SHENG, Phys. Rev. Lett. **37**, 1059 (1976).
[8] P. RICHETTI, L. MOREAU, P. BAROIS, AND P. KÉKICHEFF, Phys. Rev E **54**, 1749 (1996).
[9] K. KOČEVAR, R. BLINC, AND I. MUŠEVIC, Phys. Rev. E **62**, R3055 (2000).
[10] K. KOČEVAR AND I. MUŠEVIC, Phys. Rev. E **65**, 021703 (2002).
[11] B. ZAPPONE, *Films nanométriques de cristeaux liquides étudiés par mesure de force SFA et AFM*, PhD. Thesis, University of Bordeaux, France (2004).
[12] F. BISI, E.G. VIRGA, AND G.E. DURAND, Phys. Rev. E **70**, 042701 (2004).
[13] N. SCHOPOHL AND T.J. SLUCKIN, Phys. Rev. Lett. **59**, 2582 (1987).
[14] P. PALFFY-MUHORAY, E.C. GARTLAND, AND J.R. KELLY, Liq. Cryst. **16**, 713 (1994).
[15] P.G. DE GENNES AND J. PROST, *The Physics of Liquid Crystals* (Clarendon Press, Oxford 1993), pp. 56-57.
[16] P. KAISER, N. WIESE, AND S. HESS, J. Non-Equilib. Themodyn. **17**, 153 (1992).
[17] F. BISI, E.C. GARTLAND, R. ROSSO, AND E.G. VIRGA, Phys. Rev. E **68**, 021707 (2003).
[18] S. KRALJ, E.G. VIRGA, AND S. ŽUMER, Phys. Rev. E **60**, 1858 (1999).
[19] H.J. COLES, Mol. Cryst. Liq. Cryst. Lett. **49**, 67 (1978).
[20] R. BARBERI, F. CIUCHI, G.E. DURAND, M. IOVANE, D. SIKHARULIDZE, A.M. SONNET, AND E.G. VIRGA, Eur. Phys. J. E **13**, 61 (2004).
[21] K. MIYANO, Phys. Rev. Lett. **43**, 51 (1979).
[22] E.C. GARTLAND AND E.G. VIRGA, *in preparation* (2005).
[23] J. L. ERICKSEN, Arch. Rational Mech. Anal. **9**, 371 (1962).
[24] A.M. SONNET, P.L. MAFFETTONE, AND E.G. VIRGA, J. Non-Newtonian Fluid Mech. **119**, 51 (2004).

[25] MATLAB is a registered trademark of The MathWorks, Inc. `http://www.mathworks.com`.
[26] See `http://sourceforge.net/projects/auto2000`.
[27] J.N. ISRAELASCHVILI AND D. TABOR, Proc. R. Soc. London A, **331**, 19 (1972).
[28] J.N. ISRAELASCHVILI AND G.E. ADAMS, J. Chem. Soc. Faraday Trans. I **74**, 975 (1978).
[29] J.N. ISRAELASCHVILI, *Intermolecular and surface forces* (Academic Press, London, 1992).
[30] G. BARBERO, Z. GABBASOVA, AND M.A. OSIPOV, J. Phys. II **1**, 691 (1991).
[31] M.A. OSIPOV AND S. HESS, J. Chem. Phys. **99** (5), 4181 (1993).
[32] M. CAGNON AND G. DURAND, Phys. Rev. Lett. **73**, 3556 (1993).
[33] J.S. FOSTER AND J.E. FROMMER, Nature (London) **333**, 542 (1988).
[34] J.K. SPONG, H.M. MIZES, J.R.L.J. LACOMB, M.M. DOVEK, J.E. FROMMER, AND J.S. FOSTER, Nature (London) **338**, 137 (1989).
[35] D.P.E. SMITH, J.K.H. HÖRBER, G. BINNIG, AND N. NEJOH, Nature (London) **344**, 641 (1990).
[36] B. JÉRÔME, Rep. Prog. Phys. **54**, 391 (1991).
[37] B. ZAPPONE, PH. RICHETTI, R. BARBERI, R. BARTOLINO, AND H.T. NGUYEN, Phys. Rev E **71**, 041703 (2005).

MODELLING LINE TENSION IN WETTING

RICCARDO ROSSO*

Abstract. Line tension can be viewed as the analogue, for three-phase contact, of surface tension. However, obtaining a coherent picture from the different avenues followed to model line tension is much harder than the analogous operation for surface tension. This essentially reflects the extreme sensitivity of line tension to the details of the model employed. Line tension has an impact on the equilibrium and stability of fluid droplets laid on a rigid substrate, in the presence of a vapor phase. In particular, the sign of line tension is a critical issue, that gave rise to conflicting interpretations. Here, we review the approaches to line tension from microscopic to macroscopic scales, stressing the mathematical problems involved. We also illustrate a stability criterion for wetting functionals to clarify the rôle of the sign of line tension. As an application, we discuss how stability of liquid bridges near the wetting or the dewetting transition mirrors the scaling laws for surface and line tension.

Key words. Surface tension; Line tension; wetting.

AMS(MOS) subject classifications. 74A50, 82-02.

1. Introduction. In the past three decades line tension has been one of the most studied and controversial issues in the wetting science, both at the theoretical and at the experimental level. To some extent, line tension can be studied by the methods also employed for its more renowned relative, the surface tension. In fact, line tension was first introduced by Gibbs in a footnote on p. 288 of his seminal paper *On the equilibrium of heterogeneous substances* [1] as the analogue, for three-phase contact along a line, of surface tension for two-phase contact: "These lines (*i.e.* contact lines) might be treated in a manner entirely analogous to that in which we have treated surfaces of discontinuity". However, a few pages later, Gibbs remarked that "We may here add that the linear tension there mentioned (*i.e.* line tension) may have a negative value" (*cf.* p. 296 of [1]). As we shall see, this comment contains the basic ingredient of most debates concerning line tension.

After Gibbs, line tension received less attention than surface tension essentially because its effects can be appreciated at much smaller length scales. Careful experiments are required to detect the tiny effects due to line tension and the experimental outcomes often led to contradictory results. Indeed, line tension measurements can be found in the literature which differ from one another even by five orders of magnitude. Moreover, no consensus was reached on the sign of line tension that oscillated from positive to negative according to the experiment: the papers [2] and [3] contain lists of references to relevant experimental results on line tension. Apart from poor techniques, discrepancies are mainly due to the fact that

*Dipartimento di Matematica, Istituto Nazionale di Fisica della Materia, Università di Pavia, Via Ferrata 1, 27100 Pavia, Italy (riccardo.rosso@unipv.it).

line tension measurements are indirect, as they are obtained through the values of parameters somehow related to line tension *via* theoretical predictions. For instance, line tension can be computed by use of the generalized Young equation which predicts a dependence of the contact angle on the size of sessile droplets. In this case, line tension measurements inherit the intrinsic difficulties to obtain an unambiguous value for the contact angle, which is sensitive to substrate's roughness and heterogeneity, a fact that led experimentalists to study liquid-fluid-vapor systems instead of solid-fluid-vapor systems: for an updated account on the main experimental techniques, we refer the reader to Section 5 of [4].

Also theoretical predictions differ if different avenues are followed to account for Gibbs's view of line tension as a free-energy excess. The crucial step is the strategy adopted to model the distortions of sessile droplets near the three-phase region where, unavoidably, several approximations are required. It has been appreciated for a while that line tension is greatly affected by these approximations and that different outcomes are obtained, depending on the neglected factors. For instance, gravity is often neglected [5, 6]; on exploring intermolecular interactions, the rôle of short-range forces is stressed [7, 8] at the expense of long-range forces which, in turn, are embodied in a nonlocal approach [5] where, on the contrary, multi-body effects are not taken into account.

Once a value for the line tension has been obtained from a microscopic approach based either on statistical mechanics [9] or on some phenomenological model that somehow summarizes intermolecular interactions (see *e.g.* [6, 8, 10]), it can be inserted into a macroscopic free-energy functional. The simplest model to study the equilibrium of a droplet $\mathcal{B}$ made of incompressible fluid laid on a rigid and homogeneous substrate, in the absence of body forces, is based on the free-energy functional

$$\mathcal{F}[\mathcal{B}] = \gamma \int_{\mathcal{S}} \mathrm{d}a + (\gamma - w) \int_{\mathcal{S}_*} \mathrm{d}a + \tau \int_{\mathcal{C}} \mathrm{d}s\,, \tag{1.1}$$

where the boundary $\partial\mathcal{B}$ of the droplet has been split as $\partial\mathcal{B} = \mathcal{S} \cup \mathcal{S}_*$. The *free* surface $\mathcal{S}$ is the portion of $\partial\mathcal{B}$ in contact with the vapor phase (V) and the *adhering* surface $\mathcal{S}_*$ is the portion of $\partial\mathcal{B}$ in contact with the substrate (S) (*see* Fig. 1). The positive constant $\gamma \equiv \gamma_{LV}$ is the surface tension between the liquid droplet (L) and the vapor phase, the positive constant $w := \gamma + \gamma_{SL} - \gamma_{SV}$, referred to as the *adhesion potential*, accounts for the interactions between the droplet and the substrate and can be expressed in terms of the surface tensions γ, γ_{SV} and γ_{SL} where the last two are referred to the substrate-vapor, and to the substrate-liquid interfaces. Taking $\gamma - w$ as the interfacial energy on the substrate means that $w > 0$ for an adhesive substrate. Finally, the constant τ is the line tension associated with the contact line $\mathcal{C}$ where three phases coexist in equilibrium. In (1.1) a is the area measure on $\partial\mathcal{B} = \mathcal{S} \cup \mathcal{S}_*$, and s is the arc-length along the contact line $\mathcal{C}$. At this stage, the sign of line tension raises a basic question. In fact, a

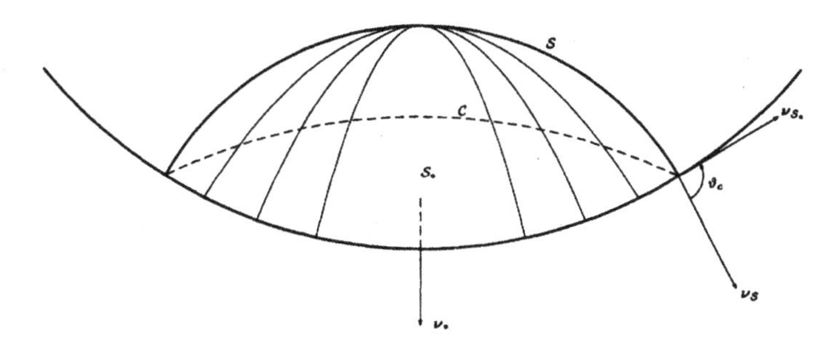

FIGURE 1. *Sketch of a drop deposited on a curved solid substrate. The boundary of the drop is composed of the free surface S and the adhering surface S_*. The contact line C is the common border of S and S_*. The contact angle ϑ_c is the angle between the conormals ν_S and ν_{S_*} of C, viewed as a curve on S and S_*, respectively.*

negative line tension makes the contact line unstable against perturbations with a short wavelength (*see* [2, 11]) since, in that case, the term

$$\tau \int_C \mathrm{d}s\,, \tag{1.2}$$

makes the free-energy functional (1.1) unbounded from below, and variational problems ill-posed [12]. As an aside, it is interesting to note that objections to negative values of τ raised on thermodynamic grounds [11] were rejected since "in the three-phase equilibrium the contact line cannot pucker to increase its length without at the same time changing the areas of the two-phase interfaces –which are of positive tension– in such a way as to increase the free energy of the whole system" (p. 237 of [13], see also the similar argument invoked by Solomentsev and White [6]). Although appealing, this counter-objection does not address the issue from the mathematical point of view, since it is not the amplitude of the perturbation, but its wavelength to induce instability, and indeed the perturbations could be concentrated so as to make the variation in the interfacial area negligible with respect to the length of the perturbed contact line. To tackle this problem, a possible strategy pursued in [12] relies on relaxation techniques to regularize the behaviour of (1.2).

Alternatively, one could assume that the functional (1.2) poorly accounts for line tension effects and that further terms, presumably dependent on the curvature of C, should be added to get rid of the unboundedness of (1.2), along the lines of Boruvka and Neumann's generalized theory of capillarity [14]. Finally, as a third way, it is possible to note [15, 16] that the ratio $|\tau|/\gamma$ between the line and the surface tension of the droplet introduces a natural length below which the model based on (1.1) is unreliable. Hence, if the wavelength of a destabilizing mode is shorter than $|\tau|/\gamma$, the instability would have a mathematical meaning with no physical counterpart since it would be effective at length scales that lie outside the realm of the model. In this case, a cutoff should be imposed to avoid too wiggly perturbations. In this way, conditionally stable equilibria would exist even for negative values of τ, *provided* that its magnitude $|\tau|$ is not too high.

The ratio $|\tau|/\gamma$ and the contact angle can be treated as independent on temperature only far from either the wetting or the dewetting transition, where the droplet either wets completely the underlying substrate, or is separated from the substrate by a vapor layer. When these transitions are approached, $|\tau|/\gamma$ and the contact angle are correlated *via* their dependence on temperature. This, in turn, takes us into another controversial topic, that is, the behavior of line tension at the wetting transition. While surface tension vanishes at the transition (*see* Chapter 9 of [13]), several scaling laws have been proposed for line tension, predicting all kind of behaviors. For instance, Widom and Clarke [17] predicted a vanishing line tension at wetting, while in [18] Szleifer and Widom predicted a possible divergence of line tension, though their conclusions were affected by numerical uncertainties: for an overview on this topic, *see* Section 1 of [19]. Such undulating predictions mirror once again the sensitivity of line tension to the microscopic description of the three-phase line.

The plan of this paper is the following. In Section 2 we start at the macroscopic level by building a wetting functional more general than (1.1) to cope with problems posed by technological applications of wetting. Then, we discuss the effects of line tension on the equilibrium equations for a droplet. Remaining at the macroscopic level, however, does not yield information on surface and line tension which are only parameters of the model. Hence, in Section 3 we review the main microscopic approaches to surface tension both at, say, the *phenomenological* level –where the details of intermolecular interactions are accounted for by use of a phenomenological potential – and at the *fundamental* level, where the tools of statistical mechanics are employed. The digression on surface tension is expedient to the treatment of line tension that, to some extent, follows a parallel avenue, and forms the content of Section 4. In Section 5 we are ready to study the effects of line tension on the stability of equilibria. We will outline the basic features of the general stability criterion introduced in [15, 20], and discuss in detail our approach to negative line tension illustrating by example how the cutoff on the wavelength of the perturbations comes into play. In Sections 6 and 7 we apply the stability criterion introduced in Section 5 to explore the consequences of a class of scaling laws on the stability of liquid bridges, close to the wetting and the dewetting transitions. Finally, a closing section summarizes the paper.

2. Line tension effects on equilibria. The renewed interest in wetting phenomena is due to the emergence of problems in different areas which call for a theoretical explanation. Among these are the effects of both material inhomogeneities and geometric microstructures of the substrate supporting the liquid drop $\mathcal{B}$ (*see* [21, 22]). Applications of these effects range from botany to the coating of fibers. To the former category belongs the Lotus effect, that is, the property endowed by the leaves of some plants to repel virtually any liquid [23]. Here, the geometric microstructure of the leaves seems to play a major rôle in the phenomenon [24, 25]. In the latter category, it is the substrate's curvature that strongly influences the stability of the sit droplets [26, 27]. Besides interfacial interactions, the droplet $\mathcal{B}$ can be subject to *body* forces with free-energy density f that could depend on the position in space. Apart from gravity, at the macroscopic level f could also account for contributions due to diluted interactions between the drop and the substrate, in the spirit of [28]. An appropriate generalization of (1.1) to account also for the effects just alluded to is

$$\mathcal{F}[\mathcal{B}] = \mathcal{F}_b[\mathcal{B}] + \mathcal{F}_s[\mathcal{S}] + \mathcal{F}_a[\mathcal{S}_*] + \mathcal{F}_\ell[\mathcal{C}]\,. \tag{2.1}$$

The first term is the bulk energy

$$\mathcal{F}_b[\mathcal{B}] := \int_{\mathcal{B}} f \mathrm{d}v\,. \tag{2.2}$$

The second term is the interfacial energy of the free surface $\mathcal{S}$:

$$\mathcal{F}_s[\mathcal{S}] := \gamma \int_{\mathcal{S}} \mathrm{d}a\,. \tag{2.3}$$

Hence, we are supposing that the "tangible surface of discontinuity" (p. 769 of [29]) $\mathcal{S}$ is the surface of tension introduced by Gibbs (see Section 3). The next term,

$$\mathcal{F}_a[\mathcal{S}_*] := \int_{\mathcal{S}_*} (\gamma - w)\mathrm{d}a\,, \tag{2.4}$$

describes the adhesive properties of the substrate. Finally,

$$\mathcal{F}_\ell[\mathcal{C}] := \int_{\mathcal{C}} \tau \mathrm{d}s \tag{2.5}$$

is the free energy of the contact line $\mathcal{C}$, where τ can be interpreted as a line tension. At variance with (1.1), in (2.4) and (2.5) we assume that both w and τ depend on the position on the substrate, so that wetting on inhomogeneous substrates can be studied: for an account on theoretical investigations on this topic *see* [22, 30, 31]. Moreover, when both w and τ are constant, arbitrary geometric microstructures are still possible in the substrate. To arrive at the equilibrium equations for the free surface $\mathcal{S}$ and the contact line $\mathcal{C}$, we perturb the points p on a putative equilibrium configuration into points p_ε, according to

$$p \mapsto p_\varepsilon := p + \varepsilon \boldsymbol{u} \tag{2.6}$$

where $\boldsymbol{u}$ is a smooth vector field. Since the droplet has to *glide* on the substrate, $\boldsymbol{u}$ must obey the constraint

$$\boldsymbol{u} \cdot \boldsymbol{\nu}_* = 0 \quad \text{on } \mathcal{S}_* \tag{2.7}$$

where $\boldsymbol{\nu}_*$ denotes the unit normal of $\mathcal{S}_*$ oriented outward to $\mathcal{B}$ (*see* Figure 1). Moreover, since all admissible variations must preserve the volume of the droplet, $\boldsymbol{u}$ is also constrained to satisfy

$$\int_{\mathcal{S}} u da = 0\,, \tag{2.8}$$

where $u := \boldsymbol{u} \cdot \boldsymbol{\nu}$ is the component of $\boldsymbol{u}$ along the unit normal $\boldsymbol{\nu}$ to $\mathcal{S}$. Referring the interested reader to Section 2 of [20] for the details of the computations, by setting the first variation of (2.1) equal to zero, we arrive at the following equilibrium equations:

$$\begin{cases} \gamma H + f = \lambda & \text{on } \mathcal{S} \\ \gamma \cos\vartheta_c + \gamma - w + \nabla_s \tau \cdot \boldsymbol{\nu}_{\mathcal{S}_*} - \tau\kappa_{g*} = 0 & \text{along } \mathcal{C}\,, \end{cases} \tag{2.9}$$

where H is the total curvature of $\mathcal{S}$, that is, twice its mean curvature, λ is the Lagrange multiplier associated with the constraint (2.8), κ_{g*} is the geodesic curvature of $\mathcal{C}$, imagined as a curve on the substrate $\mathcal{S}_*$, ∇_s is the surface-gradient operator, and ϑ_c is the contact angle, that is, the angle between the conormals $\boldsymbol{\nu}_{\mathcal{S}}$ and $\boldsymbol{\nu}_{\mathcal{S}_*}$ to $\mathcal{C}$, viewed as a curve on $\mathcal{S}$ and $\mathcal{S}_*$, respectively. Equation $(2.9)_1$ is the standard Laplace equation. If gravity is negligible and no other bulk effects are incorporated in the description so that $f = 0$, $(2.9)_1$ predicts that the equilibrium shape of the free surface $\mathcal{S}$ has to be a surface with constant mean curvature, making straight cylinders and spheres admissible solutions.

Equation $(2.9)_2$ is the generalised Young equation. The neutral attribute *generalised* appended to $(2.9)_2$ reminds us that several contributors worked on it. The generalization $(2.9)_2$ was obtained by Swain and Lipowsky [21], although Rusanov and Toshev (see §9.2 of [32]) obtained a condition similar to $(2.9)_2$, rephrased in terms of the angle between the wetted substrate and the local plane of the three-phase contact line. The equation

$$\gamma \cos\vartheta_c + \gamma - w - \tau\kappa_{g*} = 0 \tag{2.10}$$

to which $(2.9)_2$ reduces when the line tension is constant, was first introduced by Vesselovsky and Pertzov in 1936, according to [32] and [33]; it was employed by Pethica in 1961, and is sometimes referred to as the Gretz rule [34], quoting a 1966 note by Gretz [35]. Apart from priority matters, Equations $(2.9)_2$ and (2.10) are interesting since they show how line tension affects the equilibrium. It can be proved (see, *e.g.*, [36]) that, even on a flat substrate, a spherical sessile droplet can have no equilibrium configuration if the line tension is positive and large enough, and two distinct equilibria if the line tension, still positive, is smaller. On the other hand, no lack of either existence or uniqueness occurs when the line tension is negative. The equilibrium reveals a richer structure when the substrate is curved, as discussed in [37]. These features should be contrasted with the behavior in the absence of line tension, where the classical Young equation

$$\gamma \cos \vartheta_c + \gamma - w = 0 \tag{2.11}$$

replaces (2.10). In fact, when constitutive restrictions on γ and w guarantee that $\cos \vartheta_c \in [-1, 1]$, then (2.11) always possesses a unique solution. The different behavior can be explained by the presence in (2.10), as compared to (2.11), of the natural length scale τ/γ which breaks the invariance of (2.11) under rescaling. It should also be remarked that line tension effects on equilibrium are coupled with the geometric properties of the substrate through the geodesic curvature κ_{g*}. Hence, even in the presence of a non-vanishing line tension τ, no effects of τ on the equilibrium can be detected if the geometry of the substrate guarantees that $\kappa_{g*} = 0$, as happens for liquid bridges on a flat substrate [16].

3. Modelling surface tension. The macroscopic model (2.1) does not give any information on the surface and the line tension, that should be the outcome of a microscopic analysis accounting for the intermolecular interactions of which γ and τ are macroscopic signatures. Moreover, the interfaces $\mathcal{S}$ and $\mathcal{S}_*$ defined in the macroscopic approach introduce sharp discontinuities and it is not evident why only modifications of the interface areas should be penalized in (1.1) or in (2.1), and not also bending of the interfaces. Similarly it is not evident why only the length of the contact line, and not also its curvature, should contribute to line effects in the three-phase region. To extract information on both surface and line tension out of a microscopic analysis, and to understand the mechanisms behind possible generalizations of the free-energy functional (2.1), different avenues were followed in the literature. In this section, I refer to the approaches concerning surface tension, postponing a parallel review concerning line tension to the next section.

The microscopic origin of surface tension has been recognized for a long time, seemingly since the work of Von Segner published in 1752: *see* [38] and p. 15 of Finn's book [39]. Referring to Figure 1, the molecules of the liquid droplet $\mathcal{B}$ that lie either at $\mathcal{S}$ or at $\mathcal{S}_*$ feel different molecular interactions with respect to the droplet's molecules located in the interior of $\mathcal{B}$: it is this discrepancy that justifies the introduction of surface tension.

In fact, though Gibbs knew that the properties of $\mathcal{B}$ change continuously across a *transition layer* and not abruptly, he found it easier to imagine that the two phases remain homogeneous up to a certain *dividing surface*, that is "sensibly coincident with the physical surface of discontinuity, but shall have a precisely determined position" (p. 219 of [1]). In doing so, however, the system is somehow modified, since properties like its mass, energy, and entropy do not coincide with the sum of the masses, the energies and the entropies ascribed to the ideal homogeneous phases in contact at the dividing surface. To recover a correct description, Gibbs attributed to the dividing surface the *excess* energy U –and, similarly, the excess entropy, mass, etc.– defined as the difference between the real energy of the system and the energies of the ideal homogeneous phases that replace it. Different choices for the dividing surface could be made, depending on the specific requirement we place on it, with the only restriction that it belongs to a family of surfaces *parallel* to the "tangible physical surface of discontinuity". Gibbs indeed postulated that U should depend on the dividing surface through its area A, and its principal curvatures c_1 and c_2, so that

$$U = U(S, N_i, A, c_1, c_2),$$

where S is the entropy, and N_i is the mole number of the i-th component. Then, he was able to show that, by suitably selecting the dividing surface $\mathcal{S}$, it was always possible to get rid of the contributions arising from the curvature, and so he postulated a reduced surface energy U^r given by

$$U^r = U^r(S, N_i, A) :$$

the dividing surface on which the excess energy is U^r is called the *surface of tension*. The excess energy per unit area of the surface of tension is the *surface tension* associated with the interface separating two different phases. Hovever, Gibbs argument to obtain U^r from U works only if the interface is planar or spherical, and it is approximately true when the interface slightly differs from these cases. At this point, we stress an important difference between planar and spherical dividing surfaces. While shifting the dividing surface does not modify the surface tension in the former case, as it does not modifies the area of the surface, a displacement within a spherical layer is more delicate since, when the dividing surface is shifted its area changes, inducing an effective dependence of the surface tension on the radius of the dividing surface. This dependence was studied by Tolman [40] who proved that the surface tension $\gamma(R)$ associated with a spherical dividing surface of radius R can be written, at the lowest level, as

$$\gamma(R) = \gamma_0 \left(1 - \frac{2\delta_T}{R}\right),$$

where γ_0 is the line tension of a planar dividing surface, and the microscopic length δ_T is now called the *Tolman length*: accounts of Tolman's approach can be found in Section 2.4 of [13] and in Chapter 11 of [41]. Tolman also explored the spherical interfaces in [29], where he modelled the transition layer as a spherical shell, within which he located the surface of tension, obtaining an expression for its surface tension γ which depends on the layer's structure. Additional work on this topic is due to Buff [42] who explored the dependence on curvature of the surface tension pertaining to the *equimolar* surface, defined as the dividing surface over which no excess of mass exists.

Although Gibbs did not analyze arbitrarily curved interfaces, nonetheless he suggested the lines along which concentrate the efforts: "It will be observed that the position of this surface (*i.e.* the dividing surface) is as yet to a certain extent arbitrary, but that the directions of its normals are already everywhere determined, since all the surfaces that can be formed in the manner described are evidently parallel to one another." (*see* p. 219 of [1]). Seemingly, Buff [43] was the first who dropped the restrictions on the shape of the lamellae forming the transition layer. Buff, adopting a mechanical rather than thermodynamical approach, based his analysis upon the hydrostatics equation

$$\operatorname{div}\boldsymbol{\sigma} + \boldsymbol{f} = \mathbf{0} \tag{3.1}$$

where $\boldsymbol{\sigma}$ is the stress tensor, and $\boldsymbol{f}$ is the body force acting on the system. He assumed that in the bulk of the liquid and the vapor phases the stress tensors reduce to the standard isotropic form

$$\boldsymbol{\sigma}_L = -p_L \boldsymbol{I},$$

and

$$\boldsymbol{\sigma}_V = -p_V \boldsymbol{I}$$

where $\boldsymbol{I}$ is the identity tensor, while p_L and p_V are the pressures in the bulk phases. In the transition zone, Buff assumed the stress $\boldsymbol{\sigma}$ to be transversely isotropic about the unit normal $\boldsymbol{\nu}$ common to all the lamellae of the transition layer, by setting

$$\boldsymbol{\sigma}(\lambda) = \sigma_T(\lambda)(\boldsymbol{I} - \boldsymbol{\nu}\otimes\boldsymbol{\nu}) + \sigma_N(\lambda)\boldsymbol{\nu}\otimes\boldsymbol{\nu}, \tag{3.2}$$

where λ measures the distance from the dividing surface $\mathcal{S}$, corresponding to $\lambda = 0$. The layer is parameterized by use of confocal coordinates (u, v, λ), where u and v parameterize $\mathcal{S}$. In the spirit of Gibbs's prescription, Buff extrapolated the isotropic stress tensors $\boldsymbol{\sigma}_L$ and $\boldsymbol{\sigma}_V$ up to dividing surface. By taking the inner product between (3.1) and $\boldsymbol{\nu}$, with $\boldsymbol{\sigma}$ as in (3.2), then integrating across the layer, and repeating the same procedure for $\boldsymbol{\sigma} = \boldsymbol{\sigma}_L$ and $\boldsymbol{\sigma} = \boldsymbol{\sigma}_V$, he obtained the equilibrium equation

$$\Delta p := p_L - p_V = \gamma H - (H^2 - 2K)\frac{C}{S} + g\Gamma \boldsymbol{e}_z \cdot \boldsymbol{\nu}, \tag{3.3}$$

where H and K are the total and the Gaussian curvature of $\mathcal{S}$, $\boldsymbol{f}$ has been restricted to gravity contribution, and $\boldsymbol{g} = -g\boldsymbol{e}_z$ is the gravity acceleration.

From (3.3), Buff could interpret γ as the surface tension of $\mathcal{S}$ and express it in terms of the layer parameters as

$$\gamma := \int_{-\lambda_L}^{\lambda_V} (\sigma_T - \sigma_{LV})(1 + H\lambda)\mathrm{d}\lambda\,,$$

where λ_L and λ_V are positive numbers chosen so as the surfaces at $\lambda = -\lambda_L$ and at $\lambda = \lambda_V$ lie in the bulk phases. Moreover,

$$\sigma_{LV} := \sigma_L[1 - H(\lambda)] + \sigma_V H(\lambda)$$

where the Heaviside function

$$H(\lambda) := \begin{cases} 0 & \text{if } \lambda < 0 \\ 1 & \text{if } \lambda \geq 0 \end{cases}$$

has been introduced. Finally,

$$\frac{C}{S} = \int_{\lambda_L}^{\lambda_V} (\sigma - \sigma_{LV})\mathrm{d}\lambda \tag{3.4}$$

is the bending moment associated with variations in tl
dividing surface and

$$\Gamma = \int_{-\lambda_L}^{\lambda_V} (\varrho - \varrho_{LV})\mathrm{d}\lambda\,,$$

where

$$\varrho_{LV} := \varrho_L[1 - H(\lambda)] + \varrho_V H(\lambda)$$

is the surface excess mass of the dividing surface, in terms of the bulk mass densities ϱ_L and ϱ_V. In the remaining part of [43], Buff reexamines the phenomenological theory by repeating Gibbs's approach with the important difference that a bending moment (3.4) is retained. Hence, the free energy is modified and (3.3) follows also at this level as the equilibrium condition for a droplet, thus generalizing the classical Laplace equation.

Buff's paper stimulated further generalizations of Gibbs original approach. Indeed, as noted later by Boruvka and Neumann, Buff pointed out that "the limitations of the original analysis by Gibbs center around the curvature terms in the expression for the internal energy. [...] The Gibbs approximation worked so well that nobody pursued Gibbs' comment that more rigorous treatments are possible" (p. 5464 of [14]). Boruvka and Neumann [14] however, remarked that Buff failed to make a proper distinction between intensive and extensive parameters, resulting in an incorrect generalization of Gibbs capillarity theory. Instead of using extensive quantities as Gibbs did, Boruvka and Neumann worked with volume densities of these quantities. Moreover, when they had to ascribe a surface energy to a dividing surface, they noted that any surface $\mathcal{S}$ admits three *extensive* geometric properties, that is, the area A, and the first and second integral curvatures $\mathcal{H}$ and $\mathcal{K}$ defined as

$$\mathcal{H} := \int_{\mathcal{S}} H \mathrm{d}A \qquad \text{and} \qquad \mathcal{K} := \int_{\mathcal{S}} K \mathrm{d}A$$

so that the surface energy density should depend on H and K, while the extensive surface energy U should contain A, $\mathcal{H}$ and $\mathcal{K}$ among its arguments. Besides the quest for a general approach, it seems that Boruvka and Neumann first realized that only differential *invariants* of a surface could enter in the free-energy density. By an analysis parallel to Buff's, Boruvka, Rotenberg and Neumann [44] used the hydrostatic theory of capillarity to have a further verification of Boruvka and Neumann [14] phenomenological model. The consistency of the two approaches was shown in [44] through a procedure that was later simplified in [45] and [46].

Until now the intimate details of intermolecular interactions occurring at an interface have been bypassed, by use of more or less idealized descriptions of the transition layer. Computations of the surface tension for *planar* interfaces at the microscopic level, by resorting to statistical mechanics, were pioneered by Fowler [47], and by Kirkwood and Buff [48]. A basic ingredient in Kirkwood and Buff's analysis is the pair density $\varrho^{(2)}(\boldsymbol{r}_1, \boldsymbol{r}_{12})$ which measures the average number of molecular pairs, with one molecule located in a small volume around $\boldsymbol{r}_1$, and the other in a small volume around a second point whose relative position from the former is $\boldsymbol{r}_{12}$. Hence, by assuming the planar interface at $z = 0$, it is possible to set $\varrho^{(2)} = \varrho^{(2)}(z_1, r_{12})$, where z_1 is the z-coordinate of the first molecule, and $r_{12} = |\boldsymbol{r}_{12}|$. From these premisses, Kirkwood and Buff did not determine the surface tension from its thermodynamical definition as the work needed to form a unit area of interface, but from its mechanical definition in terms of the stress transmitted across a strip of unit width, orthogonal to the dividing surface: in some sense, they remount to Laplace's original view of capillarity [49]. In fact, it is possible to arrive at the expression for γ obtained by Kirkwood and Buff also through the standard thermodynamical avenue, by computing

$$\gamma = \left(\frac{\partial \mathcal{F}}{\partial A}\right)\bigg|_{V,T,N} \tag{3.5}$$

where $\mathcal{F}$ is the Helmholtz free energy, A is the area of the dividing surface, and the subscripts mean that differentiation should be computed at constant volume V, temperature T, and mean number of particles N. It turns out that the derivation based upon (3.5) is more general than the original approach discussed in [48] since, as proven in [50], the molecular pressure tensor employed by Kirkwood and Buff lacks of uniqueness: fortunately, this feature does not affect the expression of γ. Hence, regardless of the avenue that is followed, the surface tension can be given in the form

$$\gamma = \frac{1}{4}\int_{-\infty}^{+\infty} \mathrm{d}z_1 \int \mathrm{d}r_{12} r_{12} \left(1 - 3\frac{z_{12}^2}{r_{12}^2}\right) u'(r_{12})\varrho^{(2)}(r_{12}, z_1), \tag{3.6}$$

where u is the pairwise intermolecular potential that depends only on the relative distance r_{12} between the two molecules, and a prime denotes differentiation with respect to r_{12}: for a clear derivation of (3.6) from (3.5), see Chapter 4 of [41]. Though compact, Equation (3.6) is not easy to compute, as it requires knowledge of the intermolecular potential and the two-particle density. Partial simplification can be obtained by invoking Fowler approximation, according to which the density of the the vapor phase is neglected (*see* §4.1.2 of [41]). Even in that case, however, the evaluation of γ requires a delicate machinery to deal with the two-particle correlation function. Apart from computer simulations, progress in this field was made possible by use of perturbation schemes, according to which a reference intermolecular potential is singled out while the remaining part of the potential is treated as a perturbation. Though more schemes are possible, the Weeks-Chandler-Anderson approach [51] has been the most influential and, besides several applications to bulk properties of homogeneous fluids, it was applied by Kalikmanov and Hofmans [52] to obtain an analytic expression for the surface tension of a Lennard-Jones fluid at the *planar* interface with the vapor phase: a detailed account of the perturbation approach in the statistical mechanics of fluids can be found in Chapter 5 of [41].

The second microscopic way to surface tension is based upon van der Waals theory that postulates a free energy density ψ which depends on the local mass density $\varrho(\boldsymbol{r})$ and on its squared gradient. If no external field acts on the fluid, the Helmholtz free energy $\mathcal{F}$ can be written as

$$\mathcal{F} = \int [\psi(\boldsymbol{r})]\mathrm{d}V = \int \left\{\psi(\varrho(\boldsymbol{r})) + \frac{1}{2}A(\varrho(\boldsymbol{r}))[\nabla\varrho(\boldsymbol{r})]^2\right\}\mathrm{d}V$$

where $\psi(\varrho)$ is the free energy density of a uniform fluid with density ϱ: hence, the first term in $\mathcal{F}$ is the extension to the non-uniform case of the free-energy density of a uniform fluid, while the second term accounts for spatial inhomogeneities of the density profile. It was a major achievement of [49] to prove that the coefficient $A(\varrho)$ is related to the direct correlation function. In this approach, the surface tension arises naturally as a byproduct from minimization of $\mathcal{F}$ under the constraint that the system is closed. If $\varrho(\boldsymbol{r})$ is the density profile that minimizes $\mathcal{F}$, and the interface is the plane $z = 0$, the surface tension γ_0 can be written as

$$\gamma_0 = \int_a^b A(\varrho(z))(\varrho'(z))^2\mathrm{d}z,$$

where a and b are values of the coordinate z in the bulk phases.

With the extension of van der Waals theory in [49] we are entering a broad area of intensive research in non-uniform fluids that makes use of the density functional theory. Since entering the details of this approach would take us too far, we content ourselves with a cursory mention to its existence, referring the interested reader to the reviews [53, 54, 55], and to Chapter 9 of [41].

4. Modelling line tension. It is now time to see how the strategies employed to study the transition layer and to obtain information on the surface tension work when applied to a three-phase contact region, with special emphasis on the predictions concerning line tension. As remarked in the Introduction, the line tension was introduced by Gibbs [1] in the same spirit as surface tension. At a formal level, Gibbs prescriptions were generalized by Boruvka and Neumann in [14], who considered a line energy excess depending also on the geodesic and the normal curvatures of the contact line $\mathcal{C}$, thought of as a curve belonging to any of the three coexisting phases. Moreover they also included a dependence on the contact angles between the intervening phases. This theory takes care of the deficiencies of (1.2) by adding curvature terms to penalize from the beginning the onset of wiggly destabilizing modes. However, it is difficult to manage this theory in its full generality. On the other hand, we also mention that the use of free-energy densities depending on curvature has been questioned by Sagis and Slattery who, in the continuum theory developped in [56, 57, 58], claimed that no contributions in either the surface- or the line-energy density depending on curvatures should arise at all. They formally proved that the presence of curvature terms would be incompatible with the fulfillment of the entropy inequality which, in the spirit of Coleman and Noll theorem [59], acts as a constraint on constitutive assumptions. As far as I know, no attempt in confirming or disproving the results of Sagis and Slattery has been made.

The theoretical approaches to line tension can be divided [4] into two categories, as local and non-local. In a local approach the intermolecular forces are accounted for only in the narrow region where the three interfaces meet together, while in a non-local approach intermolecular forces are also considered beyond that region and, moreover, effects due to the global shape of the droplet are incorporated in the treatment. It should be noted, however, that several assumptions are made in both approaches which make straightforward comparisons difficult. For instance, effects of gravity can be neglected; the droplet can be imagined in equilibrium with a thin fluid film or not; predictions are often made for droplets of infinite size, more similar to a wedge, and so on. It is important to realize this, since different models often lead to drastically different conclusions concerning line tension.

All microscopic models consider the distortions of the droplet's shape in the three-phase region as the key factor governing the line tension, as

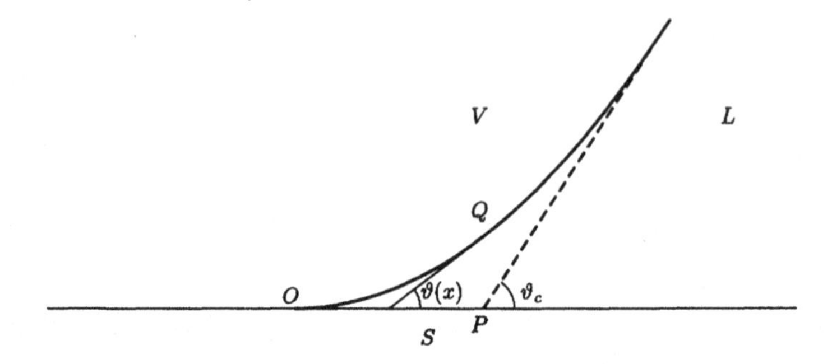

FIGURE 2. *Distortion suffered by the liquid droplet L, here sketched as an infinite wedge, in the contact region, where it meets both the substrate S and the vapor phase V. Macroscopically, the droplet is seen to meet the substrate at P, hence forming a macroscopic contact angle ϑ_c. However, at any point $Q \equiv (x, h(x))$ lying on the real, distorted profile of the droplet, the microscopic contact angle is $\vartheta(x)$.*

stressed by [6], [10], and [58]. This makes the analysis more difficult than the parallel one performed for surface tension, as was clearly stated by Buff and Saltsburg [60] who extended the capillary approach to three-phase contact. In [60] they noted that in the transition region "the distribution of matter becomes more diffuse and the physical basis for parallel dividing surfaces breaks down." (p. 27 of [60]). To overcome the difficulty they extrapolated not only bulk energies, as done to define surface tension, but also surface energies, ascribing the excess energy to the contact line, according to Gibbs's prescription.

The distortions of the droplet's shape lead, for instance, to distinguish between a *macroscopic* contact angle obtained by extrapolating the macroscopic shape of the drop up to the substrate, and the *microscopic* contact angle, bound to follow the distortion of the droplet's profile up to the substrate (Figure 2). Sagis and Slattery [58] use a perturbative approach to arrive at the droplet's shape by distinguishing an outer problem, where the profile is dictated by Laplace equation and so the effects of intermolecular forces in the three-phase region are neglected, from an inner problem, where these forces enter the equilibrium equation obeyed by the profile through a suitable potential that incorporates, besides gravity, the effects of van der Waals-London forces, while double-layer, steric, and structural forces are neglected. Not surprisingly, the perturbation parameter involves the strength of long-range intermolecular forces. The geometry of the three-phase region is carefully analyzed by Babak [10] from a slightly different perspective, as he is more concerned with situations in which the droplet is in equilibrium with either a thin fluid film or a thick fluid layer. Consequently, he considers three regions to be matched together by resort to suitable boundary conditions: first, there is a *Laplace* (outer) *region*, where the profile, in the absence of gravity, is a sphere. The rôle of intermolecular forces is particularly relevant in the *transition region* that mediates between the Laplace region and a further *flat region*, where the drop sits upon the film or the layer. In the transition zone, the contact angle gradually passes from its macroscopic value to zero. Using a different approach, Solomentsev and White [6] do not assume in advance that the droplet is in equilibrium with a fluid film on the substrate, and use the profile of the Laplace region as an asymptotic shape. In passing, we note that in all these approaches the solid substrate on which the droplet lies is planar.

Solomentsev and White [6] locally modelled the profile of an infinitely large drop by its height $h(x)$ over the substrate, of which x is the abscissa. The interaction energy between the vapor and the fluid phase close to the substrate is modelled by resorting to the Derjaguin approximation, according to which the interaction energy per unit length $E_{SLV}(h(x))$ is in fact the interaction energy between "plane parallel half-spaces of substrate and vapor phases, separated by a distance h of the liquid phase" (p. 123 of [6]). Embodied in $E_{SLV}(h(x))$ are contributions due to van der Waals, steric, and structural forces, besides possible terms due to density variations. However, use of the Derjaguin approximation limits predictions to small values of the macroscopic contact angle ϑ_c.

In these approaches, the way to line tension is twofold. From one side [10], [58], the balance equations obtained at the microscopic level are recast in a form which allows identification with the macroscopic definition of line tension based on the functional (1.2). On the other hand, it is possible to determine the droplet's profile as the minimizer of the Helmholtz free energy F per unit length of the contact line that models the free energy stored in the three-phase region [6]:

$$F = F_0 + \int_{x_0}^{\infty} [\gamma_{LV} + E_{SLV}(h(x))](1 + h'^2(x))^{1/2} \mathrm{d}x + x_0(\gamma_{SV} - \gamma_{SL}) \, .$$

Here, F_0 contains all contributions independent of the droplet's shape, and x_0 is the value of x at O, where the real contact line, that can be appreciated only at the microscopic level, meets the substrate (*see* Fig. 2). By minimizing F, Solomentsev and White obtained that the microscopic contact angle should vanish at $x = x_0$. Curiously, we mention that [58] attributed discrepancies between their theoretical model and experimental results to the *assumption* of vanishing microscopic angle. When the droplet's profile $h(x)$ has been obtained from this variational analysis, the line tension τ is defined as the energy difference (per unit length of the contact line) between the microscopic and the macroscopic profiles and can be calculated as

$$\tau = \frac{\gamma_{LV}}{\cos\vartheta_c} \times \int_{x_0}^{+\infty} \left\{ \frac{E_{SLV}(h(x))}{\gamma_{LV}} \left(2 + \frac{E_{SLV}(h(x))}{\gamma_{LV}} \right) + (x_P - x_0) \sin^2 \vartheta_c \right\} \mathrm{d}x \, ,$$

where x_P is the value of x at which the extrapolated contact line meets the substrate. Solomentsev and White obtained negative values for τ by taking, as an illustration of their method,

$$E_{SLV}(h) = -\frac{A_{SLV}}{12\pi(h+h_0)^2},$$

where $A_{SLV} > 0$ is the Hamaker constant (*see* Chapt. 11 of [61]), and h_0 is a suitable cutoff length that measures the range of the forces involved in the interaction. It should be noted that, although this simplified example does not account for steric and structural forces, it maintains reference to the multi-body effects associated with the dispersion forces. It should also be remarked that [6] was also concerned with the finite-size effects of the droplet showing that the infinite-drop model is inadequate only when the radius R of a spherical drop is comparable with h_0. The approach of [6] retains many features in common with the *interface displacement* models which were discussed by de Gennes (*see* §II.D of [62]) and then applied extensively to study line tension (see, *e.g.*, [7, 8, 63]). Here, the free energy per unit length of the contact line of an infinite droplet sitting on a flat substrate and in equilibrium with a thin liquid film of width h_1 is taken as

$$\Omega[h] := \int \left[\frac{\gamma_{LV}}{2}\left(\frac{\mathrm{d}h(x)}{\mathrm{d}x}\right)^2 + V(h(x))\right]\mathrm{d}x\,. \tag{4.1}$$

The quadratic term in $\mathrm{d}h(x)/\mathrm{d}x$ penalizes changes in the interface area and so accounts for distortions near the contact region, while $V(h(x))$ is the *interface potential* which embodies intermolecular interactions: the equilibrium profile $h_e(x)$ is the minimizer of (4.1). The line tension is then obtained by replacing $h_e(x)$ into (4.1), and can be recast as

$$\tau = \sqrt{2\gamma_{LV}}\int_{h_1}^{+\infty}\left[\sqrt{V(h)}-\sqrt{E}\right]\mathrm{d}h\,,$$

where $E := \lim_{h\to+\infty} V(h)$. In [8], τ was computed for systems with long-range forces such that either

$$V(h) = E + \beta(h^{-3} - ah^{-2}) \tag{4.2}$$

with β and a positive constants, or

$$V(h) = E + \beta(ah^{-2} - 2h^{-3} + h^{-4})\ : \tag{4.3}$$

while for (4.2) the line tension is always negative, for the choice in (4.3) line tension turns from negative to positive on decreasing the macroscopic contact angle.

As already remarked, Solomentsev and White [6] did not assume the existence of a thin liquid film in equilibrium with the droplet and, in turn, they obtained a necessary condition for this to happen. However, when such a film exists, the energy density $E_{SLV}(h)$ is related to the *disjoining pressure* $\Pi(h)$ of a film of height h through

$$\Pi(h) := -\frac{\partial E_{SLV}(h)}{\partial h}.$$

The effect on line tension due to the profile of the disjoining pressure isotherm Π against h is the major concern of Babak's generalized theory [10] which aims at mediating between two extreme models that view the adsorbed film in equilibrium with the droplet either as an idealized surface of zero thickness, or as a layer of thickness H_f. Babak's theory points out the crucial rôle played by the profile of $\Pi(h)$ to distinguish between the two cases and, more importantly, it relates the predictions on line tensions to the qualitative features of the function $\Pi(h)$, in particular to the presence of maxima and minima.

The approaches illustrated so far are local, as they neglect nonlocal contribution to liquid-liquid interactions that have been incorporated by Getta and Dietrich [5, 64], who were motivated by apparent deficiencies in the interface displacement model, when applied to structured substrates. Starting form a density functional theory, Getta and Dietrich [64] concluded that in most situations predictions based on interface displacement models are in good agreement with the nonlocal theory which, on the contrary, requires careful numerical analysis. In fact, resort to nonlocal models seems mandatory only when the interface profiles are highly curved. It should also be stressed that multi-body effects are neglected within a nonlocal theory, while they are embodied in the simpler, phenomenological interface displacement models (*see* [6, 7]).

To close this section, we recall that the approach to line tension *via* statistical mechanics, in the spirit of Kirkwood and Buff [48], was pursued by Tarazona and Navascués [9]. As they stressed, "The three-phase contact line is a very inhomogeneous region [...]. The topology of the density in this region can be very complicated and, at present, it prevents us from trying to determine the molecular distribution functions" that enter the expression of the line tension (*cf.* p. 3116 of [9]). In fact, even restricting attention to a straight contact line, they had to resort to a suitable form of Fowler's approximation to calculate the line tension of a Lennard-Jones fluid.

5. Line tension effects on stability. A wealth of mathematical literature exists on the equilibrium of sessile droplets and, in general, on equilibrium capillary surfaces [39]. It is however crucial to ascertain the stability of equilibrium configurations, since only stable or, at least, metastable equilibria can be observed. Studies on the stability of fluid surfaces appear less frequently, especially when constraints are imposed on the free-energy functional, that must be obeyed up to second order in the perturbation parameter ε. The special role played by the contact line can be understood by recalling a classical result in the theory of minimal surfaces. In fact, Barbosa and Do Carmo [65] proved that the area functional (2.3) is locally stable against perturbations confined in a sufficiently small region of the equilibrium surface that does not involve the contact line, supposed as fixed. Similarly, turning attention to capillary surfaces, the local stability of spheres was proved in [66] by limiting attention to perturbations that, again, do not move the contact line. An important contribution to this topic was the paper by Sekimoto, Oguma, and Kawasaki [67] who studied the stability of several wetting morphologies, for a planar and homogeneous substrate, without line tension. In fact, for small contact angles, they proved that the sessile droplet is stable against all perturbations that move the contact line, apart from a uniform translation. Taking small contact angles allowed the authors to draw their conclusion from the study of a linear partial differential equation. Among the morphologies explored in [67] is the fluid ridge, where a liquid is drawn along a solid wedge. This geometric setting was studied by Roy and Schwartz [68] who generalized the problem by considering a cylindrical droplet laid on a cylindrical substrate. The analysis employed is confined to a two-dimensional problem, and cannot be extended to less symmetric shapes. In [68] however, a systematic exploration of the role played by the substrate's curvature was made for the first time. Along a different avenue, Lenz and Lipowsky [69] obtained a general stability criterion, for arbitrary wetting morphologies, in the presence of a structured substrate. In [69] the substrate is flat and line tension is absent, so that no effect of curvature is at play. In [20], we derived a general stability criterion for the wetting functional (2.1) which hence incorporates the effects of inhomogeneous, arbitrarily curved substrates, in the presence of line tension. Before reviewing the contents of [20], it is worth digressing slightly to put the results in the right perspective.

In the mathematical literature, a clear distinction is made between two concepts: *stability* and *minimality*. In fact, given a functional $\mathcal{F}$ defined on a Banach space $\mathbb{B}$, a point $P \in \mathbb{B}$ is stable for $\mathcal{F}$ if, for all $Q \in \mathbb{B}$, the function $f(\varepsilon) := \mathcal{F}(P + \varepsilon Q)$ attains a minimum at $\varepsilon = 0$. A point $P \in \mathbb{B}$ is a strict local minimum for $\mathcal{F}$ in $\mathbb{B}$ if there is $\varepsilon > 0$ such that $\mathcal{F}(P) < \mathcal{F}(Q)$ for all $Q \neq P$ in $\mathbb{B}$ such that $||P - Q||_{\mathbb{B}} < \varepsilon$, where $|| \cdot ||$ is the norm in $\mathbb{B}$. Considerable literature exists on the stability of capillary surfaces lying on homogeneous substrates, mostly in the absence of gravity: see *e.g.* [70, 71]. The question about strict local minima in capillarity theory has been treated in a series of papers by Vogel [72, 73, 74], who proposed minimality criteria arising from the requirement that the second variation $\delta^2\mathcal{F}$ of the free-energy functional $\mathcal{F}$ be *strongly* positive, that is, there exists a positive constant k such that $\delta^2\mathcal{F} > k||h||_{\mathbb{B}}$, for all $h \in \mathbb{B}$ (*see* p. 100 of [75]). Invoking strict positivity of the second variation, Vogel could elude the difficulty pointed out by Finn [76] who remarked that the mere requirement that the second variation of a functional be positive cannot guarantee by itself that the energy actually attains a minimum.

In [20], we systematically addressed the stability of the functional (2.1), leaving untouched the issue as to whether our stability criterion was related to the minimality of (2.1). To arrive at a correct expression for the second variation of (2.1), the crucial step was to properly take care of the constraints on the system, up to second order in ε. Requiring (2.7) and (2.8) suffices to obey the constraints on the system up to first order in ε but, in general, use of a single vector field $\boldsymbol{u}$ could lead to inconsistencies when the constraints are enforced to second order, as required in a stability analysis. Following Peterson [77, 78], instead of (2.6) we used the following mapping

$$p \mapsto p_\varepsilon := p + \varepsilon \boldsymbol{u} + \varepsilon^2 \boldsymbol{v} \tag{5.1}$$

to perturb the shape $\mathcal{B}$, where $\boldsymbol{v}$ is a smooth vector field. Imposing the volume constraint up to second order in ε requires, besides (2.8), the fulfillment of

$$\int_{\mathcal{S}} (u \operatorname{div}_s \boldsymbol{u} - \boldsymbol{u} \cdot \boldsymbol{a} + 2v) \mathrm{d}a = 0 , \tag{5.2}$$

where $\operatorname{div}_s \boldsymbol{u} := \operatorname{tr} \nabla_s \boldsymbol{u}$ is the surface divergence of $\boldsymbol{u}$, $\boldsymbol{a} := (\nabla_s \boldsymbol{u})^{\mathsf{T}} \boldsymbol{\nu}_*$, and $v := \boldsymbol{v} \cdot \boldsymbol{\nu}$. Similarly, contact between the droplet and the substrate is preserved up to the required order if

$$\boldsymbol{v} \cdot \boldsymbol{\nu}_* = -\frac{1}{2} \boldsymbol{u} \cdot (\nabla_s \boldsymbol{\nu}_*) \boldsymbol{u} \qquad \text{on } \mathcal{S}_* \tag{5.3}$$

is obeyed, in addition to (2.7). From (5.3), it follows that, in general, were the substrate curved and had we chosen $\boldsymbol{v} \equiv \mathbf{0}$, it would be impossible to ensure contact between the droplet and the substrate, up to terms quadratic in ε. After painstaking computations, in [20] we arrived at the following expression for the second variation [79] of (2.1):

$$\begin{aligned} \delta^2 \mathcal{F} &= \gamma \int_{\mathcal{S}} \left\{ |\nabla_s u|^2 + (2K - H + (\gamma^{-1}) \partial_\nu f) u^2 \right\} \mathrm{d}a \\ &\quad + \int_{\mathcal{C}} \{ \tau u_{s*}'^2 - \gamma \beta u_{s*}^2 \} \mathrm{d}s , \end{aligned} \tag{5.4}$$

where:

$$\begin{aligned} \gamma\beta : &= \tau(K^* + \kappa_g^{*2}) + (\nabla_s w \cdot \boldsymbol{\nu}_{\mathcal{S}_*}) - (\nabla_s^2 \tau) \boldsymbol{\nu}_{\mathcal{S}_*} \cdot \boldsymbol{\nu}_{\mathcal{S}_*} + \kappa_g^* \nabla_s \tau \cdot \boldsymbol{\nu}_{\mathcal{S}_*} \\ &\quad + \gamma [H^* \sin \vartheta_c + H \cos \vartheta_c \sin \vartheta_c + \kappa_g \sin^2 \vartheta_c]. \end{aligned} \tag{5.5}$$

In (5.4) and (5.5) H^* and K^* are the total and the Gaussian curvatures of $\mathcal{S}_*$, $\partial_\nu f := \nabla f \cdot \boldsymbol{\nu}$ is the normal derivative of f on $\mathcal{S}$, κ_g^* is the geodesic curvature of $\mathcal{C}$, thought of as a curve on $\mathcal{S}_*$, the field u_{s*} is related to u by

$$u = \sin \vartheta_c u_{s*}, \tag{5.6}$$

and a prime denotes differentiation with respect to the arc-length s of $\mathcal{C}$. It is worth noting that $\delta^2\mathcal{F}$ is independent of the field $\boldsymbol{v}$. This is in fact obtained by resorting to (5.2), (5.3), and to the equilibrium equations (2.9). We can now derive from equation (5.4) a criterion for the local stability of the equilibrium configurations of (2.1), where local stability means strong positiveness of $\delta^2\mathcal{F}$, in the same sense previously recalled, but here relative to the L^2-norm over $\mathcal{S}$. This is indeed equivalent to subjecting $\delta^2\mathcal{F}$ to the further, non-homogeneous constraint

$$\int_{\mathcal{S}} u^2 \mathrm{da} = 1\,, \tag{5.7}$$

in addition to (2.8) and requiring $\delta^2\mathcal{F}$ to attain a positive minimum on the set of functions u satisfying (5.7). As is standard in such a case (see Sect. VI.1 of [80]), we define the functional

$$\begin{aligned} F[u] \;:=\; & \frac{1}{2}\int_{\mathcal{S}}\{|\nabla_s u|^2 + \alpha u^2\}\mathrm{da} + \lambda\int_{\mathcal{S}} u\mathrm{da} - \\ & -\frac{1}{2}\mu\int_{\mathcal{S}} u^2\mathrm{da} + \frac{1}{2}\int_{\mathcal{C}}\{\xi u_{s*}'^2 - \beta u_{s*}^2\}\mathrm{ds}\,, \end{aligned} \tag{5.8}$$

where λ and μ are Lagrange multipliers associated with (2.8) and (5.7),

$$\alpha := 2K - H + \partial_\nu f\,, \tag{5.9}$$

and the ratio

$$\xi := \frac{\tau}{\gamma}\,, \tag{5.10}$$

sets a natural length scale of the problem. Like τ, ξ can be of either sign. The equilibrium equations associated with (5.8) are

$$\Delta_s u - \alpha u - \lambda + \mu u = 0 \qquad \text{on } \mathcal{S}\,, \tag{5.11}$$

$$\nabla_s u \cdot \boldsymbol{\nu}_{\mathcal{S}} - \chi(\xi(\chi u)')' - \beta\chi^2 u = 0 \qquad \text{along } \mathcal{C}\,, \tag{5.12}$$

where $\chi := 1/(\sin\vartheta_c)$. By standard arguments [20], it is possible to prove that the least value of μ for which there is a solution to Equations (5.11) and (5.12) is also the minimum of $\delta^2\mathcal{F}$ on the constraints (2.8) and (5.7). We thus conclude that an equilibrium configuration of a droplet with energy represented as in (2.1) is locally stable whenever the least eigenvalue μ in (5.11) is positive.

We applied the stability criterion to find out how a line tension of either sign affects the stability of droplets laid upon a substrate. In [16] and [36], where the substrate was assumed flat and homogeneous, we focussed attention on liquid bridges and spherical sessile droplets. In [37] we explored the effects of the substrate's curvature by studying a spherical

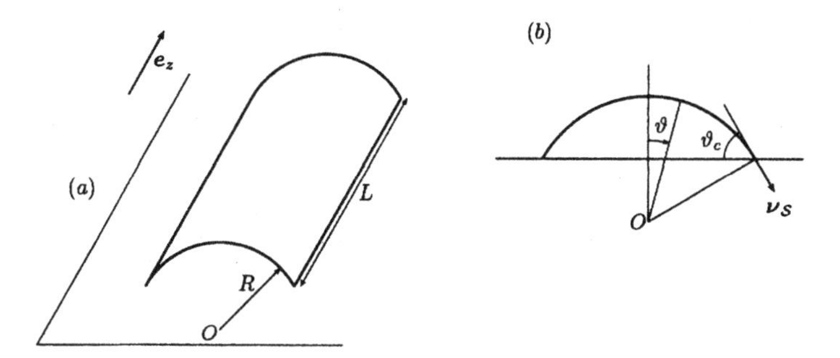

FIGURE 3. *(a). Sketch of a liquid bridge lying on a flat, homogeneous substrate. The bridge is a straight circular cylindrical sector. (b). Every planar cross-section of the bridge orthogonal to* $\mathbf{e}_z$ *is a circular sector of radius* R*, centered at the point* O*. The contact angle* ϑ_c *is the angle between the substrate and the bridge's tangent plane along the contact line* C*. The points on the bridge's profile are parameterized by the* z*-coordinate along* $\mathbf{e}_z$ *and by the polar angle* ϑ*. Since we imagine the liquid bridge as infinite along* $\mathbf{e}_z$*,* L *is interpreted as the typical length associated with the distortion* u*.*

sessile droplet on a spherical, homogeneous substrate. In the sequel, we shall stress the results of [16] that are expedient to the computations of Sections 6 and 7. The geometrical setting is sketched in Figure 3 where a liquid bridge is modelled as a straight cylinder of radius R. Since both w and τ were assumed constant, we are focussing attention on the functional $\mathcal{F}$ in (1.1). It is easy to prove from (2.10) that line tension does not affect the equilibrium, since $\kappa_{g*} = 0$. Once cylindrical polar coordinates (ϑ, z) are introduced, the eigenvalue problem (5.11)-(5.12) reduces to

$$\frac{1}{R^2}\frac{\partial^2 u}{\partial \vartheta^2} + \frac{\partial^2 u}{\partial z^2} + (\mu + \frac{1}{R^2})u + \lambda = 0, \tag{5.13}$$

$$\frac{1}{R}\sin^2\vartheta_c \left.\frac{\partial u}{\partial \vartheta}\right|_{\vartheta=\vartheta_c} - \xi \left.\frac{\partial^2 u}{\partial z^2}\right|_{\vartheta=\vartheta_c} - \frac{1}{R}\sin\vartheta_c\cos\vartheta_c u(\vartheta_c, z) = 0\,. \tag{5.14}$$

We imagine the liquid bridge as infinite along $\mathbf{e}_z$, and we introduce the typical length L associated with the distortions the bridge undergoes. Thus, it is natural to require that

$$u(\vartheta, 0) = u(\vartheta, L) = 0\,, \quad \forall \vartheta \in [0, \vartheta_c]\,. \tag{5.15}$$

Limiting attention to perturbations that do not introduce further symmetry breaking in the problem, we also impose the requirement

$$\left.\frac{\partial u}{\partial \vartheta}\right|_{\vartheta=0} = 0 \quad \forall z \in [0, L]\,. \tag{5.16}$$

We look for solutions to Eqs. (5.13) and (5.14) in the form

$$u(\vartheta, z) = u_0(\vartheta) + \sum_{n=1}^{\infty} \sin\left(\frac{2n\pi}{L}z\right) u_n(\vartheta)\,. \tag{5.17}$$

so that (5.15) is automatically satisfied. Attention is restricted to even modes to satisfy automatically the incompressibility condition (2.8), which now reads

$$\int_0^L \mathrm{d}z \int_0^{\vartheta_c} u(\vartheta, z)\mathrm{d}\vartheta = 0\,.$$

Note that the addition of odd modes does not modify the stability threshold [16]. Furthermore, the z-independent mode $u_0(\vartheta)$ compatible with (5.15) is $u_0(\vartheta) \equiv 0$, whence $\lambda = 0$ in (5.13) follows. By setting

$$\varrho_n := \left(\frac{2\pi n R}{L}\right)^2 \tag{5.18}$$

the projection of (5.13) onto the eigenspaces corresponding to different values of n gives rise to the following set of boundary value problems, with $n \in \mathbb{N}$, $n \geq 1$:

$$\begin{cases} \ddot{u}_n(\vartheta) + \sigma_n u_n(\vartheta) = 0 \qquad \forall \vartheta \in (-\vartheta_c, \vartheta_c) \\ \sin^2 \vartheta_c \dot{u}_n(\vartheta_c) + [\frac{\xi}{R}\varrho_n - \sin\vartheta_c \cos\vartheta_c] u_n(\vartheta_c) = 0 \\ \dot{u}_n(0) = 0\,, \end{cases} \tag{5.19}$$

where

$$\sigma_n := \mu R^2 + 1 - \varrho_n\,, \tag{5.20}$$

and the condition $(5.19)_3$ reflects the requirement (5.16). According to the value of σ_n, Equation $(5.19)_1$ is solved by

$$\begin{cases} u_n(\vartheta) = A \cosh\sqrt{-\sigma_n}\vartheta \qquad \text{if } \sigma_n < 0 \\ u_n(\vartheta) = B \qquad \text{if } \sigma_n = 0 \\ u_n(\vartheta) = C \cos\sqrt{\sigma_n}\vartheta \qquad \text{if } \sigma_n > 0\,, \end{cases}$$

referred to as the *hyperbolic*, the *linear*, and the *circular* modes. The constants A, B, C are irrelevant in the sequel, as they are fixed by (5.7).

The stability of linear modes was readily addressed by proving that

$$\begin{array}{l} \text{if } \zeta(\frac{R}{\xi}, \vartheta_c) > 1 \text{ linear modes are stable;} \\ \text{if } 0 < \zeta(\frac{R}{\xi}, \vartheta_c) < 1 \text{ linear modes are unstable,} \end{array} \tag{5.21}$$

where

$$\zeta(\frac{R}{\xi}, \vartheta_c) := \frac{R}{\xi} \sin\vartheta_c \cos\vartheta_c\,. \tag{5.22}$$

Ascertaining the stability of circular and hyperbolic modes was more involved as it required solving the transcendental equation

$$\varrho_n = \frac{R}{\xi} \sin\vartheta_c \left[\cos\vartheta_c + \frac{\sin\vartheta_c}{\vartheta_c} x_n \tan x_n\right] =: f_c(x_n)\,, \tag{5.23}$$

with $x_n := \sqrt{\sigma_n}\vartheta_c$, for circular modes, and the transcendental equation

$$\varrho_n = \frac{R}{\xi}\sin\vartheta_c\left[\cos\vartheta_c - \frac{\sin\vartheta_c}{\vartheta_c}x_n\tanh x_n\right] =: f_h(x_n) \tag{5.24}$$

for hyperbolic modes, where now $x_n := \sqrt{-\sigma_n}\vartheta_c$. By (5.20) and the definitions of x_n, it follows that the pairs (x_n, ϱ_n) for which

$$\varrho_n = 1 \mp \left(\frac{x_n}{\vartheta_c}\right)^2, \tag{5.25}$$

correspond to $\mu = 0$, for either circular $(-)$ or hyperbolic modes $(+)$. Hence, the parabolas (5.25) can be referred to as the marginal parabolas for the corresponding class of modes. Then, we study the intersections in the set

$$\mathcal{Q} := \{(x_n, \varrho_n), x_n \geq 0, \varrho_n \geq 0\}.$$

between $f_c(x_n)$ $-f_h(x_n)-$ and the marginal parabola for circular $-$ hyperbolic$-$ modes. As discussed in [16], any marginal parabola divides $\mathcal{Q}$ into two sets: $\mathcal{S}_c$ and $\mathcal{U}_c$ for circular modes, $\mathcal{S}_h$ and $\mathcal{U}_h$ for hyperbolic modes. The modes associated with the sets $\mathcal{S}_c$ and $\mathcal{S}_h$ correspond to $\mu > 0$ and so are stable, whereas modes associated with $\mathcal{U}_c$ and $\mathcal{U}_h$ are unstable. It is then crucial to see whether the intersections between a marginal parabola and $f_c(x_n)$ or $f_h(x_n)$, depending on the mode under study, lie in a stable set or not. The case where there are no intersections needs further study. In fact, if the graphs of $f_c(x_n)$ or $f_h(x_n)$ belong to either $\mathcal{S}_c$ or $\mathcal{S}_h$, the modes are stable whereas, if they lie within $\mathcal{U}_c$ or $\mathcal{U}_h$ they are unstable, regardless of the values of ϱ_n. As proved in [16], different regimes occur, according to the value of $\zeta(\frac{R}{\xi}, \vartheta_c)$, whether it exceeds 1 or not. The typical stability diagrams for a positive line tension are reproduced in Figure 4 for a contact angle $\vartheta_c < \pi/2$, and in Figure 5 for $\vartheta_c \in (\pi/2, \pi)$. Here, the value $\overline{\varrho}$ of ϱ_n at marginal stability is plotted against R/ξ. Figure 4 shows the stabilizing effect of a positive line tension since the larger is ξ, the smaller is $\overline{\varrho}$, and in the limit as $\xi \gg R$ the size L of the destabilizing modes diverges. When $\vartheta_c \in (\pi/2, \pi)$, the line tension maintains a stabilizing effect, but there always survive unstable modes over a size L sufficiently large: precisely circular modes such that

$$\overline{\varrho} \leq \varrho^0 := 1 - \left(\frac{\pi}{2\vartheta_c}\right)^2$$

always induce instability. As proved in [16], the classical Rayleigh instability, occurring at $\overline{\varrho} = 1$ when $\vartheta_c = \pi/2$ is recovered in the limit where $R/\xi \gg 1$.

A different scenario occurs when the line tension is negative. The stability diagram in Figure 6 shows that conditionally stable liquid bridges

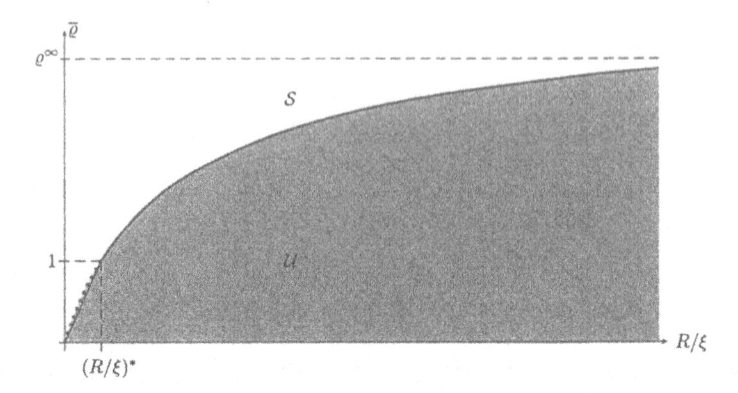

FIGURE 4. *In this stability diagram we plotted the value $\overline{\varrho}$ of ϱ_n at marginal stability against the dimensionless ratio R/ξ, for $\vartheta_c = 35°$. When a pair $(R/\xi, \varrho_n)$ lies in the set $\mathcal{S}$, the liquid bridge is stable against all modes, whereas when a pair $(R/\xi, \varrho_n)$ lies in the set $\mathcal{U}$, the liquid bridge is unstable. The analysis in [16] proves that, when $R/\xi < (R/\xi)^* := 1/(\sin\vartheta_c \cos\vartheta_c)$, linear, circular, and hyperbolic unstable modes coexist whereas, when $(R/\xi) > (R/\xi)^*$, only hyperbolic unstable modes survive. When $(R/\xi) < (R/\xi)^*$ the straight-line segment $\varrho_n = (R/\xi)\sin\vartheta_c\cos\vartheta_c$ (solid-thin line) marks the onset of instability for both linear and hyperbolic modes. However, the circular modes (dotted line) impose a strictier requirement on the stability of liquid bridges. In the limit where $(R/\xi) \gg 1$ the instability region is bounded by the line $\overline{\varrho} = \varrho^\infty := 1 + (\ell/\vartheta_c)^2$, where ℓ is the unique positive root of $\ell \tanh \ell = \vartheta_c \cot \vartheta_c$.*

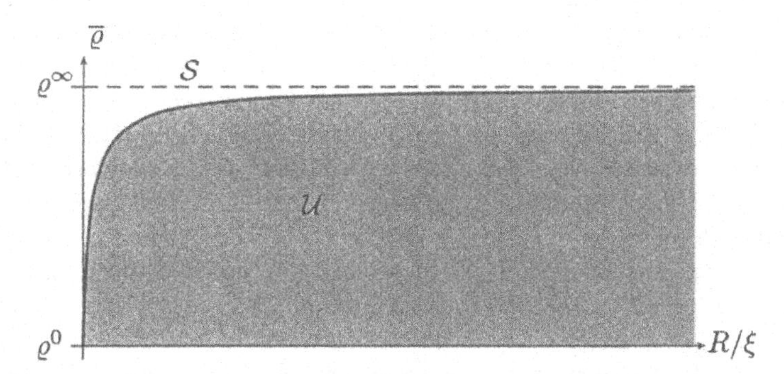

FIGURE 5. *The value $\overline{\varrho}$ of ϱ_n at marginal stability is plotted against the dimensionless ratio R/ξ, for $\vartheta_c = 125°$. When a pair $(R/\xi, \varrho_n)$ lies in the set $\mathcal{S}$, the liquid bridge is stable against all the modes we have examined, whereas when a pair $(R/\xi, \varrho_n)$ lies in the set $\mathcal{U}$, the liquid bridge is unstable against circular modes. Along the curve of marginal stability $\overline{\varrho}$ ranges between ϱ^0 and $\varrho^\infty \equiv 1 - (\ell/\vartheta_c)^2$, where ℓ is now the unique positive root of $\ell \tanh \ell = -\vartheta_c \cot \vartheta_c$. Hence, regardless of the value of R/ξ, unstable liquid bridges always exist, provided ϱ_n is chosen sufficiently small.*

exist also when the line tension is negative. The intuitive expectation that they would be totally unstable is valid only when the absolute value of the line tension is sufficiently large, that is, for $(R/|\xi|) < (R/|\xi|)_c$. For $(R/|\xi|) > (R/|\xi|)_c$, alongside the lower limit of stability $\overline{\varrho}$ which is related to the classical Rayleigh's instability, there is an upper limit ϱ^* on ϱ_n

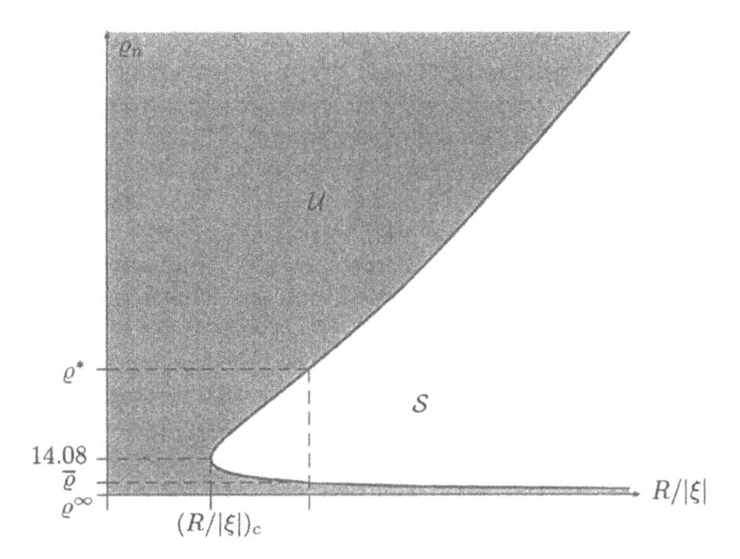

FIGURE 6. *When $R/|\xi| > (R/|\xi|)_c \simeq 20.36$, there are two values $\overline{\varrho}$ and ϱ^* of ϱ_n that correspond to marginal stability, here plotted for $\vartheta_c = 35°$. When a pair $(R/|\xi|, \varrho_n)$ lies in the set S, the liquid bridge is stable against hyperbolic modes, whereas the liquid bridge is unstable against hyperbolic modes when a pair $(R/|\xi|, \varrho_n)$ lies in the set $\mathcal{U}$: both circular and linear modes are uneffective in this case. Clearly, the values of $(R/|\xi|)_c$, $\overline{\varrho}$, and ϱ^* depend on the contact angle ϑ_c. When $\vartheta_c = 35°$ the common value of $\overline{\varrho}$ and ϱ^*, when $(R/|\xi|) = (R/|\xi|)_c$ is $\varrho_n \simeq 14.08$. Finally, the asymptotic value ϱ^∞ formally coincides with that pertaining to Figure 4.*

and, hence, on the mode index n. This means that highly wiggly modes make eventually the liquid bridge unstable; however, the corresponding wavelength can be so short to render these modes physically irrelevant, as is definitely the case for $R/|\xi|$ sufficiently large. When $R/|\xi| < (R/|\xi|)_c$, all existing modes are unstable; this means that a large, negative line tension causes instability of all liquid bridges. A similar result occurs also when $\vartheta_c \in (\pi/2, \pi)$.

Hence, we clearly find a scenario coherent with the unboundedness from below of (1.2) but, instead of seeking a different mathematical framework to replace (1.2) with functionals that are bounded from below, as done in [12], we suggest a different avenue by saying that a limitation on the admissible perturbations should be introduced, since those with a too short wavelength would induce a fictitious instability, occurring at length scales where the model based on (1.1) is doubtful and where, moreover, other stabilizing mechanisms could be at work, especially those related to possible curvature effects on the line tension. Similar results have also been obtained for sessile droplets on either a flat [36] or a curved substrate [37]. However, a quantitative selection rule that could rule out wiggly modes is still lacking.

6. Wetting transition. We just learnt that the ratio R/ξ plays a major role in the stability of a liquid bridge. However, the procedure followed in Section 5 must be modified near the wetting or the dewetting transition since in that case the dependence of R/ξ and the contact angle ϑ_c on the absolute temperature T is relevant: in fact, near the transitions we have $R/\xi = (R/\xi)(T)$ and $\vartheta_c = \vartheta_c(T)$. This suggests to assume that, when $\vartheta_c \ll 1$ or $\vartheta_c \approx \pi$, $R/\xi = (R/\xi)(\vartheta_c)$, where the dependence on ϑ_c is obtained by inverting the function $\vartheta_c(T)$ and replacing the result into $(R/\xi)(T)$. The reader concerned with the somehow related field of phase-ordering kinetics is referred to [81], where recent developments on the determination of growth laws describing the time dependence of characteristic length scales have been reviewed. Using a local stability analysis near the wetting or the dewetting transition might seem inappropriate, since most wetting transitions are first-order transitions. However, it was recently shown experimentally [82] that light alkanes on water undergo, under certain circumstances, a second-order (or critical) wetting transition besides a first-order transition. Furthermore, the behavior of line tension at wetting can be affected by the order of the wetting transition [83], and metastable states play a relevant rôle in the nucleation of wetting layers in the proximity of a first-order transition [84, 85, 86]. Hence, we think it appropriate to use the local stability analysis illustrated in Section 5 also close to wetting transitions.

The wetting transition separates partial wetting, in which the contact angle of a droplet lies in $(0, \pi)$ from complete wetting in which $\vartheta_c = 0$, and so a layer of fluid separates the vapor from the solid phase. We assume that the line tension is positive, and postulate the following scaling law in the limit where $\vartheta_c \ll 1$:

$$\frac{R}{\xi} \sim a\vartheta_c^{\alpha}, \tag{6.1}$$

for unknown parameters $a > 0$ and α. Here, we aim at exploring to which extent a scaling law like (6.1) influences the stability of a liquid bridge. To keep our approach as simple as possible, we perform the computations for liquid bridges.

In [16] we showed that the circular modes at marginal stability are the pairs (x_n, ϱ_n) obtained by searching in $\mathcal{Q}$ the solutions of the equation

$$1 - \left(\frac{x_n}{\vartheta_c}\right)^2 = f_c(x_n), \tag{6.2}$$

which gives the intersections between $f_c(x_n)$ as defined in (5.23) and the corresponding $(-)$ marginal parabola (5.25). As remarked in the caption of Figure 4, circular modes cannot induce instability when the function $\zeta(\frac{R}{\xi}, \vartheta_c)$ in (5.22) exceeds 1. Since, by (6.1), $\zeta \simeq a\vartheta_c^{\alpha+1}$, we conclude that circular modes do not induce instability at wetting, when $\alpha + 1 < 0$. On the contrary, when $\alpha+1>0$, we repeat the same approach as in [16] to seek the smallest positive root $\overline{x}_n$ of Eq. (6.2) such that $\overline{x}_n \sim b\vartheta_c^\beta$, with $b>0$ and β a non negative number. The method of dominant balance (see *e.g.* Chapter 3 of [87]) applied to Eq. (6.2) leads to

$$1-b^2\vartheta_c^{2(\beta-1)}=a\vartheta_c^{\alpha+1}\left[1-\frac{\vartheta_c^2}{2}+b^2\vartheta_c^{2\beta}\right],$$

which is inconsistent, if $\beta\neq 1$. In fact, in the case $0\leq\beta<1$, the dominant balance would give $0>-b^2\vartheta_c^{2(\beta-1)}=a\vartheta_c^{\alpha+1}>0$, which is a contradiction. On the other hand, were $\beta>1$, the dominant balance would be $1=a\vartheta_c^{\alpha+1}$ which, again is not compatible with $\alpha+1>0$. Thus, we conclude that $\beta=1$ and $1-b^2=a\vartheta_c^{\alpha+1}$ which holds, provided that $b=1$. By (5.25), we conclude that circular modes induce instability whenever

$$\varrho_n\leq a\vartheta_c^{\alpha+1}\ll 1\,, \tag{6.3}$$

and so most liquid bridges are stable against circular modes. To gain further insight into the behavior of $\overline{x}_n$, we assume $\overline{x}_n\sim\vartheta_c(1+d\vartheta^\delta)$, for unknown coefficients d and $\delta>0$. It is not difficult to prove that a consistent balance follows provided that $d=-a/2$ and $\delta=\alpha+1$.

Finally, when $\alpha=-1$, the balance holds if $\beta=1$ and $b=\sqrt{1-a}$, and we conclude that liquid bridges are unstable against circular modes if

$$\varrho_n\leq a<1\,. \tag{6.4}$$

To detect instabilities induced by hyperbolic modes, we repeat a procedure similar to that followed for circular modes, by solving the equation

$$1+\left(\frac{x_n}{\vartheta_c}\right)^2=a\vartheta_c^{\alpha+1}\left[1-\frac{\vartheta_c^2}{2}-(1-\frac{\vartheta_c^2}{6})x_n\tanh x_n\right] \tag{6.5}$$

to find the intersection between $f_h(x_n)$ defined in (5.24) and the corresponding marginal parabola in (5.25). When $\alpha+1>0$, $\zeta(R/\xi,\vartheta_c)<1$ holds and, as explained in the caption of Figure 4, liquid bridges are unstable against hyperbolic modes when

$$\varrho_n\leq a\vartheta_c^{\alpha+1}\ll 1\,,$$

that is, the same threshold as for circular modes. When $\alpha+1\leq 0$, we set $\overline{x}_n\sim b\vartheta_c^\beta$ into Eq. (6.5) where $b>0$, while β can be any real number. In the sequel, we will skip the details of the analysis when similar to those illustrated in Section 3A of [16].

- $\alpha=-1$. Here we arrive at $\beta=1$, and $b=\sqrt{a-1}$. Were $a<1$, we should have $\zeta(\frac{R}{\xi},\vartheta_c)<1$, falling in a case already shown. Thus, we conclude that instability occurs when

$$\varrho_n\leq a\,.$$

- $\alpha \in (-3,-1)$. The dominant balance $b^2\vartheta_c^{2(\beta-1)} = a\vartheta_c^{\alpha+1}$ leads to $b = \sqrt{a}$ and $\beta = \frac{\alpha+3}{2} \in (0,1)$; correspondingly, the instability threshold has still the form

$$\varrho_n \leq a\vartheta_c^{\alpha+1}.$$

- $\alpha = -3$. In this case β vanishes and b is the unique positive root of the transcendental equation $b^2 = a(1-b\tanh b)$. Instability occurs when

$$\varrho_n \leq b^2\vartheta_c^{-2}.$$

- $\alpha < -3$. It is possible to show that the assumption $\overline{x}_n \sim b\vartheta_c^\beta$, $\beta \neq 0$, leads to an inconsistent balance. Setting $\beta = 0$ yields a consistent balance, provided that b is the unique positive root of the transcendental equation

$$b\tanh b = 1. \tag{6.6}$$

By inserting the ansatz $\overline{x}_n \sim b + c\vartheta_c^\gamma$, with $\gamma > 0$, into Eq. (6.5), and using Eq. (6.6), we arrive at

$$1 + b^2\vartheta_c^{-2} + 2bc\vartheta_c^{\gamma-2} + c^2\vartheta_c^{2(\gamma-1)} = a\vartheta_c^{\alpha+1}\left\{ -\frac{\vartheta_c^2}{3} - bc\vartheta_c^\gamma \right\}. \tag{6.7}$$

A consistent dominant balance $b\vartheta_c^{-2} = -ac\vartheta_c^{\alpha+1+\gamma}$ requires $\gamma < 2$, whence $c = -b/a$, and $\gamma = -(\alpha+3) \in (0,2)$ follow. Hence, we can account for scaling laws (6.1) with $\alpha \in (-5,-3)$. When $\alpha \leq -5$, we have to set $\gamma = 2$ and $c = -\frac{1}{3b}$, so that the coefficient in front of $\vartheta_c^{\alpha+3}$ on the right-hand side of (6.7) vanishes. Moreover, we need further refinement of the putative behavior of $\overline{x}_n$ by setting $\overline{x}_n \sim b - (3/b)\vartheta_c^2(1 + d\vartheta_c^\delta)$, and then iterating the procedure. Further refinements are needed as soon as $\alpha = -7, -9$, and so on. Though involved, this procedure always yields the same instability threshold

$$\varrho_n \leq b^2\vartheta_c^{-2}$$

regardless of the value of $\alpha < -3$. It should be noted that this procedure works under the assumption that no corrections arise due to further terms in the expansion of $\frac{R}{\xi}$. However, were these corrections taken into account, the threshold on ϱ_n would not be affected.

Finally, Equation (5.21) guarantees that linear modes can induce instability when $\alpha + 1 \geq 0$. Precisely, if $\alpha = -1$ instability occurs when $\varrho_n \leq a < 1$, while, if $\alpha > -1$, instability occurs when $\varrho_n \leq a\vartheta_c^{\alpha+1}$. The table in Fig. 7 summarizes the outcome of the analysis. The remarkable

modes \ α	$(-\infty, -3)$	-3	$(-3, -1)$	-1	$(-1, +\infty)$
circular	stable	stable	stable	$\varrho_n < a < 1$	$\varrho_n \leq a\vartheta_c^{\alpha+1}$
linear	stable	stable	stable	$\varrho_n \leq a < 1$	$\varrho_n \leq a\vartheta_c^{\alpha+1}$
hyperbolic	$\varrho_n \leq b^2\vartheta_c^{-2}$, $b\tanh b = 1$	$\varrho_n \leq b^2\vartheta_c^{-2}$, $a(1 - b\tanh b) = b^2$	$\varrho_n \leq a\vartheta_c^{\alpha+1}$	$\varrho_n \leq a$, $\forall a > 0$	$\varrho_n \leq a\vartheta_c^{\alpha+1}$

FIGURE 7. *Instability thresholds when the line tension is positive. The values of ϱ_n corresponding to unstable equilibria against a specific class of modes are tabulated for all the values of α in Eq. (6.1).*

result shown here is that instability thresholds closely follow the behavior of the scaling law (6.1) only when $\alpha \geq -3$ while, when $\alpha < -3$, the thresholds saturate at $b^2\vartheta_c^{-2}$, exhibiting a universal behaviour.

As explained in [16], when the line tension τ is negative, we can limit attention to hyperbolic modes whose behavior, setting $(R/|\xi|) \sim a\vartheta_c^\alpha$ with $a > 0$, is dictated by

$$1 + \left(\frac{x_n}{\vartheta_c}\right)^2 = -a\vartheta_c^{\alpha+1}\left[1 - \frac{\vartheta_c^2}{2} - \left(1 - \frac{\vartheta_c^2}{6}\right)x_n \tanh x_n\right], \tag{6.8}$$

a slight modification of (6.5). When $\alpha \leq -3$, Eq. (6.8) has no positive root $\overline{x}_n \sim b\vartheta_c^\beta$, with $\beta > 0$. Assuming $\overline{x}_n \sim b + c\vartheta_c^\gamma$, with $\gamma > 0$ yields

$$\begin{aligned} 1 + b^2\vartheta_c^{-2} + 2bc\vartheta_c^{\gamma-2} + c^2\vartheta_c^{2(\gamma-1)} \\ = -a\vartheta_c^{\alpha+1}\Bigg[1 - b\tanh b - \frac{\vartheta_c^2}{2}\left(\frac{1 + b\tanh b}{3}\right) \\ -c\vartheta_c^\gamma[\tanh b + b(1 - \tanh^2 b)]\Bigg] \end{aligned} \tag{6.9}$$

which has the same structure as (6.7), and can be studied in a similar way. As a result, when $\alpha = -3$ two positive, finite roots $\overline{x}_n^{(1)} \sim b_1$ and $\overline{x}_n^{(2)} \sim b_2 > b_1$ exist, where b_i are the positive roots of the transcendental equation $b^2 = -a[1 - b\tanh b]$. Liquid bridges are unstable when either $\varrho_n \leq b_1^2\vartheta_c^{-2}$ or $\varrho_n \geq b_2^2\vartheta_c^{-2}$. When $\alpha < -3$ (6.8) has still a finite root $\overline{x}_n^{(1)} \sim b$, where b is the unique positive root of the transcendental equation $b\tanh b = 1$. Moreover, (6.8) has a further root $\widetilde{x}_n \sim b_1\vartheta_c^{\beta_1}$, provided that $\beta_1 = -(\alpha + 3) < 0$, and $b_1 = a$. Finally, the same argument can be used to prove that (6.8) has no positive root x_n, when $\alpha > -3$ and so, as discussed in Section 5, liquid bridges are unstable against hyperbolic

modes \ α	$(-\infty,-3)$	-3	$(-3,+\infty)$
circular	do not exist	do not exist	do not exist
linear	do not exist	do not exist	do not exist
hyperbolic	$\varrho_n \leq b^2\vartheta_c^{-2}$, $b\tanh b = 1$; $\varrho_n \geq a^2\vartheta_c^{2(\alpha+2)}$	$\varrho_n \leq b_1^2\vartheta_c^{-2}$, $\varrho_n \geq b_2^2\vartheta_c^{-2}$ $-a(1-b_i\tanh b_i) = b_i^2$, $b_1 < b_2$	unstable

FIGURE 8. *Instability thresholds when the line tension is negative. The unstable sets of ϱ_n for a specific class of modes are tabulated for all the values of α in Eq. (6.1). When $\alpha > -3$ hyperbolic modes induce instability regardless of the value of ϱ_n.*

modes, regardless of the value of ϱ_n. On the other hand, when $\alpha < 3$, liquid bridges are unstable when either $\varrho_n \leq b^2\vartheta_c^{-2}$ or $\varrho_n \geq a^2\vartheta_c^{2(\alpha+2)}$. The results we obtained are summarized in Fig. 8.

7. Dewetting Transition. The dewetting transition corresponds to the case where $\vartheta_c \simeq \pi$: it separates partial wetting from the situation in which the liquid does not wet the substrate at all. Apart from minor changes, it can be studied in the same way as the wetting transition. As usual, we first suppose that the line tension is positive. Since $\vartheta_c \simeq \pi$, we already know from the general analysis of Section 5 that only circular modes can induce instability, and so we can restrict attention to Eq. (6.2) which now, by setting $\vartheta_c = \pi - \varepsilon$, can be recast as

$$(7.1)\quad 1 - \frac{x_n^2}{\pi^2}\left(1 + \frac{2\varepsilon}{\pi}\right) = \frac{R}{\xi}\varepsilon\left[-1 + \frac{\varepsilon^2}{2} + \frac{\varepsilon}{\pi}\left(1 + \frac{\varepsilon}{\pi}\right)x_n\tan x_n\right].$$

Starting with positive line tension, we now assume

$$(7.2)\qquad \frac{R}{\xi} \sim a\varepsilon^{\alpha}$$

with $a > 0$, and look at the asymptotic behavior of the smallest positive root $\overline{x}_n$ of Eq. (7.1). As a result, $\overline{x}_n$ cannot tend to 0, since the inconsistent dominating balance $1 = -a\varepsilon^{\alpha+1}$ would follow. Similarly, assuming $\overline{x}_n \sim \ell \neq \pi/2$ would also lead to an inconsistent balance. Thus, we are left with $\overline{x}_n \to \pi/2$. The value of α affects the rate of convergence of $\overline{x}_n$ to $\pi/2$, but does not affect the instability threshold which is then independent of the scaling law (7.2). Indeed, instability occurs when $\varrho_n \leq \frac{3}{4}$.

When the line tension is negative, circular modes should satisfy the equation

$$(7.3)\qquad 1-\frac{x_n^2}{\pi^2}\left(1+\frac{2\varepsilon}{\pi}\right)=-\frac{R}{|\xi|}\varepsilon\left[-1+\frac{\varepsilon}{\pi}\left(1+\frac{\varepsilon}{\pi}\right)x_n\tan x_n\right].$$

Again, using the general case treated in Section 5 as a guide, and setting $\frac{R}{|\xi|}\sim a\varepsilon^{\alpha}$, we conclude that if $\alpha+1>0$, and so $a\varepsilon^{\alpha+1}<1$, instability occurs whenever $\varrho_n\leq a\varepsilon^{\alpha+1}$. On the other hand, if $\alpha+1<0$ Eq. (7.3) has a unique root in the interval $[0,\pi/2]$. It is not difficult to prove that if either $x_n\to 0$ or $x_n\sim\ell\neq\pi/2$, an inconsistent balance would follow so that we are led again to suppose that $x_n\to\pi/2$: from this point on, the analysis follows the same avenue as for positive line tension. The case where $\alpha=-1$ requires $\varrho_n<a$ for instability to occur, regardless of the value of $a>0$.

By use of (5.21), it is easy to prove that linear modes cannot induce instability when $\alpha<-1$. On the contrary, when $\alpha=-1$ they induce instability for liquid bridges such that $\varrho_n<a<1$, whereas they are harmless when $a>1$. Finally, when $\alpha>-1$ instability against linear modes occurs when $\varrho\leq a\vartheta_c^{\alpha+1}$.

As to hyperbolic modes, we have to study the equation

$$(7.4)\qquad 1+\frac{x_n^2}{\pi^2}\left(1+\frac{2\varepsilon}{\pi}\right)=a\varepsilon^{\alpha+1}\left[1-\frac{\varepsilon^2}{2}+\frac{\varepsilon}{\pi}\left(1+\frac{\varepsilon}{\pi}\right)x_n\tanh x_n\right]$$

which, again, gives rise to two different regimes according to whether $\alpha+1<0$ or $\alpha+1>0$. In the former case $a\varepsilon^{\alpha+1}>1$ and Eq. (7.4) has a unique root. By supposing $x_n\sim b\varepsilon^{-\beta}$, for positive constants b and β, the dominant balance yields

$$\frac{b^2}{\pi^2}\varepsilon^{-2\beta}=a\varepsilon^{\alpha+1}\left[1+\frac{b}{\pi}\varepsilon^{1-\beta}\right]$$

and three cases arise.

- $\alpha=-3$: we need to impose $\alpha+1=-2\beta=-2$, so that $\beta=1$, while b is the unique positive root of the equation $b^2/[\pi(b+\pi)]=a$. Whenever $\varrho_n\geq(b/\pi)^2\varepsilon^{-2}$, liquid bridges are unstable.
- $\alpha\in(-3,-1)$: the dominant balance now reads

$$\frac{b^2}{\pi^2}\varepsilon^{-2\beta}=a\varepsilon^{\alpha+1}$$

  from which we obtain $b=\pi\sqrt{a}$ and $\beta=-[(1+\alpha)/2]\in(0,1)$. Instability occurs when $\varrho_n\geq a\varepsilon^{\alpha+1}$.
- $\alpha<-3$: the relevant dominant balance is

$$\left(\frac{b}{\pi}\right)^2\varepsilon^{-2\beta}=a\varepsilon^{\alpha-\beta+2}$$

modes \ α	$(-\infty,-3)$	-3	$(-3,-1)$	-1	$(-1,+\infty)$
circular	$\varrho_n \leq \frac{3}{4}$	$\varrho_n \leq \frac{3}{4}$	$\varrho_n \leq \frac{3}{4}$	$\varrho_n < a$, $\forall a > 0$	$\varrho_n \leq a\vartheta_c^{\alpha+1}$
linear	stable	stable	stable	$\varrho_n \leq a < 1$	$\varrho_n \leq a\vartheta_c^{\alpha+1}$
hyperbolic	$\varrho_n \geq a\varepsilon^{2(\alpha+2)}$	$\varrho_n \geq \left(\frac{b}{\pi}\right)^2 \varepsilon^{-2}$, $\frac{b^2}{\pi(\pi+b)} = a$	$\varrho_n \geq a\varepsilon^{\alpha+1}$	$\varrho_n \geq a$, if $a > 1$, unstable, if $a \in (0,1)$	unstable

FIGURE 9. *Instability thresholds are tabulated, for all the values of α in Eq. (7.2), and for circular, linear, and hyperbolic modes. Here, we are near the dewetting transition and the line tension is negative.*

whence $b = \pi\sqrt{a}$ and $\beta = -(\alpha+2) > 1$ follow. The bridges satisfying $\varrho_n \geq a\varepsilon^{2(\alpha+2)}$ are unstable.

When $\alpha = -1$, instability occurs for $\varrho_n \geq a$, if $a > 1$, whereas all liquid bridges are unstable against hyperbolic modes if $a \in (0,1)$. Finally, when $\alpha > -1$ the same analysis just performed makes us sure that Eq. (7.4) has no roots, and so all hyperbolic modes induce instability. Fig. 9 summarizes the outcomes just illustrated. In general, while linear modes, when they exist, induce an instability threshold that mirrors the scaling law chosen in (7.2), for circular and hyperbolic modes this is true only up to a certain extent: $\alpha \geq -1$ for circular modes, $\alpha \leq -1$ for hyperbolic modes. Otherwise, the modes induce an instability threshold independent of the choice made in (7.2).

8. Conclusions. In this paper, we reviewed some topics concerning line tension modelling in the wetting science. We stressed how line tension affects both the equilibrium and the stability of a sessile droplet laid on a rigid substrate. In particular, we reviewed recent work on the stability of liquid bridges that confirmed the crucial role of the sign of line tension on stability. In fact, when line tension is negative, destabilizing modes always exist. However, the unstable modes involve wiggly corrugations of the contact line that manifest themselves at a length-scale that could be smaller than the length-scale at which the model we adopted is reliable. This calls for an effective criterion that can rule out from the beginning the perturbations that are too wiggly to be physically meaningful: as far as I know, no precise, quantitative criterion has been proposed. The computations illustrating the behaviour of liquid bridge close to either the wetting or the dewetting transition reveal that information on the stability of liquid bridges could be used to infer the relation between line tension, surface tension and the contact angle close to the transition. As a result, we proved that the instability thresholds follow the scaling law only up to a certain extent, while they exhibit a universal behaviour when the exponent α in (6.1) ranges suitable intervals.

REFERENCES

[1] J.W. Gibbs, *On the equilibrium of heterogeneous substances*, In *The Collected Papers of J. Willard Gibbs* Vol. I, 55–353. London: Yale University Press, 1957.

[2] D.J. Steigmann and D. Li, *Energy-minimizing states of capillary systems with bulk, surface, and line phases*, IMA J. Appl. Math. **55**: 1–17, 1995.

[3] J. Drelich, *The significance and magnitude of the line tension in three-phase (solid-liquid-fluid) systems*, Colloid Surface A **116**: 43–54, 1996.

[4] A. Amirfazli and A.W. Neumann, *Status of the three-phase line tension*, Adv. Colloid Interf. Sci. **110**: 121–141, 2004.

[5] T. Getta and S. Dietrich, *Line tension between fluid phases and a substrate*, Phys. Rev. E **57**: 655–671, 1998.

[6] Y. Solomentsev and L.R. White, *Microscopic drop profiles and the origins of line tension*, J. Colloid Interf. Sci. **218**: 122–136, 1999.

[7] J.O. Indekeu, *Line tension near the wetting transition: results from an interface dispacement model*, Physica A **183**: 439–461, 1992.

[8] H. Dobbs, *The elasticity of a contact line*, Physica A **271**: 36–47, 1999.

[9] P. Tarazona and G. Navascués, *A statistical mechanical theory for line tension*, J. Chem. Phys. **75**: 3114–3120, 1981.

[10] V.G. Babak, *Generalised line tension theory revisited.*, Colloid Surface A **156**: 423–448, 1999.

[11] D. Li and D.J. Steigmann, *Positive line tension as a requirement of stable equilibrium*, Colloids Surface A **116**: 25–30, 1996.

[12] G. Alberti, G. Bouchitté, and P. Seppecher, *Phase transition with the line-tension effect*, Arch. Ration. Mech. Anal. **144**: 1–46, 1998.

[13] J.S. Rowlinson and B. Widom, *Molecular Theory of Capillarity*, Dover, Mineola, 2002.

[14] L. Boruvka and A.W. Neumann, *Generalization of the classical theory of capillarity*, J. Chem. Phys. **66**: 5464–5476, 1977.

[15] R. Rosso and E.G. Virga, *A general stability criterion for wetting*, Phys. Rev. E **68**: 012601, 2003.

[16] ———, *Sign of line tension in liquid bridge stability*, Phys. Rev. E **70**: 031603, 2004.

[17] B. Widom and A.S. Clarke, *Line tension at the wetting transition*, Physica A **168**: 149–159, 1990.

[18] I. Szleifer and B. Widom, *Surface tension, line tension, and wetting*, Mol. Phys. **75**: 925–943, 1992.

[19] J.O. Indekeu, *Line tension at wetting*, Int. Journ. Mod. Phys. B **8**: 309–345, 1994.

[20] R. Rosso and E.G. Virga, *Local stability for a general wetting functional*, J. Phys. A: Math. Gen. **37**: 3989–4015, 2004. *Corrigendum*, ibid. **37**: 8751, 2004.

[21] P.S. Swain and R. Lipowsky, *Contact angles on heterogeneous substrates: a new look at Cassie's and Wenzel's laws*, Langmuir **14**: 6772–6780, 1998.

[22] R. Lipowsky, *Structured surfaces and morphological wetting transitions*, Interface Sci. **9**: 105–115, 2001.

[23] C. Neinhuis and E. Barthlott, *Characterization and distribution of water-repellent, self-cleaning plant surfaces*, Ann. Botany-London **79**: 667–677, 1998.

[24] S. Herminghaus, *Roughness-induced non-wetting*, Europhys. Lett. **52**: 165–170, 2000.

[25] J. Bico, C. Tordeaux, and D. Queré, *Rough wetting*, Europhys. Lett. **55**: 214–220, 2001.

[26] G. McHale, N.A. Käb, M.I. Newton, and S.M. Rowan, *Wetting of a high-energy fiber surface*, J. Colloid Interface Sci. **186**: 453–461, 1997.
[27] G. McHale and M.I. Newton, *Global geometry and the equilibrium shapes of liquid drops on fibers*, Colloids Surf. A **206**: 79–86, 2002.
[28] C. Bauer, and S. Dietrich, *Shapes, contact angles, and line tensions of droplets on cylinders*, Phys. Rev. E **62**: 2428–2438, 2000.
[29] R.C. Tolman, *Consideration of the Gibbs theory of surface tension*, J. Chem. Phys. **16**: 758–774, 1948.
[30] P. Lenz and R. Lipowsky, *Morphological transitions of wetting layers on structured surfaces*, Phys. Rev. Lett. **80**: 1920–1923, 1998.
[31] R. Lipowsky, P. Lenz, and P.S. Swain, *Wetting and dewetting of structured and imprinted surfaces*, Colloid Surface A **161**: 3–22, 2000.
[32] A.I. Rusanov, *Thermodynamics of solid surfaces*, Surf. Sci. Rep. **23**: 173–247, 1996.
[33] B.V. Toshev, D. Platikanov, and A. Schedulko, *Line tension in three-phase equilibrium systems*, Langmuir **14**: 489–499, 1998.
[34] A.I. Milchev and A.A. Milchev, *Wetting behaviour of nanodroplets: the limits of Young's rule validity*, Europhys. Lett. **56**: 695–701, 2001.
[35] R.D. Gretz, *Line-tension effect in a surface energy model of a cap-shaped condensed phase*, J. Chem. Phys. **45**: 3160–3161, 1966.
[36] L. Guzzardi, R. Rosso, and E.G. Virga, *Sessile droplets with negative line tension*, To appear: 2005.
[37] L. Guzzardi and R. Rosso *Sessile droplets on a curved substrate: effects of line tension*, To appear: 2005.
[38] J. Gaydos, *Differential geometric theory of capillarity*, Colloids Surf. A **114**: 1–22, 1996.
[39] R. Finn, *Equilibrium Capillary Surfaces*, Springer, New York, 1986.
[40] R.C. Tolman, *The effect of droplet size on line tension*, J. Chem. Phys. **17**: 333–337, 1949.
[41] V.I. Kalikmanov, *Statistical Physics of Fluids*. Springer, Berlin, 2001.
[42] F.P. Buff, *The spherical interface I. Thermodynamics*, J. Chem. Phys. **19**: 1591–1594, 1951.
[43] F.P. Buff, *Curved fluid interfaces. I. The generalized Gibbs-Kelvin equation*, J. Chem. Phys. **25**: 146–153, 1956.
[44] L. Boruvka, Y. Rotenberg, and A.W. Neumann, *Hydrostatic approach to capillarity*, J. Chem. Phys. **90**: 125–127, 1986.
[45] M. Pasandideh-Fard, P. Chen, J. Mostaghimi, and A.W. Neumann, *The generalized Laplace equation of capillarity I. Thermodynamic and hydrostatic considerations of the fundamental equation for interfaces*, Adv. Colloid Interface Sci. **63**: 151–178, 1996.
[46] P. Chen, S.S. Susnar, M. Pasandideh-Fard, J. Mostaghimi, and A.W. Neumann, *The generalized Laplace equation of capillarity II. Hydrostatic and themodynamics derivations of the Laplace equation for high curvatures*, Adv. Colloid Interface Sci. **63**: 179–193, 1996.
[47] R.H. Fowler, *A tentative statistical theory of Macleod's equation for surface tension, and the parachor*, Proc. R. Soc. London A **159**: 229–246, 1937.
[48] J.G. Kirkwood and F.P. Buff, *The statistical mechanical theory of surface tension*, J. Chem. Phys. **17**: 338–343, 1949.
[49] A.J.M. Yang, P.D. Fleming III, and J.H. Gibbs, *Molecular theory of capillarity*, J. Chem. Phys. **64**: 3732–3747, 1976.
[50] P. Schofield and J.R. Henderson, *Statistical mechanics of inhomogeneous fluids*, Proc. R. Soc. London A **379**: 231–246, 1982.
[51] J.D. Weeks, D. Chandler, and H.C. Andersen, *Role of repulsive forces in determining the equilibrium structure of simple liquids*, J. Chem. Phys. **54**: 5237–5247, 1971.
[52] V.I. Kalikmanov and G.C.J. Hofmans, *The perturbation approach to the sta-*

tistical theory of surface tension: new analytical results, J. Phys.: Condens. Matter **6**: 2207–2214, 1994.

[53] R. EVANS, *Microscopic theories of simple fluids and their interfaces*. In *Liquid at Interfaces*, Proocedings of the Les Houches Summer School, Session XLVIII, J. CHARVOLIN, J.F. JOANNY, AND J. ZINN-JUSTIN, Editors: pp. 3–98, 1989.

[54] R. EVANS, *The nature of the liquid-vapour interface and other topics in the statistical mechanics of non-uniform, classical fluids*, Adv. Phys. **28**: 143–200, 1979.

[55] F.F. ABRAHAM, *On the thermodynamics, structure and phase stability of the nonuniform fluid state*, Phys. Rep. **53**: 93–156, 1979.

[56] L.M.C. SAGIS AND J.C. SLATTERY, *Incorporation of line quantities in the continuum description for multiphase, multicomponent bodies with intersecting dividing surfaces. I. Kinematics and conservation principles*, J. Colloid Interf. Sci. **176**: 150–164, 1995.

[57] ———, *Incorporation of line quantities in the continuum description for multiphase, multicomponent bodies with intersecting dividing surfaces. II. Material behavior*, J. Colloid Interf. Sci. **176**: 165–172, 1995.

[58] ———, *Incorporation of line quantities in the continuum description for multiphase, multicomponent bodies with intersecting dividing surfaces. III. Determination of line tension and fluid-solid surface tensions using small sessile drops*, J. Colloid Interf. Sci. **176**: 173–182, 1995.

[59] B.D. COLEMAN AND W. NOLL, *The thermodynamics of elastic materials with heat conduction and viscosity*, Arch. Ration. Mech. Anal. **13**: 167–178, 1963.

[60] F.P. BUFF AND H. SALTSBURG, *Curved fluid interfaces. II. The generalized Neumann formula*, J. Chem. Phys. **26**: 23–31, 1957.

[61] J. ISRAELACHVILI, *Intermolecular and Surface Forces*, Academic Press, San Diego, 1992.

[62] P.G. DE GENNES, *Wetting: statics and dynamics*, Rev. Mod. Phys. **57**: 827–863, 1985.

[63] J.O. INDEKEU AND H.T. DOBBS, *Line tension at wetting: finite-size effects and scaling functions*, J. Phys. I France **4**: 77–85, 1994.

[64] C. BAUER AND S. DIETRICH, *Quantitative study of laterally inhomogeneous wetting films*, Eur. Phys. J. B **10**: 767–779, 1999.

[65] J.L. BARBOSA AND M.P. DO CARMO, *On the size of a stable minimal surface in* $\mathbb{R}^3$, Amer. J. Math. **98**: 515–527, 1974.

[66] ———, *Stability of hypersurfaces with constant mean curvature*, Math. Z. **185**: 339–353, 1984.

[67] K. SEKIMOTO, R. OGUMA, AND K. KAWASAKI, *Morphological stability analysis of partial wetting*, Adv. Phys. **176**: 359–392, 1987.

[68] R.V. ROY AND R.W. SCHWARTZ, *On the stability of liquid ridges*, J. Fluid Mech. **391**: 293–318, 1999.

[69] P. LENZ AND R. LIPOWSKY, *Stability of droplets and channels on homogeneous and structured surfaces*, Eur. Phys. J. E **1**: 249–262, 2000.

[70] A. ROS AND R. SOUAM, *On the stability of capillary surfaces in a ball*, Pac. J. Math. **178**: 345–361, 1997.

[71] H.C. WENTE, *The stability of the axially symmetric pendant drop*, Pac. J. Math. **88**: 421–470, 1980.

[72] T.I. VOGEL, *On constrained extrema*, Pac. J. Math. **176**: 457–561, 1996.

[73] ———, *Sufficient conditions for multiply constrained extrema*, Pac. J. Math. **180**: 377–380, 1997.

[74] ———, *Sufficient conditions for capillary surfaces to be energy minima*, Pac. J. Math. **194**: 469–489, 2000.

[75] I.M. GELFAND AND S.V. FOMIN, *Calculus of Variations*, Prentice-Hall, Englewood Cliffs, 1963.

[76] R. FINN, *Editorial comment on 'On the stability of catenoidal liquid bridges' by L. Zhou*, Pac. J. Math. **178**: 197–198, 1997.

[77] M.A. PETERSON, *An instability in the red blood cell shape*, J. Appl. Phys. **57**: 1739–1742, 1985.

[78] ———, *Geometrical methods for the elasticity theory of membranes*, J. Math. Phys. **26**: 711–717, 1985.

[79] The factor $(\gamma)^{-1}$ in fornt of $\partial_\nu f$ was absent from equation (3.16) of [20]. I am grateful to Dr. GUZZARDI for pointing out this misprint.

[80] R. COURANT AND D. HILBERT, *Methods of Mathematical Physics* vol. 1, Wiley-Interscience, New York, 1953.

[81] A.J. BRAY, *Theory of phase-ordering kinetics*, Adv. Phys. **51**: 481–587, 2002.

[82] S. RAFAÏ, D. BONN, E. BERTAND, J. MEUNIER, V.C. WEISS, AND J.O. INDEKEU, *Long-range critical wetting: observation of a critical end point*, Phys. Rev. Lett. **92**: 245701, 2004.

[83] A. AMIRFAZLI, A. KESHVARZ, L. ZHANG, AND A.W. NEUMANN, *Determination of line tension for systems near wetting*, J. Colloid Interf. Sci. **265**: 152–160, 2003.

[84] D. BONN AND J.O. INDEKEU, *Nucleation and wetting near surface spinodals*, Phys. Rev. Lett. **79**: 3844–3847, 1995.

[85] R. BAUSCH AND R. BLOSSEY, *Critical droplets on a wall near a first-order wetting transition*, Phys. Rev. E **48**: 1131–1135, 1993.

[86] J.O. INDEKEU AND D. BONN, *The role of surface spinodals in nucleation and wetting phenomena*, Jour. Mol. Liq. **71**: 163–173, 1997.

[87] C.M. BENDER AND S.A. ORSZAG, *Advanced Mathematical Methods for Scientists and Engineers. Asymptotic Methods and Perturbation Theory*, Springer, Heidelberg, 1999.

VARIATIONAL PROBLEMS AND MODELING OF FERROELECTRICITY IN CHIRAL SMECTIC C LIQUID CRYSTALS*

JINHAE PARK† AND M. CARME CALDERER†

Abstract. This article deals with modeling and analysis of chiral smectic C liquid crystals and their ferroelectric phases. The polarization field plays an important role in the problem. The total energy of the smectic C* contains the Oseen-Frank free energy of the nematic, together with smectic terms quadratic in the second order gradients of the complex wave function describing smectic layering, and expression for the surface energy. In addition, the polar self-interaction is taken into account, together with the electrostatic energy associated with an external electric field. The case of spatial variable electric fields is also addressed. Stability properties of the solutions are discussed to determine the interplay between the surface and electric energy terms. The physically relevant boundary conditions of the admissible fields bring out analogies to the problems of vortex tubes and vortex sheets in fluid mechanics.

Key words. Liquid crystals, smectic C*, chiral, ferroelectric, minimizer, electric field, surface energy.

AMS(MOS) subject classifications. 49J40 (Variational Methods), 49J45 (Methods of semicontinuity; relaxation), 76A15 (Liquid crystals), 82B21 (particle systems, continua), 82D45 (Ferroelectrics).

1. Introduction. In this article, we study chiral smectic C liquid crystals and their ferroelectric phases, addressing the roles of the flexoelectric and spontaneous polarizations. We investigate existence of energy minimizers, their characterization, and discuss stability properties of critical points of the energy in order to examine the relationship between applied electric fields and surface energy contributions.

We will use the conventional notation C* to denote smectic C liquid crystals with chirality. Upon lowering the temperature, the orientationally ordered nematic liquid crystals develop positional ordering of the molecular centers of mass forming smectic phases. In the smectic A, molecules tend to orient themselves perpendicular to the layers. In the lower temperature smectic C phase, the average angle of molecular alignment with respect to the layer normal takes values between 20 and 35 degrees. This angular tilt is responsible for electric polarization; in smectic C, the polarization field $\mathbf{P}$ is perpendicular to both the nematic director $\mathbf{n}$ and the layer normal $\nabla\omega$. The scalar field ω denotes the phase of the complex order parameter $\psi = \rho e^{i\omega}$ that describes the layer structure: layer locations correspond to level surfaces $\omega =$ constant; ρ represents the density of molecules arranged

*This work has been partially funded by the Institute for Mathematics and its Applications (IMA).

†School of Mathematics, University of Minnesota, Minneapolis, MN 55455, USA (jinhae@math.umn.edu and mcc@math.umn.edu).

in layers.

Some smectic phases are also chiral. In addition to layer ordering, smectic C* molecules may tend to follow a helical pattern with the polarization vector field varying periodically in space. This may yield a zero average polarization. However due to an electric field effect or to surface interactions, the helix may be suppressed, inducing uniform or splayed ferroelectric phases with nonzero net polarization [22]. In display applications, the oppositely polarized ferroelectric states allow the bistable switching such as in the case of the surface stabilized ferroelectric liquid crystal (SSFLC) device [5].

The total energy of the problem consists of several parts: the Oseen-Frank free energy of the nematic, the smectic, the surface energy, the polar self-interaction energy, and the electrostatic terms associated with an applied electric field.

We model the smectic free energy according to the covariant Ginzburg-Landau form developed by Chen and Lubensky [3] that applies to both smectic types, A* and C*; such functional contains terms quadratic on the gradients of ψ, with temperature dependent coefficients. In the case of the smectic A, the energy is minimized with $\mathbf{n}$ parallel to the layer normal $\nabla\omega$, and with the wave number equal to the material parameter q. The energy functional may fail to be positive definite at temperatures below the transition from smectic A to smectic C allowing for minimizers with $\mathbf{n}$ tilted from $\nabla\omega$. For the problem to be well-posed, one approach is to allow quadratic terms on second derivatives of ψ [3, 15]. The covariant nature of such gradients brings the term $|\Delta\psi|^2$ rather than the full second derivatives matrix. However, we show coercivity of the higher order energy within the class of fields satisfying physically relevant boundary conditions on a bounded domain. One such class of boundary conditions corresponds to the layers reaching the boundary in a perpendicular fashion, with the prescribed wave number. For another class, the layers reach the boundary in a tangent. The first case corresponds to the geometry of the Clark-Lagerwall effect in ferroelectric displays [5, 11, 24].

We need the requirement that the boundary of the domain is piecewise smooth, in order to treat configurations without point defects. Moreover, we also assume cylinder-like geometries. Specifically, the domain and its boundary can be accurately described by a vortex tube and by a vortex sheet, respectively, with boundary conditions for $\nabla\omega$ being analogous to those of the vorticity field in fluid mechanics [4].

In addition to including an intrinsic cholesteric twist contribution, the Oseen-Frank energy also involves the analogous term to account for intrinsic bending not usually found in nematic modeling [7]. The latter results from the molecular structure that induces polarization and enters the model through the bending energy term $K_3|\mathbf{n}\times(\nabla\times\mathbf{n})+\mathbf{P}|^2$. Specifically, $\mathbf{P} = P_0\frac{\nabla\omega\times\mathbf{n}}{|\nabla\omega\times\mathbf{n}|}$; the significance of P_0 is discussed later. Although the interaction between $\mathbf{P}$ and an applied electric field may be large in comparison with the dielectric term of ordinary nematics, it may not be sufficiently large as to justify the inclusion of self-interacting effects. This case corresponds to flexoelectric polarization with P_0 being a material constant. Most studies of ferroelectricity of smectic C* liquid crystals are consistent with such an approach. The treatment presented in this article includes energy of self-interaction, and thus allows for spontaneous polarization; P_0 is thus taken to be a field of the problem.

We also carry out asymptotic characterizations of the minimizers as relevant energy coefficients become unbounded. For instance, the unboundedness of K_2 and K_3 as the temperature approaches the transition value to smectic C* yields helical configurations. The unboundedness of the smectic coefficients $C_a \equiv C_{\parallel} - C_{\perp}$ and D give uniform layer structures. Likewise, the unboundedness of K_1 at temperature lower than the smectic A to smectic C transition produces ferroelectric states.

We finally give a survey on stability results to see how the competing effects of the surface energy and electrostatic energy determine the transitions between helical and ferroelectric phases [8, 9, 10, 19, 20, 21].

Although the net polarization of the C^* phase is nonzero, it is not felt in the environment. This is a main difference between ferroelectricity in liquid crystals and in solids: the force on the boundary of a polarized domain is not zero, whereas no force is measured in the case of the liquid crystal. The latter is due to a rearrangement of charges (not occurring in solids due to the much more rigid structure and location ordering) on the bounding plates that exerts a force opposing that of the polarization.

Mathematical analysis of the phase transitions between chiral nematic, smectic A*, and C* liquid crystals has been carried out by Joo and Phillips [13]. For analysis of the phase transition between chiral nematic and smectic liquid crystals, with a special emphasis on the analogies of the transition between conductor and superconductor as proposed by de Gennes [7], the reader is referred to the article [1]. For comprehensive treatments of the physical phenomena and modeling of ferroelectric liquid crystals, we refer to the books by Lagerwall [14], Pikin [22], and by Musevic, Blinc and Zeks [16]. Studies of periodic ferroelectric and antiferroelectric phases and analysis of time dependent problems arising in switching are carried out by the authors [18].

2. Free energy functions of smectic materials. Equilibrium configurations of a smectic C* liquid crystal subject to an applied constant electric field $\mathbf{E}$ correspond to quadruples of fields $(\psi, \mathbf{n}, \mathbf{E}, \mathbf{P})$ such that $\psi : \Omega \to \mathcal{C}$, $\mathbf{n} : \Omega \to \mathcal{S}^2$, $\mathbf{E} : \Omega \to \mathcal{R}^3$, and $\mathbf{P} : \Omega \to \mathcal{R}^3$ that minimize the total energy,

$$\mathcal{F}(\psi, \mathbf{n}, \mathbf{P}) = \int_\Omega [F_N(\nabla\mathbf{n}, \mathbf{n}) + F_{Sm}(\nabla^2\psi, \nabla\psi, \nabla\mathbf{n}) + F_{Elec}(\mathbf{n}, \mathbf{E}, \mathbf{P}) + F_{Pol}(\mathbf{n}, \mathbf{P}, \nabla\mathbf{P}, \nabla\psi)]d\mathbf{x} + \int_{\partial\Omega} F_{Surf}(\mathbf{n}, \mathbf{P}, \nu)\, dS, \tag{2.1}$$

where F_N, F_{Sm}, F_{Surf} denote the Oseen-Frank, the smectic, the surface energy densities ([6], page 99), and F_{Elec}, F_{Pol} are energy densities associated with the electric field and polar self-interaction, respectively.

2.1. Nematic and Smectic free energy. The Oseen-Frank free energy density is given by

$$\begin{aligned} F_N &= K_1(\nabla \cdot \mathbf{n})^2 + K_2(\mathbf{n} \cdot \nabla \times \mathbf{n} + \tau)^2 + K_3|\mathbf{n} \times (\nabla \times \mathbf{n}) + \mathbf{b} \times \mathbf{n}|^2 \\ &+ (K_2 + K_4)(tr(\nabla \mathbf{n})^2 - (\nabla \cdot \mathbf{n})^2). \end{aligned} \tag{2.2}$$

where K_i, $i = 1, 2, 3, 4$ denote elasticity constants. The scalar τ represents the chiral pitch of the helical structure of the cholesteric phases, and $\mathbf{b} \times \mathbf{n}$ is an intrinsic bending stress. The vector $\mathbf{b}$ is parallel to the layer normal, $\nabla\omega$. Such a term appears only in connection with the modeling of the smectic C* since nematics with intrinsic bending have not been observed. Both quantities result from the loss of mirror symmetry of the smectic C* phases. The fourth term in F_N is a null-Lagrangian; its integral is determined by $\mathbf{n}$ on $\partial\Omega$. The free energy density associated with the smectic layering follows the covariant form presented in [3]:

$$\begin{aligned} F_{Sm} &= D(\mathbf{D}^2\psi)(\mathbf{D}^2\psi)^* + [C_{||}n_in_j + C_\perp(\delta_{ij} - n_in_j)](\mathbf{D}_i\psi)(\mathbf{D}_j\psi)^* \\ &+ r|\psi|^2 + \frac{g}{2}|\psi|^4. \end{aligned} \tag{2.3}$$

with $\mathbf{D} \equiv \nabla - iq\mathbf{n}$ and $r = a(T - T^*), a > 0$; here T denotes the (constant) temperature of the material and T^* is the transition temperature from nematic to smectic A. The model (2.3) yields the de Gennes model for smectic $A*$ when $C_{||} - C_\perp = 0$ and $D = 0$. The smectic C phase is characterized by

$$C_\perp < 0. \tag{2.4}$$

Moreover, $C_\perp \geq 0$ in the smectic A* and $C_\perp = 0$ characterizes the transition to smectic C. Equivalently, (2.3) can also be written as follows:

$$F_{Sm} = D|\mathbf{D}^2\psi|^2 + C_\perp|\mathbf{D}\psi|^2 + C_a|\mathbf{n} \cdot \mathbf{D}\psi|^2 + r|\psi|^2 + \frac{g}{2}|\psi|^4. \tag{2.5}$$

Remark. The first term in (2.3) is obtained from reference [15] and is a modification of $D_\perp(\delta_{ij} - n_in_j)(\delta_{kl} - n_kn_l)(\mathbf{D}_i\mathbf{D}_j)(\mathbf{D}_k\mathbf{D}_l)^*$ in the original Chen-Lubensky model. The purpose of introducing the new term is to obtain coercivity of the energy.

2.2. Flexoelectric and Spontaneous Polarization. The presence of the intrinsic bending in the smectic C phase causes the material to be polarized. The polarization field $\mathbf{P}$, the volume dipole density, is given by

$$\mathbf{P} = P_0\frac{\mathbf{b} \times \mathbf{n}}{|\mathbf{b} \times \mathbf{n}|}, \tag{2.6}$$

where P_0 is a material parameter. We observe that such a polarization is a secondary field determined from the geometry of the director and the layers; this type of ferroelctricity is usually known as *improper* [22]. However for some smectic materials the modulus of $\mathbf{P}$ is so large that the equivalent point charge distribution, $-\nabla \cdot \mathbf{P}$, generated by the polarization cannot be neglected. Then the energy associated with the charge distribution has to be included in the model. Although the direction of $\mathbf{P}$ is still as in (2.6), its modulus P_0 becomes a variable of the problem. In such a case, $|\mathbf{P}|$ is a variable of the problem, and we represent $\mathbf{P}$ as follows:

$$\mathbf{P} = |\mathbf{P}|\frac{\nabla\omega \times \mathbf{n}}{|\nabla\omega \times \mathbf{n}|}, \quad \text{if } \nabla\omega \times \mathbf{n} \neq 0$$

$$|\mathbf{P}| = 0, \quad \text{if } \nabla\omega \times \mathbf{n} = 0. \tag{2.7}$$

One important difference between ferroelectric smectics and solids is the freedom of the polarization field to rotate in the layer plane in the former ($\mathbf{P}$ is a Goldstone variable) as opposed to taking specific values determined by the solid lattice [14]. Because of this vectorial symmetry the energy density of the field $\mathbf{P}$ contains, together with the term $(\nabla \cdot \mathbf{P})^2$, a term of the form $|\nabla\mathbf{P}|^2$ which penalizes interfaces in the material:

$$F_{Pol} = P_0[B_1|\nabla\mathbf{P}|^2 + B_2|\nabla\cdot\mathbf{P}|^2 + a_0|\mathbf{P}|^2 + b_0|\mathbf{P}|^4$$

$$+\frac{1}{\varepsilon^2}|(|\nabla\omega \times \mathbf{n}|)\mathbf{P} - |\mathbf{P}|(\nabla\omega \times \mathbf{n})|^2], \tag{2.8}$$

where $B_1 > 0$, $B_2 > 0$, $P_0 > 0$, a_0 and b_0 are temperature dependent parameters.

The last term of the previous equation penalizes departure of $\mathbf{P}$ from the parallel direction to $\nabla\omega \times \mathbf{n}$, in the case that $\nabla\omega \times \mathbf{n} \neq 0$. One can use $|\mathbf{P} - |\mathbf{P}|\frac{\nabla\omega\times\mathbf{n}}{|\nabla\omega\times\mathbf{n}|}|^2$ as a penalty term, but it may cause a discontinuity of $\mathbf{P}$ when $\nabla\omega \times \mathbf{n} = 0$.

2.3. Anchoring Energy. The presence of a nonzero polarization induces a surface charge on the boundary which, in turn, requires an opposite charge layer in the bounding plate; likewise, an analogous layer if the liquid crystal is surrounded by fluid. This boundary effect can be taken into account through an effective anchoring energy, which depends on the polarization. We also include a quadratic term in the polarization to account for the Rapini-Pouplar anchoring energy [7, 14]:

$$E_{surf} = \int_{\partial\Omega} (\omega_p(1 - \frac{\mathbf{P}\cdot\nu}{|\mathbf{P}|}) + \omega_r(1 - \frac{(\mathbf{P}\cdot\nu)^2}{|\mathbf{P}|^2}) \tag{2.9}$$

$$+\omega_n(1 - \alpha_0(\mathbf{n}\cdot\nu)^2)\, dS, \tag{2.10}$$

where $\omega_r, \omega_n, \omega_p$ and $|\alpha_0| < 1$ are material constants, and ν denotes the unit normal to the surface.

2.4. Electrostatic energy. The electrostatic energy is given by

$$F_{\text{Elec}} = -\frac{1}{2}\mathbf{D}\cdot\mathbf{E}, \tag{2.11}$$

$$\mathbf{D} = \varepsilon\mathbf{E} + \mathbf{P}, \tag{2.12}$$

$$\varepsilon = \varepsilon_{\perp} I + \varepsilon_a \mathbf{n}\otimes\mathbf{n}, \tag{2.13}$$

where $\varepsilon, \varepsilon_{\perp}$, and ε_a denote the susceptibility tensor, dielectric permittivity, and dielectric anisotropy, respectively. Note that (2.11) can be written as,

$$\begin{aligned} F_{\text{Elec}} &= -\frac{1}{2}(\varepsilon\mathbf{E} + \mathbf{P})\cdot\mathbf{E} \\ &= -\frac{1}{2}((\varepsilon_{\perp} I + \varepsilon_a \mathbf{n}\otimes\mathbf{n})\mathbf{E} + \mathbf{P})\cdot\mathbf{E} \\ &= -\frac{1}{2}(\varepsilon_{\perp}|\mathbf{E}|^2 + \varepsilon_a(\mathbf{n}\cdot\mathbf{E})^2 + \mathbf{P}\cdot\mathbf{E}). \end{aligned} \tag{2.14}$$

In the case of variable electric fields, the total energy may be unbounded from below due to the term F_{Elec}; we then characterize equilibrium configurations as critical points of $\mathcal{E}$ subject to constraints

$$\nabla\times\mathbf{E} = 0, \quad \nabla\cdot\mathbf{D} = 0. \tag{2.15}$$

Assumptions on the constitutive parameters include

$$g > 0, \quad q \geq 0, \tau \geq 0, r < 0, D > 0, C_{\perp} < 0, C_a > 0, \tag{2.16}$$

$$c_1 \geq K_2 + K_4 \geq c_0, K_1 \geq K_2 + K_4, \tag{2.17}$$

$$K_3 \geq K_2 + K_4, 0 \geq K_4, \tag{2.18}$$

$$B_1 > 0, B_2 > 0, b_0 > 0, P_0 > 0, \tag{2.19}$$

$$\omega_p > 0, \omega_r > 0, \omega_n > 0, \tag{2.20}$$

where c_1 and c_0 are positive constants. The latter inequalities are necessary conditions to ensure coercivity of the energy [1].

3. Existence of minimizers. In this section we study existence of minimizers of the total energy discussed in the previous section. We assume that the conditions on the constitutive parameters hold. Let Ω be a simply connected domain, either bounded or confined between two plates, $\Omega \subset \{\mathbf{x} : |x| \leq L\}$. We let the boundary boundary $\partial\Omega$ be C^2-surface.

Let $\psi = \rho e^{i\omega}$. Then

$$\begin{aligned} F_{Sm} &= D|\mathbf{D}^2\psi|^2 + C_{\perp}|\mathbf{D}\psi|^2 + C_a|\mathbf{n}\cdot\mathbf{D}\psi|^2 + r|\psi|^2 + \frac{g}{2}|\psi|^4 \\ &= D[(\Delta\rho - \rho|\nabla\omega - q\mathbf{n}|^2)^2 + (\rho\nabla\cdot(\nabla\omega - q\mathbf{n}) \\ &\quad + 2\nabla\rho\cdot(\nabla\omega - q\mathbf{n}))^2] + C_{\perp}[|\nabla\rho|^2 + \rho^2|\nabla\omega - q\mathbf{n}|^2] \\ &\quad + C_a[(\nabla\rho\cdot\mathbf{n})^2 + \rho^2(\nabla\omega\cdot\mathbf{n} - q)^2] + r\rho^2 + \frac{g}{2}\rho^4. \end{aligned} \tag{3.1}$$

Prior to addressing the question of existence of minimizers, we will discuss the type of domains and boundary conditions to consider. For this, we review the notions of *vortex sheet* and *vortex tube* in fluid mechanics and seek analogies with the present physics.

In three-dimensional flows, a vortex tube consists of a two-dimensional surface $\mathcal{S}$, nowhere tangent to the vorticity field ξ, with vortex lines drawn through each point of the bounding curve $\mathcal{C}$ of $\mathcal{S}$. One class of domains that we consider have the geometric structure of a vortex tube. A precise definition is given as follows. Let $\mathcal{S}$ be a plane surface diffeomorphic to a disc $\mathcal{S}_0$, and consider the cylinder $\mathcal{S}_0 \times \mathcal{R}$. The resulting tube is a three dimensional domain diffeomorphic to the cylinder. Moreover, we take Ω to be a portion of the vortex tube, with lateral surface Σ contained between surfaces $\mathcal{S}_1$ and $\mathcal{S}_2$, with contour curves $\mathcal{C}_1$ and $\mathcal{C}_2$, respectively. The vorticity field ξ is everywhere tangent to Σ.

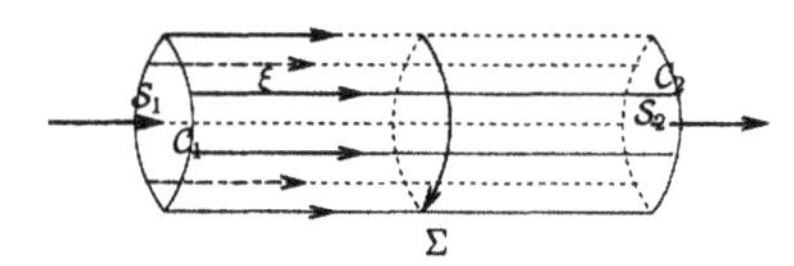

FIGURE 1. *A vortex tube*

Since in the present work we do not address defects, we construct Ω as follows. We consider a cylinder-like domain with lateral surface Σ having a nonzero tangent everywhere. Cross sections of the cylinder perpendicular to the tangent vector are two dimensional surfaces $\mathcal{S}$ diffeomorphic to a circle. The cylinder-like domain is bounded by two plane cross sections $\mathcal{S}_1$ and $\mathcal{S}_2$, with unit normal ν_1 and ν_2, respectively. So,

$$\Omega = \{\mathbf{x} \in \mathcal{R}^3 \text{ bounded by } \partial\Omega = \Sigma \cup \mathcal{S}_1 \cup \mathcal{S}_2\}. \tag{3.2}$$

The following lemma based on Gauss theorem motivates the boundary conditions taken into account.

LEMMA 3.1. *Let $\Omega \in \mathcal{R}^3$ be as previously defined. Let f be a smooth scalar function defined in Ω. Then f satisfies the following identity:*

$$\begin{aligned}\int_\Omega |\Delta f|^2 = \int_{\partial\Omega} [\nabla \cdot (\nabla f)(\nu \cdot \nabla f) - \frac{1}{2}\nabla(|\nabla f|^2) \cdot \nu]\, dS \\ + \sum_{i,j=1,2,3} \int_\Omega (\partial_i \partial_j f)^2.\end{aligned} \tag{3.3}$$

Let $k > q$ be a given constant. We require the following boundary conditions to hold on $\partial\Omega$:

$$\begin{aligned}\nabla\omega \cdot \nu = 0 \text{ on } \Sigma, \quad \nabla\omega \cdot \nu_i = k, \text{ on } \mathcal{S}_i, \; i = 1,2, \\ |\nabla\omega|^2 = k^2, \text{ on } \Sigma \cup \mathcal{S}_1 \cup \mathcal{S}_2.\end{aligned} \tag{3.4}$$

Such relations correspond to smectic layers reaching the boundary in a perpendicular manner, with a prescribed wave number k. The case with smectic layers tangent to the boundary is treated elsewhere. Moreover, for simplicity, we restrict ourselves to the case that ρ is constant (say $\rho = 1$), that is, no nematic defects are present in the sample, and rewrite the smectic energy as follows:

$$\begin{aligned} F_{Sm} &= D(\Delta\omega - q\nabla\cdot\mathbf{n})^2 + D\Big(|\nabla\omega - q\mathbf{n}|^2 + \frac{C_\perp}{2D}\Big)^2 \\ &+C_a(\nabla\omega\cdot\mathbf{n} - q)^2 - \frac{C_\perp^2}{4D^2}. \end{aligned} \tag{3.5}$$

The admissible set is defined as follows:

$$\mathcal{X} = \Big\{(\mathbf{n},\mathbf{P}) \in W^{1,2}(\Omega,\mathbf{S}^2)\times W^{1,2}(\Omega,\mathbf{R}^3) : ||P||_\infty \le P_0,\Big\} \tag{3.6}$$

$$\mathcal{H} = \{\omega \in W^{2,2}(\Omega) \mid \omega \text{ satisfies (3.4) on } \partial\Omega\}, \tag{3.7}$$

$$\mathcal{A} = \mathcal{H}\times\mathcal{X}, \tag{3.8}$$

where P_0 is the given polarization saturation constant depending on the material and temperature, and $k > q$ is a prescribed constant related to the wave number of the layers. We rewrite the energy to minimize as follows:

$$\begin{aligned} \mathcal{E} &= \int_\Omega \{F_{Sm}(\Delta\omega,\nabla\omega,\mathbf{n}) + F_N(\mathbf{P},\mathbf{n},\nabla\mathbf{n}) + F_{Pol}(\mathbf{n},\mathbf{P},\nabla\mathbf{P},\nabla\omega)\}\,d\mathbf{x} \\ &+ \int_{\partial\Omega} F_{Sur}(\mathbf{n},\mathbf{P},\nu)\,dS. \end{aligned} \tag{3.9}$$

LEMMA 3.2. *The admissible set $\mathcal{A}$ is non-empty and the inequality,*

$$\inf_{(\mathbf{n},\mathbf{P},\omega)\in\mathcal{A}} \mathcal{E}(\mathbf{n},\mathbf{P},\omega) \le M$$

holds for some $M > 0$.

We note that for all $\mathbf{n} \in \mathbf{W}^{1,2}(\Omega,\mathbf{S}^2)$, the identities

$$\begin{aligned} tr(\nabla\mathbf{n})^2 &= |\nabla\mathbf{n}|^2 - |\nabla\times\mathbf{n}|^2, \quad \text{and} \\ |\nabla\times\mathbf{n}|^2 &= |\mathbf{n}\cdot\nabla\times\mathbf{n}|^2 + |\mathbf{n}\times\nabla\times\mathbf{n}|^2 \end{aligned}$$

hold. Using these identities, we get

$$\begin{aligned} F_N(P,n,\nabla n) = &(K_1 - K_2 - K_3)(\nabla\cdot\mathbf{n})^2 + (K_2 - K)(\mathbf{n}\cdot\nabla\times\mathbf{n} + \tau)^2 \\ &+(K_3 - K)|\mathbf{n}\times\nabla\times\mathbf{n} + \mathbf{P}|^2 + (K_2 + K_4)|\nabla\mathbf{n}|^2 \\ &+K(|\mathbf{n}\cdot\nabla\times\mathbf{n} + \tau|^2 + |\mathbf{n}\times\nabla\times\mathbf{n} + \mathbf{P}|^2 - (K_2 + K_4)|\nabla\times\mathbf{n} \\ &+P_0[B_1|\nabla\mathbf{P}|^2 + B_2|\nabla\cdot\mathbf{P}|^2 + a_0|\mathbf{P}|^2 + b_0|\mathbf{P}|^4 \\ &+\frac{1}{\varepsilon^2}|(|\nabla\omega\times\mathbf{n}|)\mathbf{P} - |\mathbf{P}|(\nabla\omega\times\mathbf{n})|^2]. \end{aligned}$$

We obtain the following inequalities:

$$\begin{aligned} &K|\mathbf{n}\times\nabla\times\mathbf{n}+\mathbf{P}|^2+K(\mathbf{n}\cdot\nabla\times\mathbf{n}+\tau)^2-(K_2+K_4)|\nabla\times\mathbf{n}|^2 \\ &\geq K(\frac{1}{4}|\nabla\times\mathbf{n}|^2-2\tau^2-4|\mathbf{P}|^2)-(K_2+K_4)|\nabla\times\mathbf{n}|^2 \\ &\geq (c_1-K_2-K_4)|\nabla\times\mathbf{n}|^2-K_2(4|\mathbf{P}|^2+2\tau^2) \\ (3.10)&\geq -2K_2(2|\mathbf{P}|^2+\tau^2) \end{aligned}$$

It follows from (3.10), (3.10), and Lemma 3.1 that $\mathcal{E}$ is bounded below and coercive in $\mathcal{A}$. Therefore, we have

$$M_1 \leq \inf_{(\mathbf{n},\mathbf{P},\omega)\in\mathcal{A}} \mathcal{E}(\mathbf{n},\mathbf{P},\omega) < \infty, \text{ for some } M_1 \in \mathbf{R}.$$

Let $\{(\mathbf{n}^j,\mathbf{P}^j,\omega^j)\}$ be a minimizing sequence for $\mathcal{E}$. Since $|\mathbf{n}^j|=1$, we get

$$\begin{aligned} &\mathbf{n}^j \rightharpoonup \mathbf{n}^\infty \text{ in } W^{1,2}(\Omega), \\ &\mathbf{n}^j \to \mathbf{n}^\infty \text{ almost everywhere in } \Omega, \quad \text{and} \\ &\mathbf{P}^j \rightharpoonup \mathbf{P}^\infty \text{ in } W^{1,2}(\Omega). \end{aligned}$$

as $j\to\infty$. Furthermore, we have

$$\mathbf{n}^j\times\nabla\times\mathbf{n}^j \rightharpoonup \mathbf{n}^\infty\times\nabla\times\mathbf{n}^\infty \text{ in } L^2(\Omega),$$

$$\mathbf{n}^j\cdot\nabla\times\mathbf{n}^j \rightharpoonup \mathbf{n}^\infty\cdot\nabla\times\mathbf{n}^\infty \text{ in } L^2(\Omega), \quad \text{and}$$

$$\mathbf{n}^j\times\nabla\times\mathbf{n}^j\cdot\mathbf{P}^j \to \mathbf{n}^\infty\times\nabla\times\mathbf{n}^\infty \ \ \mathbf{P}^\infty \text{ in } L^1(\Omega)$$

as $j\to\infty$. We also show that the sequence

$$\left(\left|(|\nabla\omega^j\times\mathbf{n}^j|)\mathbf{P}^j-|\mathbf{P}^j|(\nabla\omega^j\times\mathbf{n}^j)\right|^2\right)$$

converges to

$$\left|(|\nabla\omega^\infty\times\mathbf{n}^\infty|)\mathbf{P}^\infty-|\mathbf{P}^\infty|(\nabla\omega^\infty\times\mathbf{n}^\infty)\right|^2$$

in $L^1(\Omega)$ as $j\to\infty$.

Note that for all j,

$$\int_\Omega |\nabla\omega^j|^2\,d\mathbf{x} \leq 2\int_\Omega (|\nabla\omega^j-q\mathbf{n}^j|^2+q^2)\,d\mathbf{x}, \tag{3.11}$$

and

$$\begin{aligned} \int_\Omega |\Delta\omega^j|^2\mathbf{x} &= \int_\Omega (|\Delta\omega^j-q\nabla\cdot\mathbf{n}^j+q\nabla\cdot\mathbf{n}^j|^2\,d\mathbf{x} \\ &\leq 2\int_\Omega (|\Delta\omega^j-q\nabla\cdot\mathbf{n}^j|^2+q^2(\nabla\cdot\mathbf{n}^j)^2\,d\mathbf{x} \end{aligned} \tag{3.12}$$

hold. From (3.11), (3.12), and Lemma 3.1, we get

$$||\omega^j||_{W^{2,2}(\Omega)} \leq R, \text{ for any } j$$

for some $R > 0$, and thus

$$\omega^j \rightharpoonup \omega^\infty \text{ in } W^{2,2}(\Omega), \text{ as } j \to \infty.$$

Since $\omega_p(1 - \frac{\mathbf{P}\cdot\nu}{|\mathbf{P}|}) + \omega_r(1 - \frac{(\mathbf{P}\cdot\nu)^2}{|\mathbf{P}|^2}) + \omega_n(1 - (\mathbf{n}\cdot\nu)^2)$ is continuous, it follows that $\mathcal{E}$ is weakly lower semicontinuous, that is,

$$\mathcal{E}(\mathbf{n}^\infty, \mathbf{P}^\infty, \omega^\infty) \leq \lim_{j\to\infty} \mathcal{E}(\mathbf{n}^j, \mathbf{P}^j, \omega^j).$$

Therefore we have the following theorem:

THEOREM 3.1. *There exists a minimizing triple* $(\mathbf{n}, \mathbf{P}, \omega)$ *of the energy functional* $\mathcal{E}$ *in* $\mathcal{A}$.

Remark. The proof extends with no additional difficulty to the case that the variable ρ is included in the model.

4. Asymptotic form of the energy minimizers. In this section, we aim at providing a classification of the energy minimizers previously obtained. In particular, we wish to identify the smectic layer geometry and identify parameter conditions leading to helical director configurations in the bulk with zero average polarization, as well as those giving homogeneous ferroelectric states. For this, we will consider a rectangular domain between two parallel plates:

$$\Omega = \{\mathbf{x} = (x, y, z) : 0 < y, z < L,\ 0 < x < d.\},$$

for fixed $0 < L, 0 < d$. Let $\mathbf{i}, \mathbf{j}$ and $\mathbf{k}$ denote the corresponding orthonormal system of vectors.

4.1. Helical energy minimizers. We determine the structure of the energy minimizers $(\mathbf{n}, \omega, \mathbf{P})$ when K_2 and K_3 as well as the smectic coefficients dominate over the polar energy and surface energy parameters, and $C_\perp < 0$. Such a situation arises at temperatures below the threshold of the smectic A to smectic C transition yielding helical configurations of $\mathbf{n}$ and $\mathbf{P}$. It is well known that in the higher temperature transition from nematic to smectic A, K_2 becomes unbounded and the smectic coefficient $C_\perp \geq 0$.

We take the admissible set such that

$$k = q\sqrt{\frac{|C_\perp|}{2Dq^2} + 1}. \tag{4.1}$$

We consider admissible fields such that $\mathbf{n}$ makes a constant angle α with the layer normal vector $\nabla\omega$. We take α such that

$$\tan\alpha = \sqrt{\frac{|C_\perp|}{2Dq^2}}. \tag{4.2}$$

Specifically, we let

$$\begin{cases} \mathbf{n}_0 = (a\cos\frac{\tau z}{a^2}, a\sin\frac{\tau z}{a^2}, c), \\ a = \sin\alpha \neq 0, \ c = \cos\alpha \neq 0, \ \frac{a^2}{c^2} = \frac{|C_\perp|}{2Dq^2}, \\ \mathbf{P}_0 = \frac{c\tau}{a}(-\sin\frac{\tau z}{a^2}\mathbf{i} + \cos\frac{\tau z}{a^2}\mathbf{j}), \\ \omega_0 = kz, \ k = \frac{q}{c}, \ \nu = \mathbf{i}. \end{cases} \tag{4.3}$$

A simple calculation gives

$$\nabla\omega_0 \cdot \mathbf{n}_0 = q, \ \Delta\omega_0 = 0, \ |\nabla\omega_0 - q\mathbf{n}_0|^2 = \frac{|C_\perp|}{2D},$$
$$\nabla\cdot\mathbf{n}_0 = 0, \ \mathbf{n}_0\cdot\nabla\times\mathbf{n}_0 + \tau = 0, \ |\mathbf{n}_0 \times (\nabla\times\mathbf{n}_0) + \mathbf{P}_0| = 0,$$
$$\nabla\cdot\mathbf{P}_0 = 0, \ |\nabla\mathbf{P}_0| = \frac{c\tau^2}{a^3}.$$

We observe that the quantity $\tan\alpha = \sqrt{\frac{|C_\perp|}{2Dq^2}}$ is of the order of $\tan\frac{\pi}{6}$ according to experimental measurements of the director tilt angle. This together with available information on the wave number q in the smectic A phase allows us to determine the relative value of the smectic parameters $|C_\perp|$ and D.

The total energy of (4.3) is given by

$$\mathcal{E}_0 = L^2 d[-\frac{C_\perp^2}{4D^2} + (K_2+K_4)\frac{\tau^2}{a^2} + p_0\frac{c^2\tau^2}{a^2}(B_1\frac{\tau^2}{a^4}$$
$$+a_0 + b_0\frac{c^2}{a^2}\tau^2)] + L(C_P\omega_p + C_r\omega_r + C_n\omega_n),$$

where C_r, C_p, C_n are expressions involving a, c, q and τ.

We estimate the energies:

$$E_{surf} \geq -(2|\omega_p| + 2|\omega_r| + |\omega_n|(1+|\alpha|)L^2,$$
$$\int_\Omega (F_{Sm} + F_N + F_{Pol})\,d\mathbf{x} \geq -[\frac{C_\perp^2}{4D^2} + b_0\frac{(a_0p_0 - 4c_1)^2}{4p_0} + 2c_1\tau^2]dL^2.$$

Letting $(\mathbf{n}, \mathbf{P}, \omega)$ denote an energy minimizer, we get

$$\begin{aligned} 0 &\leq \mathcal{E}(\mathbf{n},\mathbf{P},\omega) + (2|\omega_p| + 2|\omega_r| + |\omega_n|(1+|\alpha|)L^2 \\ &+[\frac{C_\perp^2}{4D^2} + b_0\frac{(a_0p_0-4c_1)^2}{4p_0} + 2c_1\tau^2]dL^2. \\ &\leq \mathcal{E}_0 + (2|\omega_p| + 2|\omega_r| + |\omega_n|(1+|\alpha|)L^2 \\ &+[\frac{C_\perp^2}{4D^2} + b_0\frac{(a_0p_0-4c_1)^2}{4p_0} + 2c_1\tau^2]dL^2 \equiv \bar{\mathcal{E}}_0. \end{aligned} \tag{4.4}$$

Note that the quantity on the right hand side of the inequality is independent of $D, C_\perp, K_1, K_2$ and K_3, with the only K_i constants appearing as the sum $K_2 + K_4$. The following estimates follow from (4.4).

THEOREM 4.1. *Let $q > 0, \tau > 0$ be fixed. Suppose that the constitutive parameters satisfy assumptions (2.16) and (2.20). Suppose that $K_2, K_3 \geq 4c_1$ and $0 < \frac{|C_\perp|}{2Dq^2} \leq 1$. If $(\mathbf{n}, \mathbf{P}, \omega)$ is a minimizer of $\mathcal{E}$, then we get*

$$||(\nabla\omega \times \mathbf{n})|\mathbf{P}| - \mathbf{P}(|\nabla\omega \times \mathbf{n}|)||_{2,\Omega}^2 \leq \epsilon^2 \bar{\mathcal{E}}_0, \tag{4.5}$$

$$||\nabla \mathbf{n}||_{2,\Omega} \leq \frac{\bar{\mathcal{E}}_0}{K_2 + K_4}, \tag{4.6}$$

$$||\mathbf{n} \times \nabla \times \mathbf{n} + \mathbf{P}||_{2,\Omega}^2 \leq \frac{\bar{\mathcal{E}}_0}{\min\{K_2, K_3\}}, \tag{4.7}$$

$$||\mathbf{n} \quad \nabla \times \mathbf{n} + \tau||_{2,\Omega}^2 \leq \frac{\bar{\mathcal{E}}_0}{\min\{K_2, K_3\}}, \tag{4.8}$$

$$|||\frac{1}{q}\nabla\omega - \mathbf{n}|^2 - \frac{|C_\perp}{2Dq^2}||_{2,\Omega}^2 \leq \frac{\bar{\mathcal{E}}_0}{Dq^2}, \tag{4.9}$$

$$||\frac{1}{q}\nabla\omega \cdot \mathbf{n} - 1||_{2,\Omega}^2 \leq \frac{\bar{\mathcal{E}}_0}{C_a}, \quad \textit{and} \tag{4.10}$$

$$||\nabla \mathbf{P}||_{2,\Omega} \leq \frac{\bar{\mathcal{E}}_0}{B_0 P_0}. \tag{4.11}$$

Next, we proceed to take limits in (4.5)-(4.10). We use the following representation for $\mathbf{n}$:

$$\mathbf{n} = \sin\theta\cos\phi\mathbf{i} + \sin\theta\sin\phi\mathbf{j} + \cos\theta\mathbf{k},$$

where $\phi = \phi(x, y, z)$ and $\theta = \theta(x, y, z)$ are functions resulting from energy minimization.

THEOREM 4.2. *Suppose that the hypotheses of the previous theorem hold. Then the energy minimizing fields $(\mathbf{n}, \mathbf{P}, \omega)$ satisfy the following limiting relations:*

$$\lim_{C_a \to \infty} \nabla\omega \cdot \mathbf{n} = q, \tag{4.12}$$

$$\lim_{|C_\perp| \to \infty} |\nabla\omega| = q\sqrt{\frac{|C_\perp|}{2Dq^2} + 1}, \tag{4.13}$$

$$\lim_{\epsilon \to 0} \mathbf{P} = -\cot\alpha\tau\frac{\mathbf{n} \times \mathbf{k}}{|\mathbf{n} \times \mathbf{k}|}, \quad \cot\alpha = \sqrt{\frac{2Dq^2}{|C_\perp|}}, \tag{4.14}$$

$$\lim_{K \to \infty} \mathbf{n} \times \nabla \times \mathbf{n} + \mathbf{P} = 0, \tag{4.15}$$

$$\lim_{K \to \infty} \mathbf{n} \cdot \nabla \times \mathbf{n} + \tau = 0, \tag{4.16}$$

where $K = \min\{K_2, K_3\}$. *Furthermore*

$$\lim_{|C_\perp|\to\infty} \omega = \left(q\sqrt{\frac{|C_\perp|}{2Dq^2}} + 1\right)z, \tag{4.17}$$

$$\lim_{K\to\infty} \mathbf{n} = \sin\alpha\cos\frac{\tau}{a^2}z\mathbf{i} + \sin\alpha\sin\frac{\tau}{a^2}z\mathbf{j} + \cos\alpha\mathbf{k}. \tag{4.18}$$

Proof. From the geometry of the domain and the boundary conditions, it follows that $\nabla\omega = |\nabla\omega|\mathbf{k}$, which together with (4.9) and (4.10) yield (4.12), (4.13), and (4.17). It now follows from relation (4.12) and (4.13) that $\theta = \alpha$ is the constant given by (4.2). These together with (4.5) yield $\mathbf{P} = |\mathbf{P}|\frac{\mathbf{n}\times\mathbf{k}}{|\mathbf{n}\times\mathbf{k}|}$. Combining this equation with (4.15) and (4.16) gives $\phi = \frac{\tau}{a^2}z$ in (4.18). This also yields equation (4.14). □

Note that from the property $\lim_{D\to\infty}(\Delta\omega - q\nabla\cdot\mathbf{n}) = 0$, it follows that the limiting director field has zero divergence, in agreement with (4.18).

4.2. Ferroelectric energy minimizers. In the previous theorem, the elasticity constants K_2 and K_3 become unbounded with respect to the parameters of the polarization contribution to the energy. We will show that the ferroelectric configurations

$$\mathbf{n} = \pm\sin\alpha\mathbf{j} + \cos\alpha\mathbf{k}, \quad \mathbf{P} = \pm P_0\mathbf{i}, \quad P_0 = \sqrt{\frac{|a_0|}{b_0}} \tag{4.19}$$

with α the constant in (4.14), are limits of minimizers at the limit of K_1 large, and when the polar coefficients ω_p and ω_r dominate over the twist and bending elasticity constants, K_2 and K_3. This situation occurs at temperatures lower than those of the helical regime. The role of the surface energy is also relevant in such a case.

Next, we take the following set of admissible fields to determine the ferroelectric limits:

$$\mathbf{n} = \pm\sin\alpha\mathbf{j} + \cos\alpha\mathbf{k}, \ \mathbf{P} = \pm\sqrt{\frac{|a_0|}{b_0}}\mathbf{i}, \tag{4.20}$$

with $0 < \alpha < \frac{\pi}{2}$, and ω as in (4.14) and (4.17), respectively. We find that the energy $\mathcal{E}_1$ corresponding to such fields is:

$$\begin{aligned}\mathcal{E}_1 = L^2[(K_2\tau^2 + \frac{|a_0|}{b_0}K_3)d + 2[\omega_p + \omega_r + \omega_n - (\frac{|a_0|}{b_0})^2\omega_r \\ -\alpha_0\omega_n\sin^2\alpha)].\end{aligned} \tag{4.21}$$

Replacing $\mathcal{E}_0$ with $\mathcal{E}_1$, the estimates of Theorem 4.1 hold. These allow us to establish the following asymptotic limits of minimizers.

$$\nabla\cdot\mathbf{n} = 0, \quad \text{as} \quad K_1 \to \infty, \tag{4.22}$$

$$|\nabla\omega| = k \quad \text{as} \quad |C_\perp| \to \infty. \tag{4.23}$$

Letting $D \to \infty$ and taking (4.22) into account, it follows that $\Delta\omega = 0$. This together with the boundary conditions on $\partial\Omega$ gives $\nabla\omega = (0,0,k)$, with k as in (4.1). Moreover, letting $C_a \to \infty$ gives $\cos\alpha = \frac{q}{k}$, and $\mathbf{P} = \sqrt{\frac{|a_0|}{b_0}}\frac{\mathbf{k}\times\mathbf{n}}{|\mathbf{k}\times\mathbf{n}|}$ results by letting $\epsilon \to 0$ and using the expression for $\nabla\omega$. Finally, letting $\omega_r \to \infty$ gives $\phi = \pm\frac{\pi}{2}$.

1. The limiting fields $(\mathbf{n}, \mathbf{P}, \omega)$ given by (4.17) and (4.19) satisfy the Euler-Lagrange equations with the prescribed boundary conditions.

2. Likewise, $(\mathbf{n}, \mathbf{P}, \omega)$ as in (4.17) and (4.18) solve the Euler-Lagrange equations at the limit $|C_\perp| \to \infty$.

5. Applied constant electric fields and boundary conditions. In this section, we present a summary of stability results about the nature of the equilibrium minimizers with respect to the parameters of the problems, boundary conditions, surface energy, and applied electric field. We also assume that the layer structure is prescribed with $\rho = 1$ and ω as in the previous section. We consider the parallel plate domain with periodic boundary conditions,

$$\begin{aligned}
&\Omega = [-d, d] \times \mathbf{R}^2,\\
&\mathcal{X} = \{\phi \in W^{1,2}(\Omega) \mid \phi(\pm d) = \frac{\pi}{2},\ \phi(x, y+M, z) = \phi(x,y,z),\\
&\qquad\qquad \phi(x,y,z+L) = (x,y,z),\ \text{for all } x,y,z \in \Omega\,\}.
\end{aligned}$$

One goal is to show that, although the ground state of the energy may be the helical smectic C* (i.e., the cone structure), the boundary conditions may impose a preference for uniform director configurations, labelled $\phi = \pm\frac{\pi}{2}$ in the proposed geometry.

THEOREM 5.1. *The energy functional $\mathcal{E}$ has a global minimizer $\frac{\pi}{2}$ among functions in $\mathcal{X}$.*

The analogous statement holds for $\phi = -\frac{\pi}{2}$ in $\mathcal{X}$. In the following, we assume some symmetry conditions on the plates but do not prescribe boundary values of the director on the top and bottom plates. The former are consistent with the additional assumption that $\omega_p = 0$.

Rather than prescribing boundary conditions on ϕ, we consider the effect of the surface energy ω_r. For the present purpose, we take $\omega_p = 0 = \omega_n$. We first study a special class of one-dimensional fields $\phi(z) \in \mathcal{X}_1$, where

$$\mathcal{X}_1 = \{\ \phi \in W^{1,2}([0,L]) \mid \phi(0) = \phi(L), \phi'(0) = \phi'(L)\ \}.$$

The Euler-Lagrange equations for such fields

$$\phi_{zz} + \frac{b^2\omega_r}{d(a^2K_2 + c^2K_3)}\sin 2\phi = 0, \tag{5.1}$$

$$\phi(0) = \phi(L), \phi'(0) = \phi'(L). \tag{5.2}$$

THEOREM 5.2. *Let ω_r be positive and d be such that*

$$\frac{2b^2L^2\omega_r}{(k+1)^2\pi^2} < d \leq \frac{2b^2L^2\omega_r}{k^2\pi^2} \text{ for some } k \in \mathbb{N}.$$

Then there exist at least k nonconstant solutions to (5.1) *and* (5.2) *and for such a solution ϕ,*

$$\mathcal{E}(\phi) \geq \mathcal{E}(\pm\frac{\pi}{2}).$$

Remark. This implies that $\phi = \pm\frac{\pi}{2}$ are global minimizers, in particular, for small $d > 0$. The analogous result in higher dimensions in general does not hold. Also, for $\omega_p \neq 0$, $\phi = \pm\frac{\pi}{2}$ may not be a critical point of the energy, and instead, an x-dependent field, *splayed configuration*, $\phi(x)$ will achieve the minimum energy.

Next, we show that the stability of the uniform states fails when the gap between the plates exceeds a critical value. Specifically, we study the sign of the second variation of the energy to conclude the following.

THEOREM 5.3. *There exists a critical value d_c of the domain thickness such that $\phi = \pm\frac{\pi}{2}$ is unstable for $d \geq d_c$. In the case that $K_2 = K_3$,*

$$d_c = 2\sin^2\alpha[L^2P_0^2(\cot^2\alpha)\pi\frac{\omega_r}{K} + L^2(2P_0 + \tau\sin 2\alpha \\ - 2P_0\cos^2\alpha)(2+\frac{\pi}{2})]^{-1}.$$

In order to study how the multiple effects of inter-plate distance $d > 0$, strength of the applied electric field $E > 0$, and surface energy coefficient ω_r, we address the stability of fields $\mathbf{n} = (a\cos\phi(z), a\sin\phi(z), c)$, $\phi \in W^{1,2}$, with periodic boundary conditions, under three dimensional perturbations. In addition to $\phi = \frac{\pi}{2}$, another critical point of the energy corresponds to

$$\sin\phi_c = -\frac{E}{2(\varepsilon_a E^2 - \frac{\omega_r}{d})\sin\alpha}.$$

A bifurcation analysis [17, 23] yields the following conclusion:

THEOREM 5.4. *Suppose $E > 0$. The field $\phi = \frac{\pi}{2}$ is stable under three dimensional periodic perturbations, if (ω_r, d, E) satisfy one of the following conditions:*

(1) $\varepsilon_a E^2 < \frac{\omega_r}{d} < \varepsilon_a E^2 + \frac{E}{2a}$,
(2) $\varepsilon_a E^2 - \frac{E}{2a} < \frac{\omega_r}{d} < \varepsilon_a E^2$,
(3) $\varepsilon_a E^2 - \frac{E}{2a} - \frac{K\pi^2}{L^2} < \frac{\omega_r}{d} < \varepsilon_a E^2 - \frac{E}{2a}$,
(4) $\varepsilon_a E^2 + \frac{E}{2a} < \frac{\omega_r}{d}$.

In cases (3) and (4), $\phi = \phi_c$ is also a critical point. Moreover, $\phi = \frac{\pi}{2}$ is unstable if $\varepsilon_a E^2 - \frac{E}{2a} - \frac{K\pi^2}{L^2} > \frac{\omega_r}{d}$.

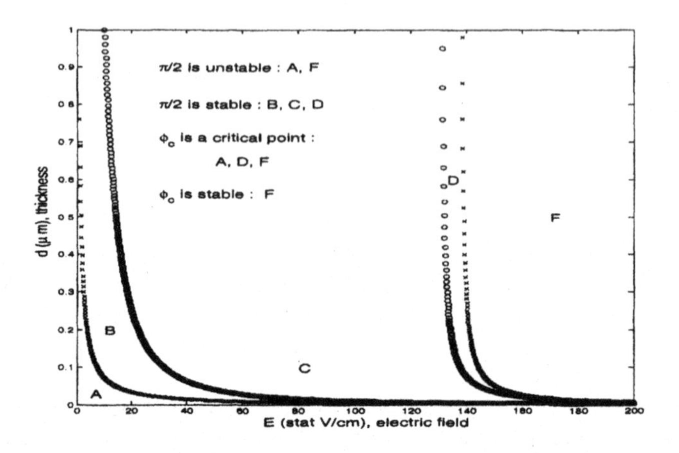

FIGURE 2. $\varepsilon_a = 0.01, a = \sin(22.5), \omega_r = 1, K = 10^{-6}, L = 10^{-3}$

Remarks.
1. If $E < 0$, then the energy of $\phi = -\frac{\pi}{2}$ always is less energy than that of $\phi = \frac{\pi}{2}$. The stability region for $-\frac{\pi}{2}$ is symmetric with respect to the line $E = 0$ in the previous diagrams.
2. Estimates of critical fields that without including dielectric effects can be found in the literature [9]. However the inclusion of such a contribution is relevant since it becomes dominant when the electric field reaches a threshold value, in which case the ferroelectric state $\phi = \frac{\pi}{2}$ becomes unstable. Although our calculations predict that the configuration with $\phi = \phi_c$ become the stable one, other ferroelectric phases may be observed in such a regime.

We conclude this section with a calculation of the correction to the threshold values that need to be made due to the field created by surface charges in polarized bulk. Such a field opposes the effect of the applied one and it rises the threshold values [14].

REMARK 5.1. *The flexoelectric polarization associated with the vector field* $\mathbf{n} = (\sin\alpha\cos\phi, \sin\alpha\sin\phi, \cos\alpha)$ *is given by* $\mathbf{P} = P_0\frac{\mathbf{k}\times\mathbf{n}}{|\mathbf{k}\times\mathbf{n}|}$. *As a result, the boundary plates sustain charges of opposite sign and density* $\mathbf{P}\cdot\mathbf{i} = -P_0\sin\phi\mathbf{i}$, *where* $\mathbf{i}$ *corresponds to the surface unit normal vector in the parallel plate geometry. The electric field created by the pair of oppositely charged plates is* $\mathbf{E}_{ion} = 4\pi P_0\sin\phi\mathbf{i}$. *In the case* $\phi = \frac{\pi}{2}$ *such a*

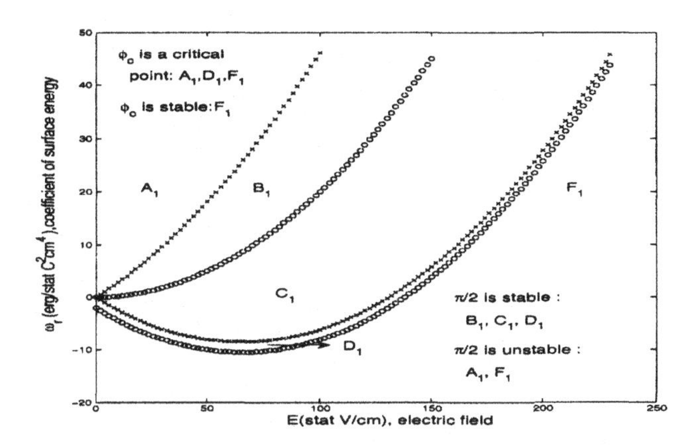

FIGURE 3. $\varepsilon_a = 0.01, a = \sin(22.5), d = 0.2, K = 10^{-6}, L = 10^{-3}$

field reduces to $\mathbf{E}_{ion} = 4\pi P_0 \mathbf{i}$, *which is opposite to the applied field* $\mathbf{E} = -E\mathbf{i}$, $E > 0$. *Therefore, the effective field is reduced to* $\mathbf{E} - \mathbf{E}_{ion}$, *which leads to an increase of the threshold values by the amount* $|\mathbf{E}_{ion}| = 4\pi P_0$.

6. Variable electric fields. As mentioned in [12], the total energy including electric potential is not bounded below. In order to overcome, we minimize it with Maxwell's equations for $\mathbf{D}$ and $\mathbf{E}$

$$\nabla \cdot \mathbf{D} = 0, \quad \nabla \times \mathbf{E} = \mathbf{0}.$$

Using an electric potential φ and imposing boundary conditions for φ, we consider the following equation,

$$\left\{ \begin{array}{c} -\nabla \cdot (\alpha_0 I + \alpha_a \mathbf{n} \otimes \mathbf{n}) \nabla \varphi = \nabla \cdot \mathbf{P} \text{ in } \Omega \\ \varphi = \varphi_0 \text{ on } \Gamma_D, \\ \frac{\partial \varphi}{\partial \nu} = 0 \text{ on } \Gamma_N, \end{array} \right. \tag{6.1}$$

where $\varphi_0 \in H^{\frac{1}{2}}(\Gamma_D)$ and $\partial\Omega = \Gamma_D \cup \Gamma_N$.

Let

$$H^1_{\varphi_0} = \{\varphi \in H^1(\Omega) : \varphi = \varphi_0 \text{ on } \Gamma_D\},$$
$$\mathcal{A}^* = \mathcal{A} \times H^1_{\varphi_0},$$

where $\mathcal{A}$ is defined in (3.8).

Define

$$\mathcal{W}^*(\mathbf{n}, \mathbf{P}, \omega, \varphi) = \mathcal{E}(\mathbf{n}, \mathbf{P}, \omega) + \int_\Omega F_{Elec}\, d\mathbf{x}.$$

For any $(\mathbf{n}, \mathbf{P}, \omega) \in \mathcal{A}$, the fundamental theory of elliptic PDEs shows that (6.1) has a unique solution which we denote by $\Phi_{\varphi_0}(\mathbf{n}, \mathbf{P})$, and thus $\Phi_{\varphi_0}(\mathbf{n}, \mathbf{P})$ is the unique minimizer of $\int_\Omega F_{Elec}\, d\mathbf{x}$ in $H^1_{\varphi_0}$.

Substituting $\Phi_{\varphi_0}(\mathbf{n}, \mathbf{P})$ for φ in $\mathcal{W}$, we define $\mathcal{W}$ by

$$\mathcal{W}(\mathbf{n}, \mathbf{P}, \omega) = \mathcal{W}^*(\mathbf{n}, \mathbf{P}, \omega, \Phi_{\varphi_0}(\mathbf{n}, \mathbf{P})).$$

The following theorem uses arguments similiar to those in [12].

THEOREM 6.1. *For any* $\varphi_0 \in H^{\frac{1}{2}}(\Gamma_D)$ *and* $\min\{K_2, K_3\} \geq 4c_1$, *there exists a triple* $(\mathbf{n}, \mathbf{P}, \omega)$ *which minimizes* $\mathcal{W}$ *on* $\mathcal{A}$. *Furthermore, a quadruple* $(\mathbf{n}, \mathbf{P}, \omega, \varphi)$ *is a critical point of* $\mathcal{W}^*$,

$$\delta\mathcal{W}^*(\mathbf{n}, \mathbf{P}, \omega, \varphi) = 0 \textit{ on } \mathcal{A}^*$$

if and only if

$$\varphi = \Phi_{\varphi_0}(\mathbf{n}, \mathbf{P}), \textit{ and } \delta\mathcal{W}(\mathbf{n}, \mathbf{P}, \omega) = 0 \textit{ on } \mathcal{A}.$$

The boundary condition φ_0 can be considered as an external field. In the case that $\varphi_0 = 0$, φ is due to material polarization only which brings out nonlocal energy. The polarization $\mathbf{P}$ causes a point charge density $\rho_p = \nabla \cdot \mathbf{P}$ and a surface charge density of charges $\sigma = \mathbf{P} \cdot \nu$. These charge densities may induce a field in the whole space. In fact, one needs to find an electric potential $\varphi \in W^{1,2}(\mathbf{R}^3)$ satisfying

$$-\nabla \cdot ((\alpha_0 I + \alpha_a \mathbf{n} \otimes \mathbf{n})\nabla\varphi) = \nabla \cdot \mathbf{P} \text{ in } \Omega, \tag{6.2}$$

$$\Delta\varphi = 0 \text{ in } \mathbf{R}^3 - \Omega, \tag{6.3}$$

$$[(\alpha_0 I + \alpha_a \mathbf{n} \otimes \mathbf{n})\nabla\varphi - \varepsilon_0\nabla\varphi] \cdot \nu = \mathbf{P} \cdot \nu \text{ on } \partial\Omega, \tag{6.4}$$

where ν is the normal to the boundary and ε_0 is the dielectric coefficient of the environment. It can be shown [2] that $\varphi|_\Omega$ belongs to $W^{2,p}(\Omega)$ for any $\mathbf{P} \in W^{1,p}(\Omega)$. In other words, there exists a constant C such that

$$||\varphi||_{W^{2,p}(\Omega)} \leq C||\mathbf{P}||_{W^{1,p}(\Omega)}.$$

We shall discuss further details for nonlocal energy associated with the polarization $\mathbf{P}$ in our forthcoming papers.

7. Conclusions. We studied modeling and analysis of static configurations of smectic C* liquid crystals with a special emphasis on polarization and ferroelectricity. The energy that we analyze contains quadratic terms on the second derivative of the complex wave function. We show that the total energy is lower semicontinuous and coercive within the class of fields satisfying physically relevant boundary conditions. These include the cases where the layer structure reaches the boundary of the domain either in a perpendicular or parallel fashion , with prescribed wave number. The problem bears a strong resemblance to the analysis of vortex tubes and vortex sheets in fluid mechanics. We studied the asymptotic properties of the energy minimizers as the parameters of the energy become unbounded upon the temperature reaching transition values from smectic A* to smectic C*, and lower temperature limits. We discuss some stability results to help interpreting the role of boundary conditions and applied electric field effects. Finally, we considered the case such that the electric field is variable. The problem reduces to finding critical points of the energy subject to the time-independent Maxwell's equations.

REFERENCES

[1] P. Bauman, M. C. Calderer, C. Liu, and D. Phillips, *The phase transition between Chiral and Smectic A* liquid crystals*, Arch. Ration. Mech. Anal., 165 (2002), pp. 161–186.

[2] G. Carbou, *Regularity for critical points of a nonlocal energy*, Calc. Var. Partial Differential Equations, 5 (1997), pp. 409–433.

[3] J. Chen and T. Lubensky, *Landau-Ginzburg mean-field theory for the nematic to smectic C and nematic to smectic A liquid crystal transistions*, Phys. Rev. A, 14 (1976), pp. 1202–1297.

[4] A. Chorin and J. Marsden, *A Mathematical Introduction to Fluid Mechanics*, Springer-Verlag, New York, 1984.

[5] N. A. Clark and S. T. Lagerwall, *Submicrosecond bistable electro-optic switching in liquid crystals*, Appl. Phys. Lett., 36 (1980), pp. 899–901.

[6] P. Collings and J. Patel, eds., *Handbook of Liquid Crystal Research*, Oxford University Press, New York, 1997.

[7] P. G. de Gennes and J. Prost, *The Physics of Liquid Crystals*, Clarendon Press, 1993.

[8] M. Glogarova, J. Fousek, L. Lejcek, and J. Pavel, *The structure of ferroelectric liquid crystals in planar and its response to electric fields*, Ferroelectrics, 58 (1984), pp. 161–178.

[9] M. Glogarova, L. Lejcek, and J. Pavel, *The influence of an external electric field on the structure of chrial Sm C* liquid crystals.*

[10] M. Glogarova and J. Pavel, *The structure of chiral Sm C* liquid crystals in planar samples and its change in an electric field*, J. Physique, 45 (1984), pp. 143–149.

[11] M. A. Handschy and N. A. Clark, *Structures and responses of ferroelectric liquid crystals in the syrface-stabilized geometry*, Ferroelectrics, 59 (1984), pp. 69–116.

[12] R. Hardt, D. Kinderlehrer, and F. H. Lin, *Existence and partial regularity of static liquid crystal configurations*, Comm. Math. Phys., 105 (1986), pp. 547–570.

[13] S. Joo and D. Phillips, *Phase transitions between chiral nematic, smectic A*, and C* liquid crystals.* Preprint, 2004.

[14] S. T. Lagerwall, *Ferroelectric and Antiferroelectric Liquid Crystals*, Wiley-VCH, 1999.

[15] I. Lukyanchuk, *Phase transition between the cholesteric and twist grain boundary C phases*, Phys. Rev. E, 57 (1998), pp. 574–581.

[16] I. MUŠEVIČ, R. BLINC, AND B. ŽEKŠ, *The Physics of Ferroelectric and Antiferroelectric Liquid Crystals*, World-Scientific, Singapore, New Jersey, London, Hong Kong, 2000.
[17] L. NIRENBERG, *Topics in Nonlinear Functiuonal Analysis*, AMS/CIMS, New York, 2001.
[18] J. PARK AND M. CALDERER, *Phase transitions between ferroelectric and antiferroelectric liquid crystal phases: static and dynamical problems.* Preprint, 2004.
[19] J. PAVEL, *Behaviour of thin planar Sm C* samples in an electric field*, J. Physique, 45 (1984), pp. 137–141.
[20] J. PAVEL AND M. GLOGAROVA, *The effect of biasing electric field on relaxations in FLC investigated by the dielectric and optical methods*, Ferroelectrics, 121 (1991), pp. 45–53.
[21] J. PAVEL, M. GLOGAROVA, AND S. S. BAWA, *Dielectric permittivity of ferroelectric liquid crystals influenced by a biasing electric field*, Ferroelectrics, 76 (1987), pp. 221–232.
[22] S. PIKIN, *Structural Transformations in Liquid Crystals*, Gordon and Breach Science Publishers, New York, 1991.
[23] P. H. RABINOWITZ, *Some global results for nonlinear eigenvalue problems*, J. Func. Anal., 7 (1971), pp. 487–513.
[24] T. P. RIEKER, N. A. CLARK AND S. LAGERWALL, *Chevron local layer structure in surface stabilized ferroelectric smectic-C cells*, Phys. Rev. Lett., 59 (1987), pp. 2658–2672.

STRIPE–DOMAINS IN NEMATIC ELASTOMERS: OLD AND NEW

ANTONIO DESIMONE* AND GEORG DOLZMANN†

Abstract. Formation of stripe–domains has often been observed in nematic elastomers, starting from the pioneering work of Finkelmann and coworkers. One of the possible interpretations of this phenomenon is to view it as a material instability driven by energy minimization. This approach, first proposed by Warner and Terentjev, has been quite helpful in the analysis of stretching experiments of thin sheets, and in the modelling of soft elasticity associated with stripe–domain formation. Recently, complex stripe–domain patterns have been observed in nematic gels undergoing the isotropic–to–nematic transition while being confined by two glass plates. We suggest that, once again, energy minimization can be seen as the driving mechanism for the formation of the observed patterns.

Key words. Polymers, Nematic elastomers, Domain patterns, Nonlinear elasticity, Nonconvex problems in the Calculus of Variations, Relaxation, Numerical simulation of microstructures.

AMS(MOS) subject classifications. 74N15, 74B20, 49J45.

1. Introduction. Nematic elastomers consist of networks of cross–linked polymeric chains, each of which contains nematic rigid rod–like molecules (either as main–chain segments or as side attachments). These are called nematic mesogens: their tendency to align below a critical transition temperature promotes the formation of nematic order. Contrary to nematic liquids, however, the orientational degrees of freedom of the mesogens are coupled to the translational degrees of freedom of an underlying elastic solid (the rubbery polymer network). This coupling makes nematic elastomers very interesting as a model physical system, and it is also at the root of their technological interest as materials for actuators and for (mechanically tunable) optical devices. The recent monograph [1], and the extensive list of references cited therein give a comprehensive account of the history of the synthesis of liquid crystal elastomers, of the envisaged applications, and of the models proposed to explain their fascinating properties.

Nematic elastomers display interesting material instabilities. At high temperatures, the nematic mesogens are randomly oriented due to thermal fluctuations, and nematic elastomers behave like isotropic rubbers. Upon cooling through the nematic transition temperature, the nematic mesogens align causing a distorsion of the underlying rubber network. Imagine, for simplicity, the possibility that locally the mesogens are perfectly aligned at

*SISSA–International School for Advanced Studies, Via Beirut 2-4, 34014 Trieste, ITALY.

†Department of Mathematics, University of Maryland, College Park, MD 20742, USA.

the microscopic scale. Then symmetry dictates that the polymer network deform uniaxially, with the distinguished axis parallel to the common direction of the mesogens. Moving now to a more macroscopic scale, if the mesogens align along different directions in different parts of the sample, then different states of spontaneous distorsion may coexist within the same sample, generating patterns which are reminiscent of magnetic domains in ferromagnetic materials. Since the optical properties of the material are controlled by the orientation of the nematic mesogens, which changes from domain to domain, domain patterns are observable under polarized light. Indeed, regions where the mesogens are differently oriented may appear opaque or transparent when observed under crossed polarizers, depending on the relative orientation of mesogens and polarization of light.

The mechanical response of nematic elastomers to imposed deformation may be unusually soft, when compared to the response of ordinary rubber. The existence of domain patterns, and the ease with which they can evolve, is what allows nematic elastomers to respond to imposed macroscopic deformations with negligible internal stress, whenever the imposed strains may be accommodated by simply reorienting the nematic mesogens.

2. A minimalist model. Both the occurrence of domains, in a characteristic striped texture, and the existence of 'soft' deformation modes have been observed experimentally, in several different laboratories. These phenomena have also been analyzed in great detail by several groups, using a variety of different models. Many of these were represented at the IMA workshop [2], and we refer to the papers [3]–[8] for a sample of the many existing viewpoints on the subject. By contrast, we take here a minimalist approach, and proceed with the help of the most basic model we can think of. In essence, all what we do is to accept that the symmetry–breaking phase transformation associated with the establishment of nematic order is able to produce a number of distinct, symmetry–related, spontaneously deformed states, and that these states are the minima of a quadratic free–energy density with the correct invariance properties. From here, we will try and deduce the largest possible number of rigorous consequences resulting from energy minimization, without further simplifications. Our aim is to contribute, in this way, to put into sharper focus the most generic and universal aspects of the phenomenon under study.

The first ingredient of our model is the spontaneous strain the material experiences when, starting from the isotropic parent phase, it undergoes the isotropic–to–nematic phase transformation. Assume that the nematic mesogens align, say, along the direction of a unit vector $\mathbf{n}$. Accepting the hypothesis of incompressibility, and assuming that the polymer chains stretch in the direction of $\mathbf{n}$, the resulting deformation is described by a uniaxial tensor of the form

$$\mathbf{U}_{\mathbf{n}} := a^{-1/3}\mathbf{n} \otimes \mathbf{n} + a^{1/6}(\mathrm{Id} - \mathbf{n} \otimes \mathbf{n}) \tag{2.1}$$

where $(\mathbf{n}\otimes\mathbf{n})_{ij} = \mathbf{n}_i\mathbf{n}_j$, Id is the identity, and $a < 1$ is a material parameter quantifying how much a spherical random polymer coil spontaneously extends in the direction of $\mathbf{n}$ upon the alignment of the mesogens. Using frame–indifference and material symmetry (recall that the relevant symmetry is the one of the isotropic parent phase), we deduce from (2.1) that the set of spontaneous deformations has the form

$$\mathbf{P}\mathbf{U}_{\mathbf{Q}\mathbf{n}} = \mathbf{P}\mathbf{Q}\mathbf{U}_{\mathbf{n}}\mathbf{Q}^T \tag{2.2}$$

where $\mathbf{P}$ and $\mathbf{Q}$ are arbitrary rotations. A deformation gradient $\mathbf{F}$ belongs to the set (2.2) if and only if the ordered principal stretches $\lambda_1(\mathbf{F}) \le \lambda_2(\mathbf{F}) \le \lambda_3(\mathbf{F})$ satisfy $\lambda_1(\mathbf{F}) = \lambda_2(\mathbf{F}) = a^{1/6} < 1$, $\lambda_3(\mathbf{F}) = a^{-1/3} > 1$.

The next ingredient of our model is an energy penalization on deformation gradients $\mathbf{F}$ which are not in the set (2.2). The simplest choice for a quadratic, isotropic, frame–indifferent expression is a minor modification of the classical neo–hookean expression

$$W(\mathbf{F}) = \begin{cases} \mu\left(\lambda_1^2(\mathbf{F}) + \lambda_2^2(\mathbf{F}) + a\lambda_3^2(\mathbf{F}) - 3a^{1/3}\right) & \text{if } \det\mathbf{F} = 1 \\ +\infty & \text{else.} \end{cases} \tag{2.3}$$

Here $\mu > 0$ is a material parameter giving the rubber energy scale (defining the initial shear modulus of the material), while the constraint $\det\mathbf{F} = 1$ on deformations with finite energy enforces incompressibility. Since the geometric mean of the squares of the principal stretches is not larger than the corresponding arithmetic mean, it is easy to see that (2.3) is always non–negative, and that it vanishes precisely on the set (2.2) (which we then call the set of the material's energy wells). The classical neo–hookean expression is simply obtained from (2.7) by setting $a = 1$.

Finally, denoting by $\mathbf{y}$ the map that gives the deformed configuration of a body whose reference configuration is $\mathcal{B}$, and by $\nabla\mathbf{y}$ its gradient, the total energy reads

$$E(\mathbf{y}) = \int_{\mathcal{B}} W(\mathbf{y}(\mathbf{x}))d\mathbf{x}\,. \tag{2.4}$$

We are interested in deformation maps which minimize this total energy when suitable loads or boundary displacements are prescribed.

Before proceeding with the analysis of our minimalist model, a few remarks are in order. Expression (2.3) can actually be derived from the one first proposed in [3] with the following procedure (see [9, 10]). First, assume that nematic order is adequately described by a director field $\mathbf{x} \mapsto \mathbf{n}(\mathbf{x})$. Then perform an affine transformation of the spatial variables which corresponds to using, as reference configuration, the highly symmetric stress–free configuration of the isotropic parent phase.[1] This leads to the following

[1] The reference configuration chosen in [3] is *one* of the stress–free configurations the material adopts *after* it has transformed to the nematic phase. One such configuration is the initial configuration in the stretching experiments discussed in Section 3.

simple expression for the energy density, depending on the vector field $\mathbf{n}$ and on the tensor field $\mathbf{F}$

$$W_{\mathrm{BTW}}(\mathbf{n},\mathbf{F}) = \mu\left(|\mathbf{F}|^2 - (1-a)|\mathbf{F}^T\mathbf{n}|^2 - 3a^{1/3}\right). \tag{2.5}$$

Finally, minimize the last expression with respect to $\mathbf{n}$ to obtain

$$W(\mathbf{F}) = \min_{|\mathbf{n}|=1} W_{\mathrm{BTW}}(\mathbf{n},\mathbf{F}). \tag{2.6}$$

Notice that the $\mathbf{n}$ minimizing W_{BTW} is the eigenvector of $\mathbf{F}^T\mathbf{F}$ associated with its largest eigenvalue $\lambda_3^2(\mathbf{F})$. It follows that, in our simple-minded model, the nematic degrees of freedom are slaved to the elastic ones, and they are simply described by a director field which is aligned to the direction of the largest principal stretch at every point of the sample.

Moving to the analysis of our model, its most important feature is the non–convexity of energy density (2.3) for $a < 1$. This implies that uniform configurations may have higher energy than complex domain patterns with the same average deformation. In fact, this is the mechanism proposed in [7] to explain the experimental observation that stretching a mono-domain sheet of nematic elastomer may induce spontaneous break-up into stripe-domains [14].

The non–convexity of (2.3) also explains the experimentally observed soft deformation paths. They arise as energetically optimal fine phase mixtures, with volume fractions evolving with the applied loads. These fine phase mixtures effectively accomplish a convexification of the underlying rough energy landscape. A plateau in the stress–strain response is the precisely signature of a flat portion of the energy graph resulting from the convexification of the energy. We conclude that macroscopic soft deformation paths result, and should then be computable, from suitable convex hulls of the materials energy wells. In fact, using this idea, the characterization of all soft deformation paths associated with the set (2.2) was obtained in [10].

The existence of macroscopic deformations that can be resolved at no energy cost by microscopic mixtures of deformation gradients lying on the energy wells is only part of the story. One may ask what is, in general, the minimal energetic cost of imposing an arbitrary affine deformation $\mathbf{F}$ on the boundary $\partial\Omega$ of a representative volume Ω, leaving the system free to develop fine structures in the interior of Ω whenever this is energetically advantageous. This is given by a well–defined mathematical object (see, e.g., [11]), the value at $\mathbf{F}$ of the quasi–convex envelope of the energy density W

$$W_{\mathrm{qc}}(\mathbf{F}) := \inf_{\mathbf{y}\in\mathcal{A}}\left\{\frac{1}{|\Omega|}\int_\Omega W(\nabla\mathbf{y}(\mathbf{x}))d\mathbf{x} : \mathbf{y}(\mathbf{x}) = \mathbf{F}\mathbf{x} \text{ on } \partial\Omega\right\} \tag{2.7}$$

where $|\Omega|$ is the volume of Ω and the infimum is taken over the set $\mathcal{A}$ of Lipschitz–continuous maps such that $\det \nabla \mathbf{y}(\mathbf{x}) \equiv 1$. It can be shown that the right–hand–side of (2.7) does not depend on the geometry of Ω.

An explicit formula for the quasi–convex envelope of (2.3) has been derived in [12]. For volume-preserving deformation gradients it reads

$$W_{\mathrm{qc}}(\mathbf{F}) = \begin{cases} 0 & \text{(phase L) if } \lambda_1 \geq a^{1/6} \\ W(\mathbf{F}) & \text{(phase S) if } a^{1/2}\lambda_3^2\lambda_1 > 1 \\ \mu\left(\lambda_1^2 + 2a^{1/2}\lambda_1^{-1} - 3a^{1/3}\right) & \text{(phase I) else} \end{cases} \tag{2.8}$$

while $W_{\mathrm{qc}}(\mathbf{F}) = +\infty$ if $\det \mathbf{F} \neq 1$. Here the labels L, S, and I refer to the fact that the resulting material response is liquid–like, solid–like, or of an intermediate type, see the discussion below.

The formula above gives a very precise picture of the macroscopic mechanical response resulting from our model, and of its microscopic origin. There are three regimes in (2.8), arising from the collective behavior of energetically optimal fine phase mixtures. They represent three different modes of macroscopic mechanical response, and they correspond to three different patterns of microscopic decomposition of the macroscopic deformation gradient $\mathbf{F}$. Phase L describes a liquid-like response (at least within the ideally soft approximation embedded in the expression (2.3) for the microscopic energy density; the semi–soft case is discussed in [13]). All gradients falling in this region of the phase diagram can be sustained at zero internal stress (in other words, the zero level set of W_{qc} is the set of all soft deformation paths mentioned above). To resolve microscopically the whole of phase L (in particular, to resolve the deformation gradient $\mathbf{F} = \mathrm{Id}$) it is necessary to allow for relatively complex microstructures (layers-within-layers, see the right panel in Figure 1). Phase S describes a solid-like response in which fine phase mixtures are ruled out. As a consequence, in this regime the effective macroscopic energy W_{qc} reproduces the microscopic energy W with no changes. Finally, gradients in the intermediate phase I can transmit stresses (unlike phase L) through microstructure formation (unlike phase S). The microscopic patterns required to resolve phase I have a relatively simple geometry (laminates, or simple-layers, see the left panel in Figure 1). Patterns of this kind have been frequently observed experimentally after being first reported in [14]. The first attempt to explain them through elastic energy minimization is in [7].

By decoupling the physical length scales into microscopic ones (which determine the effective energetics but are averaged out in the kinematics) and macroscopic ones (e.g., those which determine the deformed shape of a sample in a stretching experiment), expression (2.8) leads to a coarse–grained version of our model. In order to resolve macroscopic quantities, we minimize the effective energy

$$E_{\mathrm{eff}}(\mathbf{y}) = \int_B W_{\mathrm{qc}}(\mathbf{y}(\mathbf{x}))d\mathbf{x} \tag{2.9}$$

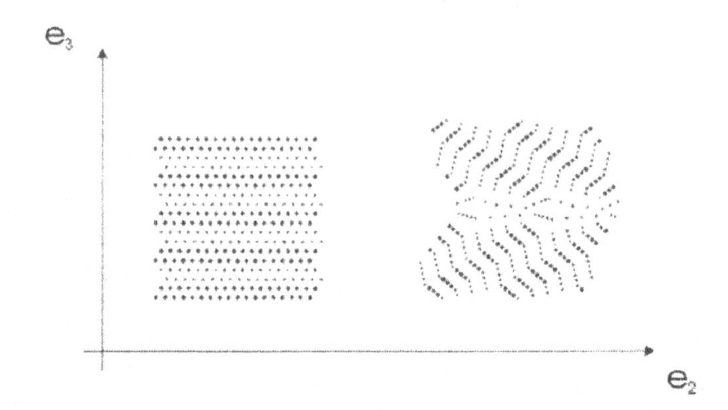

FIG. 1. *Geometry of energetically optimal microstructures: simple-layers (left) and layers-within-layers (right). The light and dark dots hint at the optical contrast these microstructure would produce in nematic elastomers under polarized light (adapted from [9]).*

subject to the boundary conditions that are relevant to the loading experiment we want to model. This can be done numerically, giving a field $\mathbf{x} \mapsto \mathbf{F}(\mathbf{x})$. Then, by locating the computed macroscopic deformation within one of the three phases L, S, I of the phase diagram, we can reconstruct, from the field of computed local deformation gradients $\mathbf{x} \mapsto \mathbf{F}(\mathbf{x})$, the corresponding field of local (energy minimizing) microstructures.

Before closing this section, one word about the length scales of the domain patterns. Energy (2.2) does not contain an intrinsic length scale. Thus, the minimization procedure (2.7) leading to W_{qc} can lead to domain patterns which are infinitely fine with respect to the size of a representative volume element, and which manifest themselves as (infinitely refining) sequences of deformation patterns driving the energy to its infimum. Clearly, in reality, physical mechanisms not active in our simple–minded model do establish a smallest characteristic length scale. Whenever the size of the sample of interest is large compared to this characteristic length, our approach can be used to gather detailed information on the macroscopic response of the sample (e.g., the force–stretch curve), and gross information on the domain patterns responsible for the macroscopic behavior. More detailed information on domain patterns requires explicit resolution of length scales. One natural possibility is to consider models containing higher order gradients (e.g., Frank-type terms in the gradient of the director, see e.g. [4]). Needless to say, since the corresponding numerical simulations will need to resolve much finer spatial scales (typically, at the sub–micron scale), a substantial increase of the computational cost for obtaining a force–stretch diagram for, say, a mm–sized sample should be expected.

3. Stripe–domain patterns: the classics. The computational approach outlined in the previous section, and based on the expression (2.7) for the energy density, has been used in [15] for the numerical simulation of stretching experiments of sheets of nematic elastomer held between two rigid clamps. The simulations are designed to reproduce the classical experimental setting of Kundler and Finkelmann [14], where stripe–domain patterns were first observed.

The specimen is a thin sheet of nematic elastomer. We choose a reference frame with axis x_1 parallel to the thickness direction. Moreover, we assume that the specimen is prepared with the director uniformly aligned along x_3, and is then stretched along x_2. By reorienting the director from the x_3 to the x_2 direction, the material can accommodate the imposed stretches without storing elastic energy. As it is well known, see e.g. [1], a uniform rotation of the director would induce large shears, which are incompatible with the presence of the clamps. Director reorientation occurs instead with the development of spatial modulations shaped as bands parallel to the x_2 axis. This is the origin of the striped texture observed in the experiments.

The numerical simulations allow us to analyze the stretching experiments in more detail. If the clamps do not allow lateral contraction, the reorientation of the director towards the direction of the imposed stretch is severely hindered. This constraint is stronger near the clamps, and it decays away from them producing two interesting effects. On one hand, the induced microstructures are spatially inhomogeneous, with director reorientation occurring more rapidly in the regions far away from the clamps. On the other hand, the stress–strain response shows a marked dependence on the geometry of the sample, with the influence of the clamps becoming less pronounced as the aspect ratio length/width increases. These effects are documented in Figure 2 and Figure 3, which show good qualitative agreement with both the experimental results from the Cavendish Laboratories [1], and with the X-ray scattering measurements in [16].

The stripe domain patterns appearing in Figure 3 are all simple laminates, either in phase L or in phase I. Focusing on the point at the center of the sample (the bottom left corner in the plots of the deformed shape), the material is in phase L as long as no force is transmitted at the clamps. The computed deformation gradient is

$$\mathbf{F}_\lambda = \begin{pmatrix} a^{1/6} & 0 & 0 \\ 0 & \lambda & 0 \\ 0 & 0 & a^{-1/6}/\lambda \end{pmatrix} \tag{3.1}$$

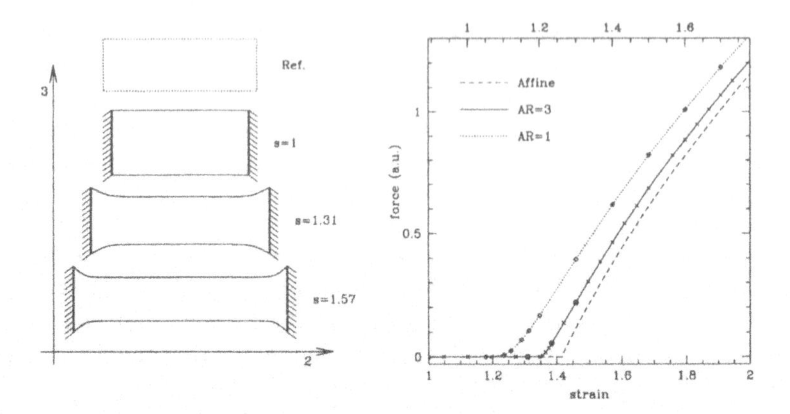

FIG. 2. *Numerical simulation of stretching experiments on thin sheets of nematic elastomers: geometry (left) and force–stretch diagrams for several aspect ratios AR (right). The panel on the left shows four configurations, namely, reference, initial, and the two at stretches s=1.31 and s=1.57 for the geometry with AR=3. On the corresponding force–stretch curve on the right panel, full dots mark the representative points of configurations shown in Figure 3 (adapted from [15]).*

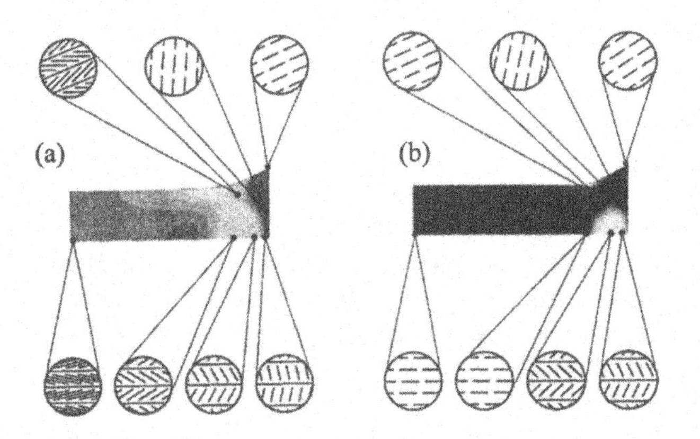

FIG. 3. *Numerical simulation of stretching experiments on thin sheets of nematic elastomers, based on the coarse–grained energy W_{qc}, at stretches s=1.31 (a), and s=1.38 (b). Only one–quarter of the sample is shown since the rest of the solution can be obtained by symmetry. The circular insets display energetically optimal microstructures at some selected locations within the sample. The sticks give the local orientation of the principal direction of maximal stretch, i.e., the orientation of the nematic director (adapted from [15]).*

with λ varying from $a^{1/6}$ to $a^{-1/3}$. This is resolved by a simple laminate in which the deformation gradient oscillates between the values

$$\mathbf{F}_\lambda^\pm = \begin{pmatrix} a^{1/6} & 0 & 0 \\ 0 & \lambda & \pm\delta \\ 0 & 0 & a^{-1/6}/\lambda \end{pmatrix} \tag{3.2}$$

in stripes perpendicular to x_3, i.e., with a geometry similar to the one shown in the left panel of Figure 1. The value of $\delta = \delta(\lambda)$ is obtained from $\delta^2 = (a^{-2/3} - \lambda^2)(1 - a^{1/3}\lambda^{-2})$, which ensures that $\mathbf{F}^{\pm}_{\lambda}$ has the characteristic principal stretches giving $W(\mathbf{F}^{\pm}_{\lambda}) = 0$. Notice that the kinematic compatibility condition $\mathbf{F}^{+}_{\lambda} - \mathbf{F}^{-}_{\lambda} = \mathbf{a} \otimes \mathbf{N}$, where $\mathbf{N}$ is the reference normal to the stripes and $\mathbf{a}$ is a shear vector, is satisfied with $\mathbf{a} = 2\delta(\lambda)\mathbf{e}_2$ and $\mathbf{N} = \mathbf{e}_3$. This guarantees the existence of a continuous map $\mathbf{y}$ such that $\nabla \mathbf{y}(\mathbf{x}) = \mathbf{F}^{\pm}_{\lambda}$ in layers with normal $\mathbf{e}_3$.

Force starts being transmitted through the sample when the deformation gradient in the central point moves to the region I of the phase diagram. The computed deformation gradient is now of the form

$$\mathbf{F}_{\lambda_1} = \begin{pmatrix} \lambda_1 & 0 & 0 \\ 0 & 1/\lambda_1\lambda_3 & 0 \\ 0 & 0 & \lambda_3 \end{pmatrix} \tag{3.3}$$

where $\lambda_3 > a^{-1/3}$ forces $\lambda_1 < a^{1/6}$. This is resolved by simple laminates similar to the ones above. The deformation gradient oscillates between the values

$$\mathbf{F}^{\pm}_{\lambda_1} = \begin{pmatrix} \lambda_1 & 0 & 0 \\ 0 & 1/\lambda_1\lambda_3 & \pm\delta \\ 0 & 0 & \lambda_3 \end{pmatrix} \tag{3.4}$$

in stripes perpendicular to x_3, and $\delta = \delta(\lambda_1)$ is computed by requiring that the principal stretches be those giving the minimal energy at given λ_1, namely, $(\lambda_1, a^{1/4}\lambda_1^{-1/2}, a^{-1/4}\lambda_1^{-1/2})$, see [15].

4. Stripe–domain patterns: recent observations. The setup of the stretching experiments described in the previous section has one peculiarity. In the initial configuration, the film thickness near the center of the sample is $a^{1/6}$ times the thickness in the reference configuration. As the film is stretched, the film thickness either stays unchanged (this is the regime given by (3.1), and it represents an unusual behavior when compared to that of a conventional rubber) or it decreases (this is the regime given by (3.3)). This implies that the smallest principal stretch can never exceed the value $a^{1/6}$. As a consequence microstructures of the layers-within-layers type are not accessible in the kind of stretching experiments described above, no matter how close the material is to the limit of ideal softness.

In the course of the workshop [2], the following recent experiment by Meyer and Meng was presented. A thin film of a soft nematic gel, confined between two horizontal glass plates, is cooled through the isotropic–to–nematic transition temperature T_{IN} while its director is kept vertical by an applied electric field. When the external field is removed, the nematic director is no longer frozen, and in–plane stripe domains develop and become visible through the glass plates.

If one focuses on points far away from the sample edges, one may think of the glass plates as infinitely extended so that, by symmetry, all the in–plane stretches are principal stretches with the same value. The constraining action of the glass plates, which do not allow the film thickness to expand, coupled with the incompressibility constraint, then results in imposing that the material does not deform at all, or

$$\mathbf{F} = \mathrm{Id} \tag{4.1}$$

at least in average, and far away from the lateral edges. This macroscopic deformation gradient can be resolved by the spontaneous deformations of the set (2.2).

To discuss the experiment more quantitatively, we fix a reference frame with axes x_1 and x_2 parallel to the mid–surface of the film, and axis x_3 along the thickness direction. The geometry of the energetically optimal microstructures resolving (4.1) entails two orders of lamination. One is along the film thickness, to accommodate the fact that at temperatures below T_{IN}, the natural thickness associated with a vertically oriented director is larger than the distance between the two glass plates. The second lamination, in the plane of the film, is the one that should be responsible for the observed contrast. Were this second order of lamination absent or, said differently, were the director to buckle while remaining in one plane (say, the x_1x_3 plane), then the film would have to contract (by the amount $a^{1/6} < 1$) in the direction of x_2. Since this contraction is incompatible with the constraints introduced by the experimental apparatus, in–plane stripe–domains are generated.

To test this hypothesis, we looked for the possibility of resolving (4.1) with exactly four deformation gradients lying in the set of zero energy spontaneous distorsions (2.2). The geometry of the construction is sketched in Figure 4. Notice that, in the layers-within-layers construction of Figure 1, fine layers are nested inside coarser layers and kinematic compatibility across the interfaces of the coarse layers holds only approximately. By contrast, in Figure 4, kinematic compatibility across all interfaces is satisfied exactly, see formulas (4.2)-(4.5) below. More in detail, we set $\eta = a^{-1/6}$ and consider the following deformation gradients

$$\mathbf{F}_{11} = \begin{pmatrix} 1 & 0 & \frac{1}{\eta} - \eta \\ \eta - \frac{1}{\eta} & 1 & 1 - \eta^2 \\ 0 & 0 & 1 \end{pmatrix} \quad \mathbf{F}_{22} = \begin{pmatrix} 1 & 0 & \frac{1}{\eta} - \eta \\ -\eta + \frac{1}{\eta} & 1 & \eta^2 - 1 \\ 0 & 0 & 1 \end{pmatrix}$$

$$\mathbf{F}_{12} = \begin{pmatrix} 1 & 0 & \eta - \frac{1}{\eta} \\ \eta - \frac{1}{\eta} & 1 & \eta^2 - 1 \\ 0 & 0 & 1 \end{pmatrix} \quad \mathbf{F}_{21} = \begin{pmatrix} 1 & 0 & \eta - \frac{1}{\eta} \\ -\eta + \frac{1}{\eta} & 1 & 1 - \eta^2 \\ 0 & 0 & 1 \end{pmatrix}$$

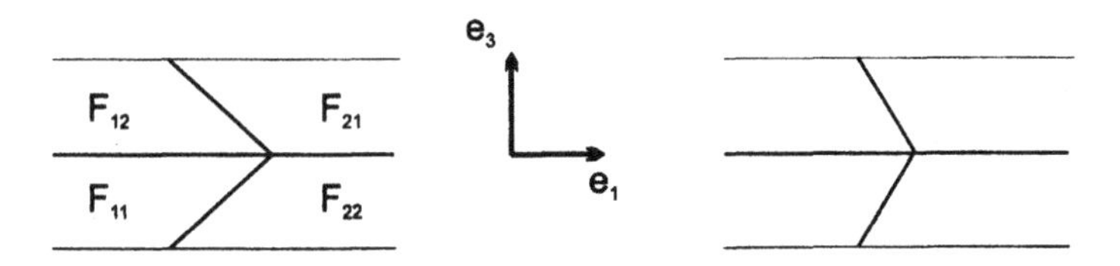

FIG. 4. *Geometry of our zero-energy four gradient construction: reference (left) and deformed (right) configurations for* $\eta = 1.1225$ *corresponding to* $a = 0.5$.

Clearly,

$$\mathbf{F}_{11} - \mathbf{F}_{12} = -2(\eta - \frac{1}{\eta}, \eta^2 - 1, 1) \otimes \mathbf{e}_3 \tag{4.2}$$

$$\mathbf{F}_{22} - \mathbf{F}_{21} = -2(\eta - \frac{1}{\eta}, 1 - \eta^2, 1) \otimes \mathbf{e}_3 \tag{4.3}$$

so that interfaces between $\mathbf{F}_{11}$ and $\mathbf{F}_{12}$, and between $\mathbf{F}_{22}$ and $\mathbf{F}_{21}$ are possible, with normal parallel to $\mathbf{e}_3$ in the reference configuration. Moreover,

$$\mathbf{F}_{21} - \mathbf{F}_{12} = -2\left(\eta - \frac{1}{\eta}\right)\mathbf{e}_2 \otimes (1, 0, \eta) \tag{4.4}$$

$$\mathbf{F}_{22} - \mathbf{F}_{11} = -2\left(\eta - \frac{1}{\eta}\right)\mathbf{e}_2 \otimes (1, 0, -\eta) \tag{4.5}$$

so that an interface between $\mathbf{F}_{21}$ and $\mathbf{F}_{12}$ is possible with reference normal parallel to $(1, 0, \eta)$, and an interface between $\mathbf{F}_{22}$ and $\mathbf{F}_{11}$ is possible with reference normal parallel to $(1, 0, -\eta)$.

We now plot the deformed configuration of the film, see Figure 4 and Figure 5, using the value $\eta = 1.1225$ corresponding to $a = 0.5$. For this purpose, it is useful to compute the deformed orientation $\mathbf{F}^*\mathbf{N}$ of each layer normal $\mathbf{N}$, where $\mathbf{F}^* = (\det \mathbf{F})\mathbf{F}^{-T}$ is the cofactor of the deformation gradient inside the layer. Since

$$\mathbf{F}^*_{11} = \begin{pmatrix} 1 & -\xi & 0 \\ 0 & 1 & 0 \\ \xi & -\Lambda & 1 \end{pmatrix} \quad \mathbf{F}^*_{22} = \begin{pmatrix} 1 & \xi & 0 \\ 0 & 1 & 0 \\ \xi & \Lambda & 1 \end{pmatrix}$$

$$\mathbf{F}^*_{12} = \begin{pmatrix} 1 & -\xi & 0 \\ 0 & 1 & 0 \\ -\xi & \Lambda & 1 \end{pmatrix} \quad \mathbf{F}^*_{21} = \begin{pmatrix} 1 & \xi & 0 \\ 0 & 1 & 0 \\ -\xi & -\Lambda & 1 \end{pmatrix}$$

where $\xi = \eta - 1/\eta$ and $\Lambda = 1 - \eta^2 + (\eta - 1/\eta)^2$, we have

$$\mathbf{F}^*_{11}\mathbf{e}_3 = \mathbf{F}^*_{12}\mathbf{e}_3 = \mathbf{F}^*_{21}\mathbf{e}_3 = \mathbf{F}^*_{22}\mathbf{e}_3 = \mathbf{e}_3 \,, \tag{4.6}$$

$$\mathbf{F}^*_{12}(1, 0, \eta) = \mathbf{F}^*_{21}(1, 0, \eta) = (1, 0, \frac{1}{\eta}) \,, \tag{4.7}$$

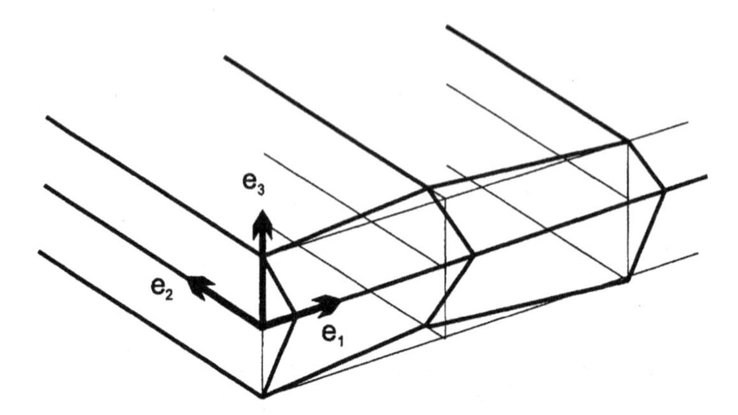

FIG. 5. *Sketch of a domain pattern resolving* $\mathbf{F} = \mathrm{Id}$. *This models the possible behavior of a film undergoing the isotropic–nematic phase transformation while being constrained between two parallel plates.*

and

$$\mathbf{F}^*_{11}(1,0,-\eta) = \mathbf{F}^*_{22}(1,0,-\eta) = (1,0,-\frac{1}{\eta}). \tag{4.8}$$

For what concerns the displacement of material particles, notice that the relative motion of the layer deformed according to $\mathbf{F}_{22}$ (respectively, $\mathbf{F}_{21}$) relative to that deformed according to $\mathbf{F}_{11}$ (respectively, $\mathbf{F}_{12}$) is a shear of amplitude $2\sqrt{1+\eta^2}(\eta - 1/\eta)$ in the $\mathbf{e}_2$ direction. This follows from formulas (4.4) and (4.5), by normalizing the modulus of the interface normals $(1,0,\pm\eta)$ to one. The resulting deformation is sketched in Figure 5, where attention should be paid to the alternating shears in the plane x_1x_2 which are responsible for the observed stripe pattern.

Finally, we give the orientations of the nematic director implied by our four–gradient construction. As noted above, within our model, the nematic director is always aligned with the direction of maximal stretch. This implies that, in the deformed configuration, the orientation $\bar{\mathbf{n}}$ of the director will be that of the eigenvector of $\mathbf{F}\mathbf{F}^T$ associated with its largest eigenvalue $\lambda_3^2(\mathbf{F})$. We thus obtain the following relations of proportionality

$$\bar{\mathbf{n}}_{11} \propto \begin{pmatrix} -\eta \\ -\eta^2 \\ 1 \end{pmatrix} \quad \bar{\mathbf{n}}_{22} \propto \begin{pmatrix} -\eta \\ \eta^2 \\ 1 \end{pmatrix}$$

$$\bar{\mathbf{n}}_{12} \propto \begin{pmatrix} \eta \\ \eta^2 \\ 1 \end{pmatrix} \quad \bar{\mathbf{n}}_{21} \propto \begin{pmatrix} \eta \\ -\eta^2 \\ 1 \end{pmatrix}.$$

A more detailed analysis of zero energy deformation patterns compatible with (4.1) and, more generally, of the stripe patterns that may arise within the confined geometry described at the beginning of this section will appear elsewhere [17].

5. Conclusions and Outlook. Nematic elastomers have been synthesized relatively recently [18]-[21]. Nevertheless, they have already attracted considerable attention in the Chemistry, Engineering, Mathematics, and Physics literature. The same is true, in particular, for the stripe–domain patterns they exhibit.

Our interest in nematic elastomers arose from the realization that the symmetry–breaking isotropic–to–nematic phase transformation which is at the root of their fascinating material instabilities has close analogies with the martensitic phase transformations exhibited by shape–memory alloys. While in the latter case, however, the underlying material symmetry is the discrete crystallographic symmetry of the austenitic parent phase, in the case of nematic elastomers the full isotropic symmetry of the high temperature amorphous polymer is available. It soon became apparent that the mathematical techniques developed for the study of displacive phase transformations in crystals are applicable to a radically different class of systems (polymers, rather than crystals) and that the simplifications accompanying the enhanced material symmetries lead to results of unprecedented completeness. One such result is the development of a combined analytical–computational approach in which the original problem is first simplified with the use of mathematical analysis, and then attacked computationally. As described above, this approach has been used with some success to simulate numerically stretching experiments of thin sheets of nematic elastomers. We believe that this combination of analysis and computation has a great potential in shedding further light on the mechanical response of nematic elastomers and, more generally, of all systems whose mechanical response is microstructure–driven.

The analogy between shape memory alloys and nematic elastomers as shape–memory polymers is fruitful. It reveals to us the underlying structure, hence the ultimate simplicity of seemingly complicated stripe–domain patterns. And yet, trivially, analogy is not identity. The devil (just like system specificity coming from the different underlying physical mechanisms) is in the details. Examples are the detailed structure of a wall between two adjacent domains, or the geometric details of how two differently oriented systems of stripe-domains meet, or the time scales with which domains appear, disappear, and respond to external fields and loads. It is through such details that the complexity of these seemingly simple domain patterns really emerges. And we should be prepared to use sharper (but, unavoidably, more complicated and system specific) models to understand them.

Acknowledgments. This paper draws on an ongoing collaboration with S. Conti (University of Duisburg-Essen) and it was written during visits of both authors to the IMA, and of the second author to SISSA.

The financial support of both Institutions is gratefully acknowledged. GD acknowledges also support through the NSF via grant DMS-0405853.

REFERENCES

[1] WARNER M. AND TERENTJEV E.M., Liquid Crystal Elastomers, Clarendon Press, Oxford 2003.
[2] IMA WORKSHOP ON MODELING OF SOFT MATTER, University of Minnesota, Minneaplis, September 2004.
[3] BLADON P., TERENTJEV E.M., AND WARNER M., 1993. Transitions and instabilities in liquid-crystal elastomers. Phys. Rev. E **47**: R3838–R3840.
[4] FINKELMANN H., KUNDLER I., TERENTJEV E.M., AND WARNER M., 1997. Critical stripe-domain instability of nematic elastomers. J. Phys. II France **7**: 1059–1069.
[5] FRIED E. AND KORCHAGIN V., 2002. Striping of nematic elastomers. Int. J. Solids Structures **39**: 3451–3467.
[6] MARTINOTY P., STEIN P., FINKELMANN H., PLEINER H., AND BRAND H.R., 2004. Mechanical properties of monodomain side chain nematic elastomers, Eur. Phys. J. E **14**: 311.
[7] VERWEY G.C., WARNER M., AND TERENTJEV E.M., 1996. Elastic instability and stripe domains in liquid crystalline elastomers. J. Phys. II France **6**: 1273–1290.
[8] WEILEPP J. AND BRAND H., 1996. Director reorientation in nematic–liquis–single–crystal elastomers by external mechanical stress. Europhys. Lett. **34**: 495–500.
[9] DESIMONE A., 1999. Energetics of fine domain structures. Ferroelectrics **222**: 275–284.
[10] DESIMONE A. AND DOLZMANN G., 2000. Material instabilities in nematic elastomers. Physica D **136**: 175–191.
[11] MÜLLER S., 1999. Variational models for microstructure and phase transitions. In: Bethuel F., Huisken G., Müller S., Steffen K., Hildebrandt S., Struwe M. (Eds.), Calculus of Variations and Geometric Evolution Problems, Lectures given at the 2nd Session of the Centro Internazionale Matematico Estivo, Cetraro 1996. Springer, Berlin.
[12] DESIMONE A. AND DOLZMANN G., 2002. Macroscopic response of nematic elastomers via relaxation of a class of SO(3)-invariant energies. Arch. Rat. Mech. Anal. **161**: 181–204.
[13] CONTI S., DESIMONE A., AND DOLZMANN G., 2002. Semi–soft elasticity and director reorientation in stretched sheets of nematic elastomers. Phys. Rev. E **60**: 61710-1–8.
[14] KUNDLER I. AND FINKELMANN H., 1995. Strain-induced director reorientation in nematic liquid single crystal elastomers. Macromol. Rapid Comm. **16**: 679–686.
[15] CONTI S., DESIMONE A., AND DOLZMANN G., 2002. Soft elastic response of stretched sheets of nematic elastomers: a numerical study. J. Mech. Phys. Solids **50**: 1431–1451.
[16] ZUBAREV E.R., KUPTSOV S.A., YURANOVA T.I., TALROZE R.V., AND FINKELMANN H., 1999. Monodomain liquid crystalline networks: reorientation mechanism from uniform to stripe domains. Liquid Crystals **26**: 1531–1540.
[17] DESIMONE A. AND DOLZMANN G., 2005. In preparation.
[18] FINKELMANN H., AND REHAGE G., 1984. Liquid crystal side chain polymers. Adv. Polymer Sci. **60/61**: 99.
[19] FINKELMANN H., GLEIM W., KOCK H.J., AND REHAGE G., 1985. Liquid crystalline polymer network - Rubber elastic material with exceptional properties. Makromol. Chem. Suppl. **12**: 49.

[20] KÜPFER J. AND FINKELMANN H., 1991. Nematic liquid single-crystal elastomers. Makromol. Chem. Rapid Comm. **12**: 717–726.
[21] ZENTEL R., 1989. Liquid crystal elastomers. Angew. Chem. Adv. Mater. **101**: 1437.

NUMERICAL SIMULATION FOR THE MESOSCALE DEFORMATION OF DISORDERED REINFORCED ELASTOMERS

DIDIER LONG* AND PAUL SOTTA*

Abstract. We study here the dynamical behavior of disordered elastic systems such as gels or filled elastomers, by dissipative molecular dynamics. We show that applied macroscopic deformations result in non-affine deformations at the scale of the filler particles. These non-affine deformations lead to slow meso-scale reorganizations, which could explain the long relaxation times measured in gels, and also in rubbers even at temperatures much above the glass transition temperature.

Key words. Statistical mechanics, Polymers, Nonlinear elasticity, Plastic materials, Materials with memory.

AMS(MOS) subject classifications. 82D60, 82C61, 74B20, 74C15, 74D10.

1. Introduction. Microscopic mechanisms at the origin of macroscopic elasticity in disordered systems [1] and in particular in systems such as gels or rubbers [2–6] have been the subject of many debates over the past 50 years. Gels or rubbers are made of cross-linked networks of polymer chains. When deforming a sample, the strands between cross-links are stretched which results in a decrease of entropy, and thus in a free energy cost. This entropic origin of the elastic properties of gels or rubbers is no longer disputed. On the other hand, a precise, microscopic description of the strand network deformation under shear has long been elusive. To overcome this difficulty, the classical models developed in the polymer literature for describing rubber elasticity have assumed affine deformation down to the strand scale. A number of experiments, such as small angle neutron scattering or light scattering experiments, have been performed in order to test this assumption, and it has been demonstrated that the deformations in gels are indeed *not* affine on this scale [4, 5, 7–10]. Many theoretical attempts have been made to go beyond the affine deformation assumption in gel or rubber elasticity. De Gennes proposed that the sol-gel transition is analogous to a percolation transition, and that the shear modulus close to the gelation critical point behaves like the electric conductivity in conductor percolation problems [11, 12]. However, Feng and Sen have shown that this is not the case when considering only central forces between cross-links [13]. The analogy can be drawn only when bending energy comes into play, an assumption for which there is no ground in gel or rubber elasticity. For the role of both stretching and bending energies, one can see e.g. the work by Arbabi and Sahimi [14]. These differences between geometric and

*Laboratoire de Physique des Solides, Universite de Paris SUD/CNRS, Bat. 510, 91405 Orsay Cedex, FRANCE.

rigidity percolation result in a higher percolation threshold and different values of the critical exponents in rigidity percolation.

The inhomogeneity in the local elastic modulus is an essential feature in describing non-affine deformation at a local scale. Finite element mapping with spring network representations is a model suitable to describe elastically inhomogeneous materials [15]. However, the dynamical behavior and large deformations have not been investigated within this framework. On the other hand, several computational studies on disordered elastic systems have addressed the question of non-affine deformations [16]. Non-affine deformation processes have been invoked to interpret yielding in colloidal gels. Brownian dynamics simulation have been performed, but such systems were not *permanent* elastic networks [17], contrarily to those of interest here. On the other hand, it has been observed recently in molecular dynamics studies that continuum elasticity breaks down at some spatial scale quite large with respect to the average interparticle distance, and this effect has been interpreted in terms of non affine displacements in the system [18, 19]. However, this study has been done in two dimensions and for very small amplitudes only.

Another important feature is the important role played by excluded volume [20, 21] and by topological constraints distinct from cross-links, that is entanglements. Indeed, polymer strands cannot cross each other, which limits their lateral spatial fluctuations and contribute to the elastic moduli of the samples. This feature makes the description of real gels or rubbers even more complicated, since a precise description of entanglements based on first principles is still lacking. Indeed, it has been demonstrated recently that topological constraints are essential in understanding the sol-gel transition [22, 23]. This can be understood qualitatively as follows: in polymer melts made of high molecular weight polymers, a plateau modulus is observed at intermediate frequencies. This is the well known result of entanglements. Upon cross-linking the melt, these entanglements are made permanent with only a small fraction of cross-links [22–25], especially in the case of very long chains. Therefore, a description of rubber or gel elasticity should take into account both entanglements and rigidity percolation, as well as their combined effects.

A further difficulty in describing gel or rubber elasticity is that these systems are intrinsically disordered and inhomogeneous, as the result of their preparation [26, 27]. The cross-linking process not only freezes in the disorder present in the melt, but it tends to enhance this disorder: during vulcanization or cross-linking, the more a region is cross-linked, the more it tends to collapse, which further enhances cross-linking in this region. This results in large scale heterogeneities which are responsible for turbidity in gels, for instance, and which have been evidenced experimentally [4]. Elastic heterogeneity may occur as well even in so called "model" networks obtained by end-to-end cross-linking of precursor chains with a controlled polymolecularity [28]. Another source of large scale disorder is due to the addition of solid inorganic particles (the so-called charges) acting as reinforcing fillers. Indeed, non-reinforced polymer matrices generally do not exhibit mechanical properties suitable for practical purposes, being too soft and fragile [29]. On the contrary, elastomers filled with carbon black or silica particles have a shear modulus much (up to 100 times) higher than that of the pure elastomer, exhibit a high dissipative efficiency, which makes them useful for damping materials, and are extremely resistant both to fracture and abrasion [29–35]. The typical diameters of these filler particles vary between 10 nm and 100 nm. Moreover, they often form fractal aggregates and/or agglomerates at even larger scales [36]. The presence of filler particle thus introduces another source of disorder on a much larger scale than the crosslinking process of pure rubbers. Deformations at the scale of polymer chains have been investigated in such systems by SANS. An average enhancement of the strain has been shown at this scale [37].

In this work, we propose a model for describing elastic properties of filled elastomers. The scale of interest here is larger than the distance between crosslinks or entanglements, which is usually considered in gels or rubbers. It is typically of order 100 nm, which corresponds to the diameter of the filler particles [29, 38, 39]. Our model must therefore be considered as a coarse-grained model aimed at describing meso-scale relaxation processes in such systems. We focus on the high temperature regime, in which the whole polymer matrix is in the rubbery state, as opposed to the lower temperature regime in which the polymer matrix is partially glassy [38–41]. In this high temperature regime, reinforcement effects are weaker than at lower temperatures, but they are still important [38, 39]. We consider therefore a rubber matrix with randomly dispersed solid particles, with strong anchorage of the matrix on the particles. We consider the matrix as being a highly cross-linked rubber, which in practical cases corresponds to a shear modulus from 10^5 Pa to 10^6 Pa. We consider typical filler volume fractions of 20 % or more. The purpose of this work is to describe the meso-scale behavior of such a system when submitted to imposed deformations. We study here both static and dynamical properties by dissipative molecular dynamics. The paper is organized as follows. We first describe in details the model and the procedure used to obtain a system at mechanical equilibrium. Then, we show that this system exhibits non-affine deformations and long time relaxations.

2. Description of the model. The basic ingredients we want to implement in the simulated system are permanent elasticity, disorder and excluded volume effects. While disorder creates a complex energy landscape with a ill-defined minimum energy state, excluded volume may introduce irreversibility in the system. The solid particles are modelized by hard spheres randomly distributed in space. The hard sphere potential is described by:

$$V_{\text{hs}}(r^*) = \begin{cases} F_{\text{hs}}^{\max} r^* + H & (r^* < r_{\min}) \\ \epsilon^* r^{*-12} & (r_{\min} < r^* < r_{\text{cut}}) \\ 0 & (r_{\text{cut}} < r^*) \end{cases} \tag{2.1}$$

where $r^* = r/\sigma$ is the reduced dimensionless distance between particles. Though we will keep it in our notations, σ should be considered to be the unit length in the problem and takes the value 1. The parameter $\epsilon^* = 1$ determines the energy scale in the system. The cutoff distance beyond which F_{hs} cancels is $r_{\text{cut}} = 2$. At this distance, the force F_{hs} is already much smaller than ϵ^*. The force is limited at short distances to prevent numerical instabilities in the initial step of the simulations, in which the centers of two particles may be very close. The constant H is chosen to insure the continuity of the potential. The force $F_{\text{hs}}(r^*)$ rises very sharply at r^* of the order one. For instance, it is already of the order 20 (in units of ϵ^*) for $r^* \approx 0.96$. This means that it is quite realistic to consider σ as the particle diameter.

To represent the effect of the rubber matrix, the particle centers are connected by harmonic springs, with an elastic interaction potential given by:

$$V_{\text{el}}(r^*) = \frac{k^*}{2}\left(\frac{r^*}{l_0^*} - 1\right)^2 \tag{2.2}$$

where $k^* = k\sigma^2 l_0^2/\epsilon^*$ is the reduced spring stiffness (in units of ϵ^*) and l_0^* is the equilibrium length of the spring (in units of σ). Thus, the particles interact with a hard-core repulsion and the elastic springs. We assume that the degrees of freedom, which are the centers of mass of the hard spheres, move relative to each other with a friction coefficient ζ. The hydrodynamic friction is computed within a mean field approximation:

$$\vec{F}_{\text{Hydro}} = -\zeta^*(\vec{v}^* - < \vec{v}^* >) \tag{2.3}$$

where $\zeta^* = \xi\sigma^2/\epsilon^* = 1$ is the reduced friction coefficient, which has the dimension of a time and thus sets the time scale in the system. $\vec{v}^*$ is the reduced velocity of a particle and $< \vec{v}^* >$ the average reduced velocity of the surrounding particles (velocities have the dimension of an inverse time). To save computational time, the hydrodynamic friction is computed with respect to the average affine deformation rate rather than to the actual motion relative to the surrounding of a particle.

Periodic boundary conditions are used in order to simulate bulk-like behavior. This means that a spring emanating from a particle close to a boundary and pointing out of the box is identified with a symmetrical one coming into the box and acting on another particle close to the opposite side of the box. To prepare the system, N particles are dispersed at random in a box of volume $V = L^3$ such that the volume fraction takes the chosen value Φ, that is:

$$L^* = \frac{L}{\sigma} = \left(\frac{\pi N}{6\Phi}\right)^{1/3} \tag{2.4}$$

The volume V of the box is kept constant throughout the simulations. The average distance between particles is:

$$a^* = \frac{a}{\sigma} = \left(\frac{\pi}{6\Phi}\right)^{1/3}. \tag{2.5}$$

For an average number of connections per particle n, the total number of springs in the system is $Nn/2$ (one spring contributes to two connections). The $Nn/2$ closest particle pairs are connected by springs. The equilibrium length l_0^* is set equal to the average distance between neighboring sites on an ordered simple cubic lattice, that is $l_0 = \frac{1}{n}(6 + (n-6)\sqrt{2})a$ (for $6 < n < 18$), where a is the lattice parameter corresponding to the volume fraction Φ, given by $a^3 = \pi/6\Phi$. This value for l_0 is quite arbitrary. The properties of the systems which will be investigated do not depend on this particular choice. Only the value of the pressure which has to be applied to maintain the volume constant, i.e. the isotropic part of the stress tensor, depends noticeably on l_0. The quantitative value of the shear modulus depends also on l_0, though much more weakly.

Thus, the parameters which are relevant to describe the system and its temporal evolution are the following ones:

- the volume fraction Φ of the solid particles
- the degree of connectivity n, defined as the average number of springs connected to one particle. Note that n does not have to be an integer.
- the spring stiffness k
- the friction coefficient ζ.

The model system considered here is disordered and only includes potentials which depend on the distances between particles, without bending energies. Since we aim at representing highly cross-linked rubber matrices, we shall only consider relatively large values of the connectivity n. Indeed, there is a threshold n_c below which the rubber matrix would become floppy, that is below which the elastic modulus μ would cancel. On the other hand, far above this threshold, the elastic modulus μ is of the order fk, where f is the number of springs per unit volume and k their stiffness. The elastic modulus should then decrease much faster on approaching n_c, then cancel at and below n_c. n_c has been estimated for some ordered lattices [42–50]. It is slightly above 6 in a simple cubic lattice, and much lower (close to 4) in a diamond lattice. We shall consider values of n larger than 8 typically in our simulations. Values of n close to or lower than 6 would correspond to solid particles imbedded in a loosely cross-linked gel or rubber, which is not our purpose here. We emphasize once again that the elastic springs in our model represent the elastic interaction between filler particles due to the elasticity of the rubber matrix as a whole, not the strands between cross-

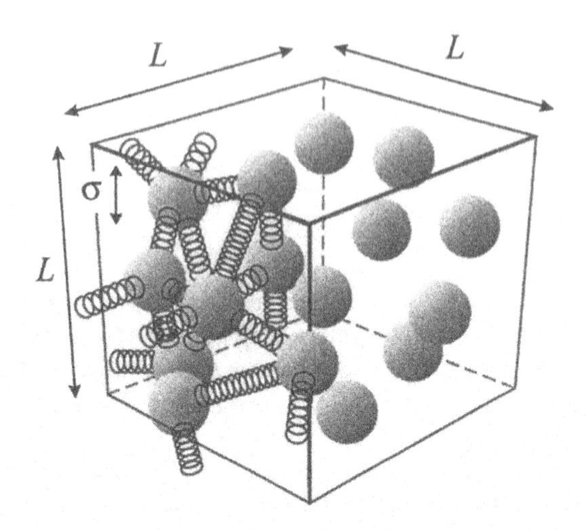

FIG. 1. *Schematics of our model for filled rubbers. Filler particles are connected by elastic springs and interact via a hard-core potential at short distances. σ is the particle diameter. For clarity, only a fraction of the springs have been drawn.*

links. Indeed, we consider the problem at the scale of the filler particles, which is much larger than the typical distance between crosslinks.

2.1. Dissipative molecular dynamics. The equations of the dissipative molecular dynamics are non inertial and include a source of dissipation in the form of a hydrodynamic friction term. The equation of motion for particle i is thus:

$$\vec{F}^i_{\rm el} + \vec{F}^i_{\rm hs} + \vec{F}^i_{\rm Hydro} = 0 \tag{2.6}$$

which gives the velocities at time t as a function of the positions $\vec{r}^i$, within the mean field approximation (see Equation (2.3)):

$$\vec{v}^i = <\vec{v}> + \frac{1}{\zeta}\left[\vec{F}^i_{\rm el} + \vec{F}^i_{\rm hs}\right] . \tag{2.7}$$

The positions and velocities are computed every time interval dt. The equations of motion are solved using the Modified Midpoint Method (MMM). The time interval dt is divided in p sub-steps such that $ddt = dt/p$, and the equations are:

$$\begin{array}{lll}
step & 0: & \vec{r}^0_i = \vec{r}^i(t) \\
step & 1: & \vec{r}^1_i = \vec{r}^0_i + \vec{v}_i(\{\vec{r}^0_j\})ddt \\
step & m+1: & \vec{r}^{m+1}_i = \vec{r}^{m-1}_{i+2}\, \vec{v}_i(\{\vec{r}^m_j\})ddt, \quad m = 1,..,p-1 \\
step & p: & \vec{r}_i(t+dt) = \frac{1}{2}(\vec{r}^{p-1}_i + \vec{r}^p_i + \vec{v}_i(\{\vec{r}^p_j\})ddt) \, .
\end{array}$$

2.2. Time scale. In the simulations, the time scale is fixed by the value of the reduced friction coefficient $\zeta^* = 1$. This time scale corresponds to the typical relaxation time of a particle. The relaxation times measured in the system should thus be compared to this typical relaxation time. Unless explicitly specified, times will be expressed in units of ζ^* denoted "sec". In real systems, the elementary time scale is determined by the rubber matrix. In the ideal case of a densely cross-linked network, and in the considered regime of high temperature, the time scale is the so-called Rouse relaxation time [2, 51]. Note however that, even in this case, real rubbers do not exhibit exponential relaxation, but display slower relaxations often described as power laws. This is thought to be a consequence of the intrinsic disorder of these systems [2, 52]. In this paper, we assume that the rubber matrix corresponds to a highly cross-linked rubber, which can be well described at long times by a single relaxation time, which is its Rouse relaxation time.

2.3. Deformation of an isotropic solid. Virial stress formulation. In an elastic isotropic body, the stress tensor $\sigma_{\alpha\beta}$ is related to the strain tensor by the constitutive equation:

$$\sigma_{\alpha\beta} = K\delta_{\alpha\beta}u_{ll} + 2\mu(u_{\alpha\beta} - \frac{1}{3}\delta_{\alpha\beta}u_{ll}) \tag{2.8}$$

where K is the bulk (compression) modulus and μ the shear modulus [53]. Consider a box of volume V^* containing N particles. The stress tensor is related to the forces exerted on the particles by the Kramers Kirkwood formula, which provides a microscopic expression for the stress tensor [51]:

$$\sigma_{\alpha\beta} = -\frac{1}{V^*}\sum_i F^i_\alpha R^i_\beta \tag{2.9}$$

where F^i_α is the α-component of the sum of the forces exerted on particle i by other particles *of the considered sample* and R^i_β is the β-component of the position of particle i. Periodic boundary conditions in the simulation box ensures that:

$$\sum_i F^i_\alpha = 0\,. \tag{2.10}$$

These boundary conditions and the fact that we consider only central forces, ensures that the total torque satisfies

$$\vec{T} = \sum_i \vec{R}^i \wedge \vec{F}^i = 0 \tag{2.11}$$

(see Appendix).

3. Preparation of the system. The purpose of the preparation steps is to obtain an equilibrated sample, in which the total force acting on each particle is zero and the stress tensor is isotropic. Just after the particles have been randomly distributed, none of these conditions is satisfied. We will have therefore to let the sample equilibrate. First, we will cancel the forces (or the velocities) on each particle, and then cancel the non-isotropic part of the stress tensor. In an infinite system, the stress tensor just after the particles have been distributed randomly should be of the form:

$$\sigma_{\alpha\beta} = -pI_{\mathrm{d}} \tag{3.1}$$

where p is the pressure, and I_{d} is the identity tensor. In a finite simulation box, $\sigma_{\alpha\beta}$ is of the form:

$$\sigma_{\alpha\beta} = -pI_{\mathrm{d}} + \tilde{\sigma}_{\alpha\beta} \ . \tag{3.2}$$

The non-isotropic part $\tilde{\sigma}_{\alpha\beta}$ is a traceless tensor. Since the torque acting on the system is zero, this tensor is symmetric [53, 54]. There is therefore five degrees of freedom to adjust in order to obtain a system with an isotropic (isostatic) stress tensor. One can show (see Appendix) that the variance of the tensor $\tilde{\sigma}_{\alpha\beta}$ scales like $N^{-2/3}$ just after the particles has been randomly distributed. In a sample with 10000 particles, this tensor is of order 0.1 and is therefore not negligible as compared to the pressure, which is of order 1.

3.1. First initialization step. This first step consists in canceling the particle velocities, or equivalently the potential forces exerted on each particle. This step is performed at constant shape. The time resolution *ddt* chosen to solve the equations of motion must be such that $vddt \ll r_0$, where v is the velocity of the particles and r_0 is the typical distance between particles. The initial velocity may be of the order 10 to 100. With $r_0 \sim 0.1$, this imposes $ddt \approx 10^{-4}$. We typically chose $dt = 10^{-3}$ and $p = 10$, which gives $ddt = dt/p = 10^{-4}$. In this step, the system is relaxed typically during 0.2 to 0.4 unit of time, which corresponds to 200 to 400 *dt* steps. At the end of this first step, the average velocity is not rigorously zero, but it has dropped to a value of the order unity at most. A predefined precision parameter value ϵ_v may be specified as well to stop the first relaxation step as soon as $< v > \leq \epsilon_{\mathrm{v}}$.

3.2. Second initialization step. At the end of the first step, the deviation of the stress tensor, $\tilde{\sigma}_{\alpha\beta}$, is still of order $N^{-1/3}$. To obtain a reference sample with isotropic stress tensor, the sample must now be deformed so as to cancel the residual stress tensor. It is important to notice that the non-isotropic stress tensor elements must be canceled with a high degree of precision (typically 10^{-4}), so that the mechanical response (such as the shear modulus) can be measured with a good accuracy. Note also that all deformations performed take place at constant volume, in order to represent the behavior of real rubbers whose bulk modulus is very large as compared to the shear modulus. The following iterative deformation process is used to cancel $\tilde{\sigma}$. At a given step, a small deformation tensor of the form

$$I_{\mathrm{d}} + d\Gamma_{\alpha\beta} = I_{\mathrm{d}} - C\delta\sigma_{\alpha\beta} \tag{3.3}$$

where $\delta\sigma_{\alpha\beta} = \sigma_{\alpha\beta} - \sigma_{\mathrm{iso}}\delta_{\alpha\beta}$ with $\sigma_{\mathrm{iso}} = \mathrm{Tr}\,\sigma_{\alpha\beta}/3$ is applied, so that the new deformation tensor $\Gamma_{\alpha\beta}(t+dt)$ is given by:

$$\Gamma_{\alpha\beta}(t+dt) = \frac{(I_{\mathrm{d}} + d\Gamma_{\alpha\beta})\Gamma_{\alpha\beta}(t)}{[\det(I_{\mathrm{d}} + d\Gamma_{\alpha\beta})]^{1/3}} \tag{3.4}$$

C is a positive number of order 0.1 typically. A new step of equilibration of the particles (as described in the first initialization procedure) is then applied for a duration Δt. This procedure is iterated until all components of $\tilde{\sigma}_{\alpha\beta}$ are zero to the prescribed precision ϵ_{σ}. Note that the dynamics of this relaxation process, namely the stress relaxation rate, is not intrinsic, but is determined by the ratio $C/\Delta t$, which has then to be optimized in order to give the best relaxation performance.

3.3. Third initialization step. Like real gels, the state of our samples depends on the way they are prepared. The purpose of this step is to remove internal instabilities which result from the preparation process, and thus to reduce the dispersion between the results of simulations performed on different samples. The samples are submitted to a few (typically 6 to 10) preliminary high amplitude elongation cycles. After performing these cycles, the second step is done again to cancel the residual stress. Non-isotropic elements of the stress are made smaller than typically 10^{-4}.

3.4. Shear experiments. After obtaining an equilibrated system along the lines described above, various experiments may be performed. The proper shear experiment is performed in the following way. The system is first sheared up to a maximum shear value γ_{max} at a constant shear rate $\dot{\gamma}$. This is done by imposing shear steps $d\gamma$ described by the deformation tensor

$$I_{\mathrm{d}} + d\Gamma_{\alpha\beta} = \begin{bmatrix} 1 & d\gamma & 0 \\ 0 & 1 & 0 \\ 0 & 0 & 1 \end{bmatrix} \tag{3.5}$$

followed by relaxation during Δt, such that $\dot{\gamma} = d\gamma/\Delta t$. This means that, during these elementary steps, the system is first affinely deformed, and then the particle positions are allowed to relax to new positions during the time Δt. A full description of the system may be stored in various states corresponding to different values of γ_{max}. After imposing a large amplitude deformation at a finite rate $\dot{\gamma}$, we let the stress relax in the sample while maintaining the system at the imposed deformation $\gamma_{\max}$. Alternatively, we may let the system come back to zero stress to achieve a large amplitude shear cycle, or perform an oscillatory shear.

3.5. Simulation times. The duration of the simulations depend on the number of time steps *dt*, or equivalently on the time over which the system is studied, given that $dt = 0.02$ has been chosen typically. In order to perform a full high amplitude shear cycle, the following steps must be achieved: first velocity and stress relaxation (3000 *dt* steps typically), initial elongation cycles (15000 *dt* steps), stress relaxation (15000 *dt* steps), shearing up to $\gamma_{\max} \approx 2$ at $\dot{\gamma} = 0.1$ (1000 *dt* steps), long time relaxation of the stress at constant $\gamma_{\max}$ (up to 50000 *dt* steps). Thus, each shear cycle performed on a new system typically corresponds to 80000 steps *dt*, or equivalently 1600 sec in terms of system unit time. The simulations have been performed on a cluster of four XEON bi-processor machines operating at 2.4 GHz. The systems simulated here have $N = 10000$ particles (which corresponds to a simulation volume of the order $21 \times 21 \times 21$ particles). This size has been chosen to insure reasonable simulation times. For this size, the performance of the machine corresponds roughly to 25 to 40 sec in terms of system time (depending on the connectivity n) per one hour computer time, which gives an idea of the overall duration of the whole experiment, that is 40 to 60 hours of computer time. On the other hand, performing an experiment (stress relaxation under large amplitude shear for example) on a system which has already been equilibrated represents about 24 hours. The results in 15 different systems (3 values of the connectivity n and 5 values of the volume fraction Φ) are presented here. In each system, the relaxation at constant shear was studied for typically 6 to 8 values of the shear amplitude. Typically two samples of each systems have been generated. All together, this corresponds roughly to 8000 hours computer time.

All the simulations presented here have been performed using the mean field approximation described in Equation (2.3), which allows to solve the dissipative molecular dynamics equations directly using Equation (2.6). In the general case where the friction term would depend on the particle velocity v_i relative to its neighbors, an (almost empty) $N \times N$ matrix would have to be inverted to solve the equations. The simulation time would then increase like N^2 instead of N. Indeed, this method has also been implemented. However, to obtain simulation times comparable to those mentioned above, the number of particles is limited to roughly 300 particles, which is obviously too small for the kind of problems addressed in this paper (non-affinity, long relaxation times due to long-ranged reorganisation) since it corresponds to $7 \times 7 \times 7$ particles only in the simulation box. On the other hand, at least in the low shear rate regime considered here, the results do not differ in a qualitative way and we believe that the essential phenomena are correctly described within the mean field approximation used here.

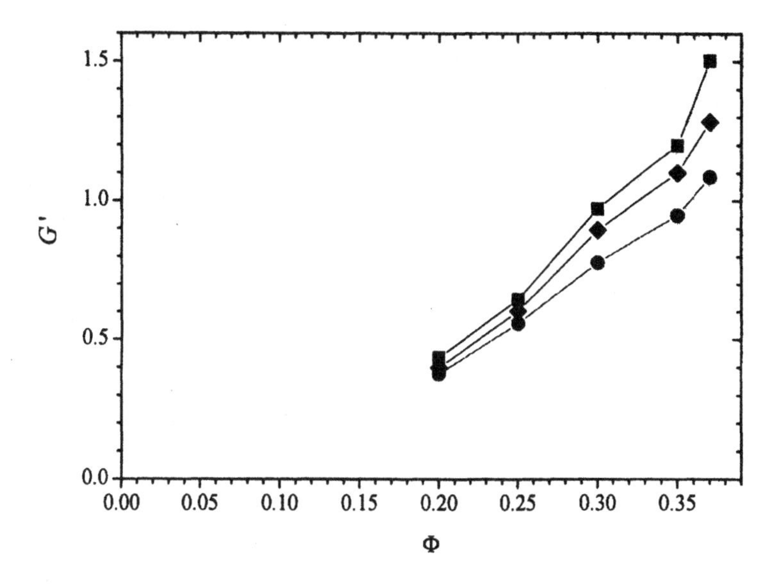

FIG. 2. *Oscillatory shear modulus, measured at an amplitude* $\gamma = 0.03$ *and frequency* $\omega/2\pi = 0.625$ sec^{-1} *as a function of the volume fraction* Φ*, for different values of the connectivity:* ■*:* $n = 10$ *;* ◆*:* $n = 9$ *;* •*:* $n = 8.5$*. The shear modulus is measured by Fourier transforming the strain and stress over 32 periods after waiting 8 periods after the beginning of the shear.*

4. Results. The main parameters describing the systems are the volume fraction Φ and connectivity n. Φ has been varied between 0.20 and 0.37 and n between 8.5 and 10. The reduced spring stiffness k^* is taken equal to 1. Systems with $N = 10000$ particles have been simulated. The size L of the simulation box varies from $L^* \approx 24.19$ (in units of σ) for $\Phi = 0.37$ to $L^* \approx 29.69$ for $\Phi = 0.20$. According to the different values of n and Φ, the spring equilibrium length varies from $l_0^* \approx 1.26$ to $l_0^* \approx 1.61$. The systems at equilibrium contain quenched forces because the spring lengths are quite widely distributed. This is a common feature of disordered systems. A fraction of springs have a length $l < l_0$, and thus are compressed. They correspond to the shell of nearest neighbors of a given particle. A fraction of springs are significantly stretched. They correspond to the shell of second nearest neighbors around a given particles. We have verified that the distribution of spring vectors is isotropic in the initial equilibrium state. In all experiments described here, the applied shear is of the type γ_{xy}, represented by a tensor with the symmetry as in Equation (3.5).

4.1. Dynamical characterization in the linear regime. In order to characterize the system in the linear regime, an oscillatory shear strain of the form $\gamma(t) = \gamma \sin \omega t$ has been applied. Both the frequency $\omega/2\pi$ and the amplitude γ may be varied. The oscillatory shear is applied during a time of the order 25 to 200 sec (in system time units), depending of the

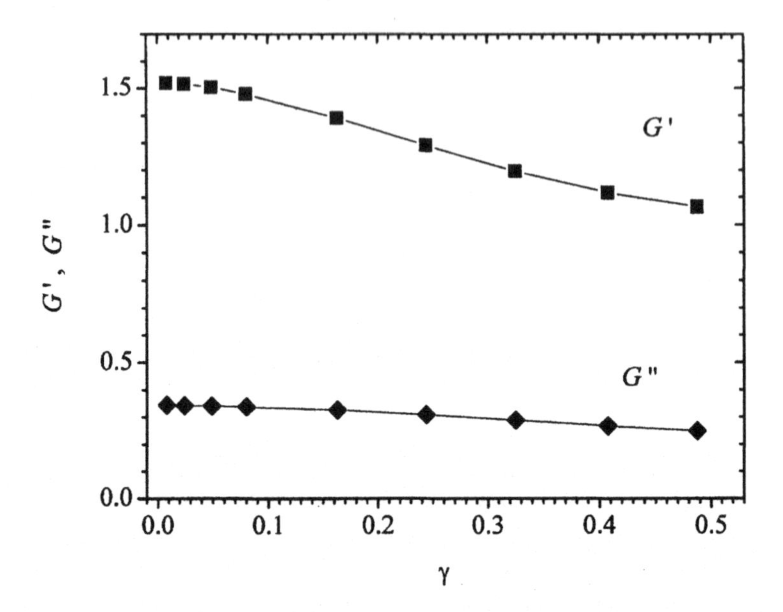

FIG. 3. *The complex shear modulus measured in an oscillatory shear* ($\omega/2\pi = 0.625$ *sec*$^{-1}$, *8 to 32 points per period, Fourier transform taken over 32 period after waiting 8 periods after the beginning of the shear) as a function of the shear amplitude, for* $n = 10$, $\Phi = 0.37$.

frequency and the amplitude, and the response is analyzed after waiting for typically 10 sec at least after the beginning of the oscillation, in order to be in a true steady state. 8 to 32 points per period are stored, depending on the amplitude. The response is analyzed by Fourier transforming both strain and stress over an integer number of periods. In this way, the complex shear modulus G', G'' may be measured as a function of the frequency ω and the amplitude γ, and the non-linear components in the response may also be detected. Note that the shear modulus μ defined by Equation (2.8) is related to the complex modulus in frequency $G(\omega) = G'(\omega) + iG''(\omega)$ by the relation $\mu = G'(\omega = 0)$ [2, 51]. Figure 2 shows the oscillatory shear modulus G' measured at an amplitude $\gamma \approx 0.03$ and frequency $\omega/2\pi = 0.625$ sec^{-1} as a function of the volume fraction Φ for different values of the connectivity n. This figure illustrates the reinforcement mechanism in the high temperature regime, as it was characterized in [38] and [39]. In these references, the shear moduli were measured in poly(ethyl acrylate) matrices filled with silica particles as a function of the frequency, temperature and particle volume fraction. In the high temperature regime ($T \gtrsim T_g + 100$ K typically), at a given (low) frequency, it was found that the modulus increases by a factor about 4 as the volume fraction increases from about 8% to about 20%. Thus, the results presented in Figure 2 are in good agreement with these observations. The complex shear modulus is plotted as a function of the shear amplitude γ in Figure 3, in the system $\Phi = 0.37$, $n = 10$. The modulus G' starts to decrease significantly at γ of the order 0.05, which can be considered as the onset of the non-linear regime.

4.2. Evolution of the energy in the system. In the disordered systems investigated, the phase space is very complicated. The system does not necessarily reach a uniquely defined minimum energy state in a finite time. To illustrate this effect of the disorder, the average energy in the system has been computed throughout the simulations. The contributions from the elastic springs and the hard-core potentials may be treated separately. The average elastic energy per spring is defined as:

$$\langle E_{\text{el}} \rangle = \frac{1}{N_s} \sum_j E_{\text{el}}^j \tag{4.1}$$

where N_s is the number of springs and the sum runs over the ensemble of springs. The average excluded volume energy is defined as:

$$\langle E_{\text{hs}} \rangle = \frac{1}{N_{\text{pair}}} \sum_m E_{\text{hs}}^m \tag{4.2}$$

where the sum runs over the ensemble of interacting pairs. The total energy per particle (including both contributions) has also been computed as:

$$\langle E_{\text{tot}} \rangle = \frac{1}{N} \left(\sum_{\text{spring}} E_{\text{el}}^j + \sum_{\text{pair}} E_{\text{hs}}^m \right). \tag{4.3}$$

Note that, with the given definitions, $\langle E_{\text{tot}} \rangle \neq \langle E_{\text{el}} \rangle + \langle E_{\text{hs}} \rangle$. The evolution of the energy throughout large amplitude stress/strain cycles is plotted in Figure 4. The cycles shown in Figure 4 consist first in shearing the system at a given shear rate $\dot{\gamma}$ up to a maximum shear γ_{max} (upwards part of the curves), and then relaxing to zero stress using the relaxation process which has been described in Section 3.2. Thus, as mentioned in Section 3.2, the relaxation times in the relaxing parts of the curves are not intrinsic here, but depend on the relaxation parameters C and Δt which have been chosen. Several observations may be drawn from Figure 4. First, the average total energy does not strictly reach a plateau in several hundreds of sec (elementary simulation time units), even though the macroscopic mechanical state of the system is stationary. Second, the energy state in which the system drops after relaxation is not unique. It depends on the mechanical history of the system, namely here, on the maximum shear γ_{max}. Third, the system drops in a state of *lower* energy after shearing at large amplitude. The tendency to decrease the energy is observed for both individual contributions (average elastic energy and average excluded volume energy). Note that this drop in energy is possible because our sample is relatively "young". But we emphasise here that even after the third preparation step, the internal energy of the sample can decrease. This is a consequence of the very complex phase space available to the fillers.

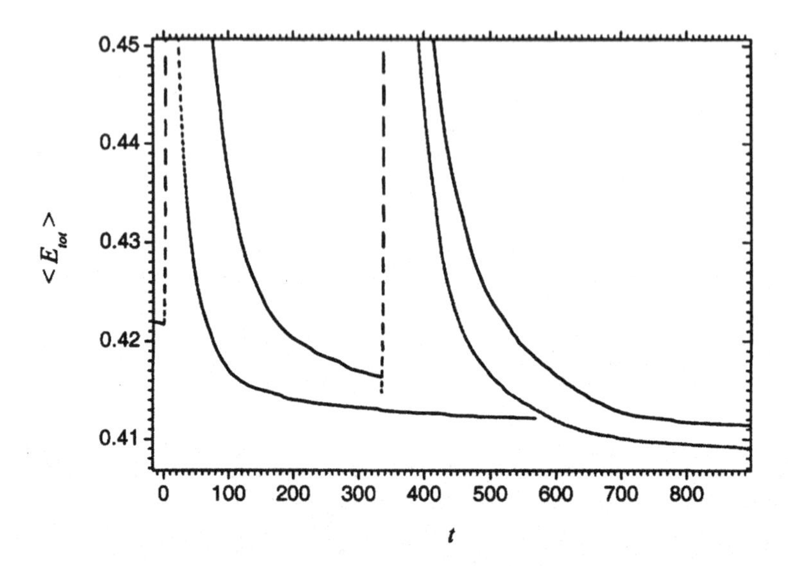

FIG. 4. *The average total energy* $< E_{tot} >$ *plotted as a function of time during large amplitude stress/strain cycles in the system* $n = 8.5$, $\Phi = 0.35$. *The abscissa is in sec (units of elementary simulation time). The time* $t = 0$ *corresponds to the starting point of the first large amplitude shear applied* ($\dot{\gamma} = 0.1$). *The two curves relaxing between* $t \approx 0$ *and* $t \approx 300$ *sec correspond to relaxing the system to zero stress after shearing at* $\gamma_{max} = 1.88$ *(top curve) and* $\gamma_{max} = 0.77$ *(bottom curve). At* $t \approx 335$ *sec, a second cycle is applied with* $\dot{\gamma} = -0.1$ *down to various maximum strain.*

4.3. Non-affine displacements. As mentioned above, most rheological models assume that the deformations are affine down to the molecular scale [2, 3, 51]. Whereas this assumption has provided a useful way of describing e.g. polymer melt dynamics, it has proven wrong and misleading in the case of systems with frozen disorder, such as gels or rubbers. We aim here at quantifying the non-affine part of the displacements, down to the scale of the filler particles. In order to quantify this deviation with respect to affine deformation, we proceed as follows. The displacements of all particles are computed with respect to a reference state. The initial state (I) is the state of the system after the preparation process. The initial equilibrium state (I) obtained after the preparation processes is taken as the reference state, characterized by the deformation tensor $\Gamma^{(I)} = I_d$. In the initial state (I) the particle i is located at position $\vec{R}_i^{(I)}$.

Then a second state (II) may be obtained, for instance after the system has been deformed at a finite rate $\dot{\gamma}$ up to a deformation $\Gamma^{(II)}$ and the particle positions have been relaxed for a long time at constant deformation $\Gamma^{(II)}$. In this second state, the particle i has a position $\vec{R}_i'^{(II)}$ which corresponds in the initial (undeformed) state to the position $\vec{R}_i^{(II)}$ given by $\vec{R}_i'^{(II)} = \Gamma^{(II)}\vec{R}_i^{(II)}$. $\vec{R}_i^{(II)}$ is the position in state (II), *referred to the*

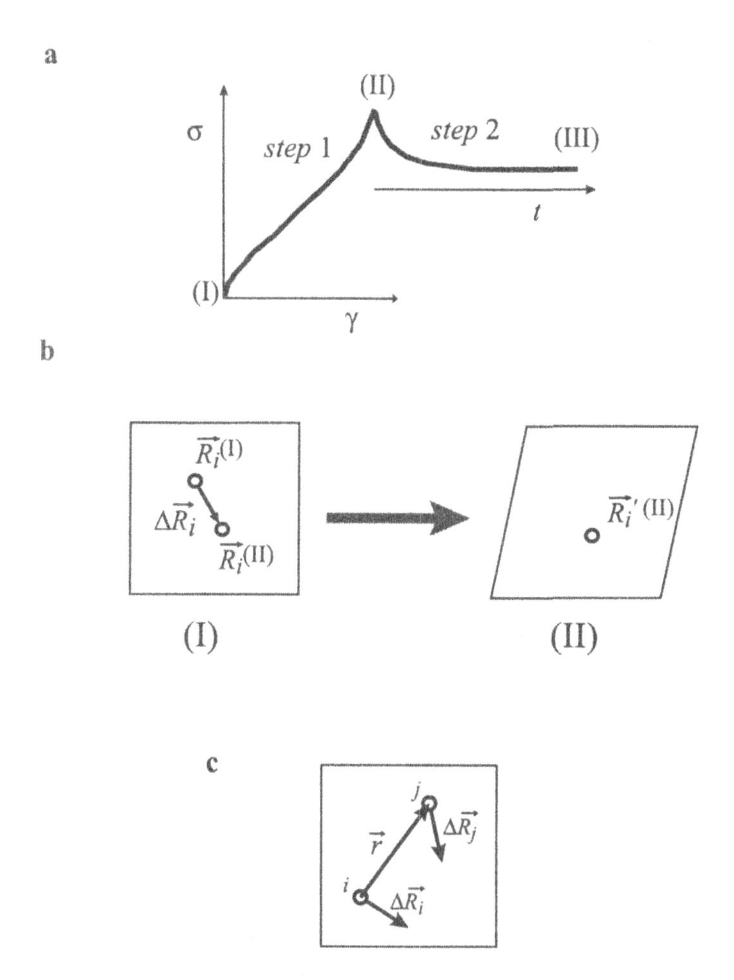

FIG. 5. *a: Schematics illustrating the type of experiment which is described in this paper. The system in the initial, reference state (I) described by the deformation tensor $\Gamma^{(I)} = I_d$ is first sheared at a given rate $\dot{\gamma}$ up to a state described by the deformation tensor $\Gamma^{(II)}$ (step 1), then the relaxation of the stress is studied as a function of time while keeping the deformation $\Gamma^{(II)}$ constant (step 2). b: The definition of the non-affine displacement vector $\Delta\vec{R}_i$, referred to the initial undeformed state (I): the position of particle i is $\vec{R}_i'^{(II)}$ in state (II), which corresponds to $\vec{R}_i^{(II)} = \Gamma^{(II)^{-1}}\vec{R}_i'^{(II)}$ in state (I). c: the definition of the correlation function of non-affine displacements.*

initial state (I). Thus, the relative displacement of particle i between the states (I) and (II), which measures the *non-affine displacement* referred to the initial undeformed state (I), is expressed directly as:

$$\Delta\vec{R}_i = \vec{R}_i^{(II)} - \vec{R}_i^{(I)} \tag{4.4}$$

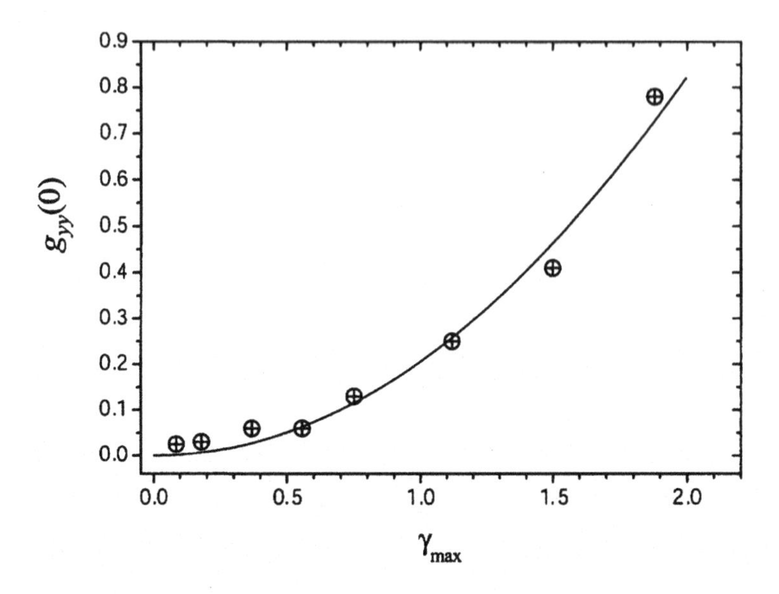

FIG. 6. *The amplitude $g_{yy}(0)$ of the correlation function of non-affine displacements in the system $n = 8.5$, $\Phi = 0.35$ sheared at $\dot{\gamma} = 0.1$, as a function of the maximum shear γ_{max}. The system is given time to relax in the imposed deformation. The continuous curve is a γ_{max}^2 law.*

This is illustrated in Figures 5**a** and **b**. Then the non-affinity may be quantified by the correlation functions

$$g_{\alpha\beta}(\vec{r}) = \langle \Delta R_{i\alpha} \Delta R_{j\beta} \rangle \tag{4.5}$$

where $\alpha, \beta = x, y, z$ and the average is done over all particle pairs (i, j) with an interparticle vector $\vec{r}$ (see Figure 5**c**). One can show (see Appendix), that for small deformations and at fixed $\vec{r}$, the correlation functions should be proportional to γ^2, where γ is the amplitude of the deformation. The amplitude $g_{yy}(0)$ of the correlation function of non-affine displacements is plotted as a function of γ_{max} in Figure 6, for $n = 8.5$ and $\Phi = 0.35$. Indeed, the variation is compatible with a γ_{max}^2 law, as illustrated by the continuous curve. The histograms for the Cartesian coordinates $(\Delta x_i, \Delta y_i, \Delta z_i)$ of the non-affine displacement vectors are plotted in Figure 7, in logarithmic scale. The distributions of non-affine displacement vectors are compatible with Gaussian distributions. They are however not isotropic. After shearing at large amplitude ($\gamma_{\text{max}} = 1.88$), the distribution widths systematically range in the order $\langle \Delta x \rangle > \langle \Delta y \rangle > \langle \Delta z \rangle$ in the various systems investigated. Histograms for the modulus of the displacement vectors $\Delta r_i = \left(\Delta x_i^2 + \Delta y_i^2 + \Delta z_i^2\right)^{1/2}$ are plotted in Figure 8. Note that the quantity $g_{xx}(0) + g_{yy}(0) + g_{zz}(0)$ is the second moment of the probability distribution $P(\Delta r)$ shown in Figure 8. The various histograms in Figure 8

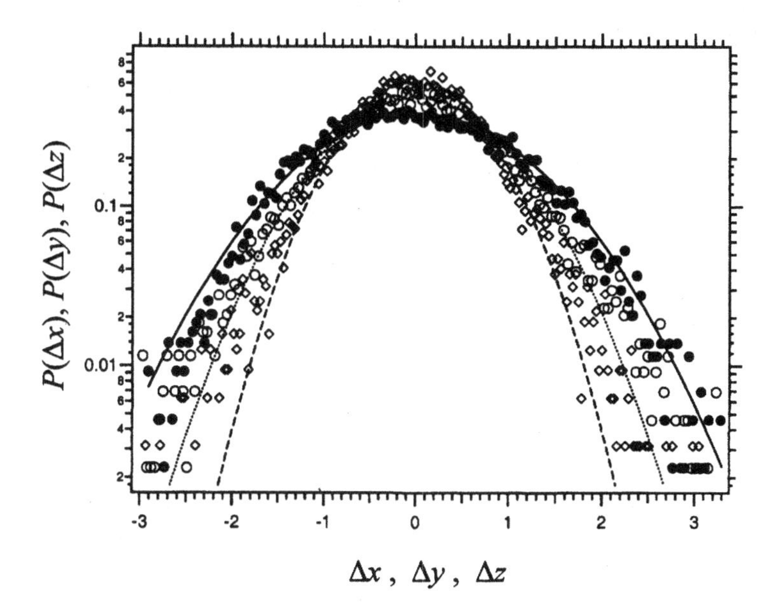

FIG. 7. *Histograms of the components Δx, Δy, Δz of the non-affine deformation vector, obtained in the system $n = 10$, $\Phi = 0.37$. Each histogram is normalized such that $\int P(\Delta\alpha)d\Delta\alpha = 1$ (with $\alpha = x$, y or z). The system is deformed at the rate at $\dot{\gamma} = 0.1$ up to $\gamma_{max} = 1.88$, then it is given time to relax internally in the imposed deformation. Gaussian fits to the histograms are shown. •: histogram for Δx (full curve: Gaussian fit) ; ◦: histogram for Δy (dotted curve: Gaussian fit) ; ⋄: histogram for Δz (dashed curve: Gaussian fit). The non-affine displacement is not isotropic.*

correspond to different values of $\gamma_{\max}$, in the system $n = 8.5$, $\Phi = 0.35$. A curve corresponding to a Gaussian distribution of the non-affine displacement vectors is shown also in Figure 8. The computed histograms are broader than the Gaussian example. This may be due to the fact that the displacements along the three directions are not independent variables. Therefore their sum may not be a Gaussian, even though they are Gaussian individually. However, when the abscissa scale Δr is normalized by the average value of the histogram (first moment of the histogram), the histograms obtained for different values of the parameters n and Φ and for different values of $\gamma_{\max}$ superpose. The average displacement $\langle\Delta r\rangle$, computed as the first moment of the histograms shown in Figure 8, is plotted in Figure 9 as a function of the volume fraction Φ, for different values of the connectivity n. Assuming that a $\gamma_{\max}^2$ law represents reasonably well the variation as a function of $\gamma_{\max}$ (see Figure 6), the values of $\langle\Delta r\rangle$ have been normalized by $\gamma_{\max}^2$ in order to compensate for the slightly different values of $\gamma_{\max}$ used in the different measurements.

The correlation functions $g_{yy}(r), g_{xx}(r), g_{zz}(r)$ and $g_{xy}(r)$ are plotted in Figure 10, in a system with $n = 8.5$, $\Phi = 0.35$ and for a final deformation

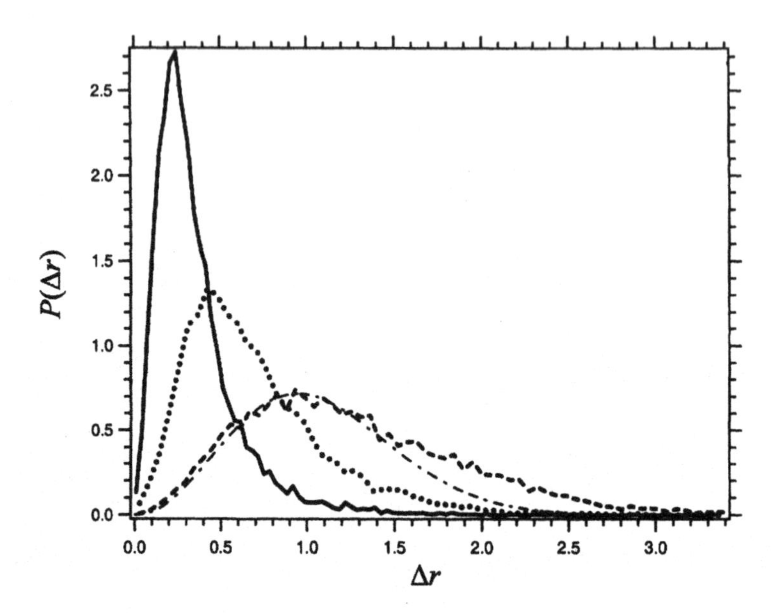

FIG. 8. *Histograms of the modulus of the non-affine deformation vector obtained in the system $n = 8.5$, $\Phi = 0.35$ which has been deformed at the rate $\dot{\gamma} = 0.1$ up γ_{max}, and after the system has been given time to relax in the imposed deformation. The abscissa is the modulus of the displacement vector $\Delta r = (\Delta x_i^2 + \Delta y_i^2 + \Delta z_i^2)^{1/2}$, the ordinate the histogram $P(\Delta r)$ normalized as $\int P(\Delta r) d\Delta r = 1$. Histograms are plotted for three different values of γ_{max}: full curve: $\gamma_{max} = 0.37$; doted curve: $\gamma_{max} = 0.745$; dashed curve: $\gamma_{max} = 1.88$. The dashed-dotted curve corresponds to a Gaussian distribution of the non-affine displacement vectors.*

of amplitude $\gamma_{\max} = 1.88$. Except for g_{xy}, these correlation functions decrease exponentially, at least over the distance which can be probed in our simulations, which is of the order 10 (that is half the dimension of the simulation box). Note that for symmetry reasons, the correlation functions $g_{xy}(r)$, as well as $g_{xz}(r)$ and $g_{yz}(r)$ are expected to be zero. Indeed, it is observed that these functions are very small in absolute value (see Figure 10). Correlation lengths ξ may be estimated from the plots of the functions $g_{\alpha\beta}(r)$ as a function of r. In a given system, the correlation length is not the same for all $g_{\alpha\beta}(r)$ functions. At high amplitudes, the correlation length of the y components ξ_{yy} is systematically the largest one. ξ_{yy} is plotted as a function of the maximum strain $\gamma_{\max}$ in Figure 11. ξ_{yy} increases significantly as $\gamma_{\max}$ increases, which means that the spatial scale of the non-affine displacements increases as well as their amplitude, though more slowly. By contrast, no systematic change of ξ_{yy} as a function of the volume fraction Φ has been measured in the systems investigated (from $\Phi = 0.20$ to $\Phi = 0.37$).

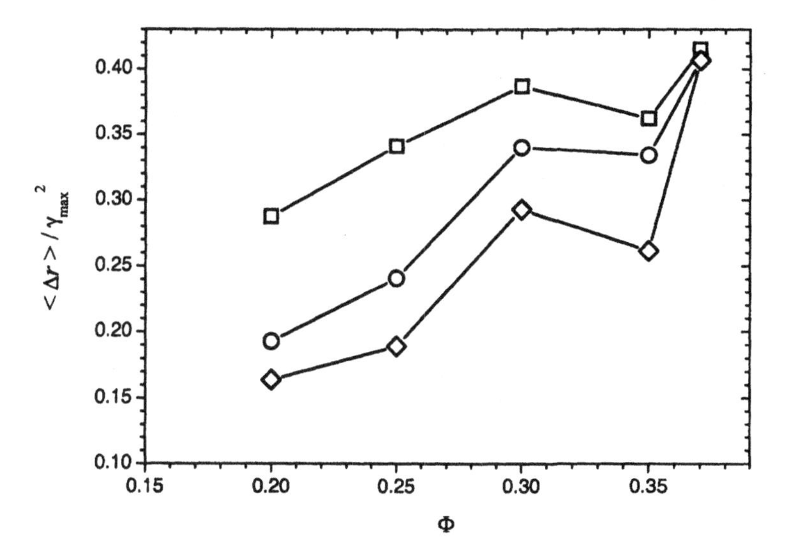

FIG. 9. *The reduced average displacement* $< \Delta r > / \gamma_{max}^2$, *computed as the first moment of the histograms (as shown in Figure 8), as a function of the volume fraction* Φ, *for different values of the connectivity* n: □: $n = 8.5$; ○: $n = 9$; ◇: $n = 10$. *All points are obtained with values of* γ_{max} *of the order 1.8 to 2.*

4.4. Stress relaxation under shear. As it is well known, gels or rubbers exhibit long relaxation times, much longer for instance than those predicted by the Rouse model, even at high temperatures [2, 52]. In the simulated filled rubbers studied here, we may expect long relaxation times as well. Indeed, due to the disorder of the systems, one may expect that the relaxation involves motions at scales larger than the filler diameters, and therefore can be much longer than unity. The relaxation times may depend on several parameters. First, the smaller the cross-link density, the longer the relaxation time, since the available phase space becomes more complex. The effect of the filler volume fraction is less clear. On one hand, the energy barriers to cross when deforming the samples become larger on increasing Φ. This effect tends to reduce the relaxation times. On the other hand, these barriers becoming higher, it takes more time to cross them back during the relaxation process. Thus, we have studied the relaxation dynamics in our systems by performing the following typical experiment. The systems are first submitted to an imposed shear deformation at a given $\dot{\gamma}$, and the subsequent intrinsic relaxation of the stress at constant deformation is measured.

We have studied the relaxation as a function of the various parameters which describe the systems. The relaxation of the stress under constant strain for different values of the strain $\gamma_{\max}$ is shown in Figure 12. $\sigma(0)$

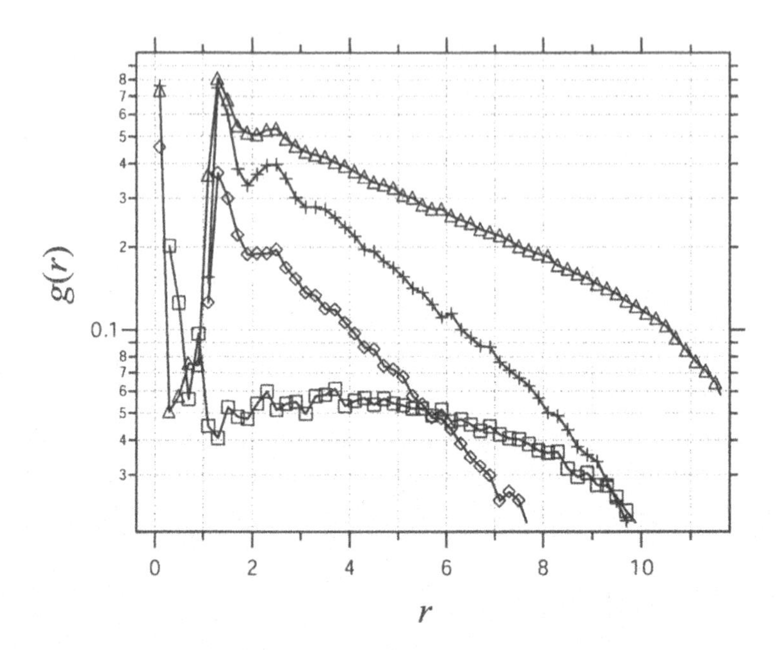

FIG. 10. *Correlation function of non-affine displacements after the system has been deformed by a large amplitude shear. r is the distance between correlated particles, the correlation function is averaged isotropically on r.* $\triangle$: $g_{yy}(r)$; $+$: $g_{xx}(r)$; $\diamond$: $g_{zz}(r)$; $\square$: $g_{xy}(r)$. $n = 8.5$, $\Phi = 0.35$, $\dot{\gamma} = 0.1$, $\gamma_{max} = 1.88$.

is the stress when the value γ_{max} has been reached. The time is measured here from that starting point. Throughout this section, the time scales are expressed in system time units (sec), determined by the reduced friction coefficient ζ^*. It appears first that the typical relaxation time is very long with respect to one, particularly at large values of the strain γ_{max}. In the top relaxation curves in Figure 12, the stress has not yet reached a plateau after more than 100 sec. The relative amount of stress relaxation is quite sensitive to the parameters of the systems. The shear stress relaxation normalized to one at the starting point, $\sigma(t)/\sigma(0)$, is plotted in Figure 13 as a function of time, for different values of the constant strain γ_{max}. The relative amount of relaxation, defined as $(\sigma(0) - \sigma_{\text{fin}})/\sigma(0)$, is plotted in Figure 14 as a function of Φ, for different values of the connectivity n and for a given applied shear γ_{max}. σ_{fin} is the stress value at the end of the measured relaxation. As mentioned above, there is some uncertainty in estimating the terminal value of the stress, which is still relaxing significantly after more than 200 sec in some cases. It is observed that the larger Φ, the larger the subsequent stress relaxation. We interpret this result as the following: at relatively short times, some fillers come into contact and interact strongly, which results in a high contribution to the stress. At subsequent times, some of these fillers may move apart in order to drop

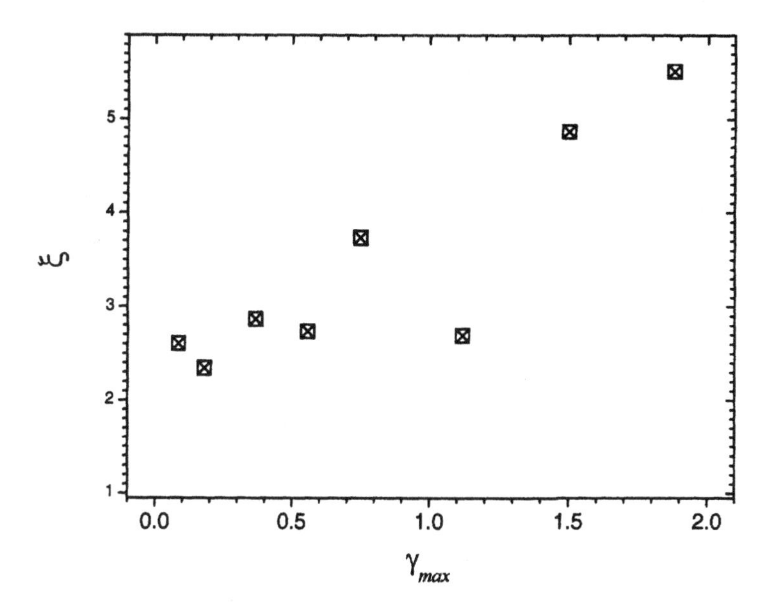

FIG. 11. *The correlation length ξ_{yy} of the correlation function of non-affine displacements $g_{yy}(0)$ in the system n = 8.5, Φ = 0.35 sheared at $\dot{\gamma}$ = 0.1 up to γ_{max}, plotted as a function of the maximum shear γ_{max}.*

in more energetically favorable position. However, this process is likely to take a long time.

Another noticeable feature is that the relaxations are non exponential. The reduced (normalized) stress $\sigma^*(t) = (\sigma(t) - \sigma_{\text{fin}})/(\sigma(0) - \sigma_{\text{fin}})$, where σ_{fin} is the final value of the stress ($\sigma_{\text{fin}} = \sigma(t = \infty)$), is plotted as a function of time in Figure 15 in the system $n = 8.5$, $\Phi = 0.35$ for different values of the applied shear strain γ_{max}. The non-exponential character of these curves is very clear. Moreover, the initial relaxation time exhibits a marked variation as a function of the strain γ: the larger γ, the longer the relaxation time, for reasons which are probably similar to those mentioned above. The dispersion of the curves at longer times comes essentially from the difficulty to determine the final stress σ_{fin}. A relaxation time τ may be defined in the usual way as the integral $\tau = \int_0^\infty \sigma^*(t)dt$. The relaxation time τ is plotted as a function of Φ in Figure 16 for different values of n, all points here being obtained at about the same value of $\gamma \approx 1.8$. Both the relaxation time τ and the relative amount of relaxation (as exemplified in Figure 13) depend on the volume fraction Φ, though not very strongly.

5. Conclusion. We have proposed a model in order to simulate the dynamical behavior of physical systems such as gels or rubbers reinforced by filler particles. This model represents a disordered elastic system made of hard spheres connected by harmonic springs. The response of this sys-

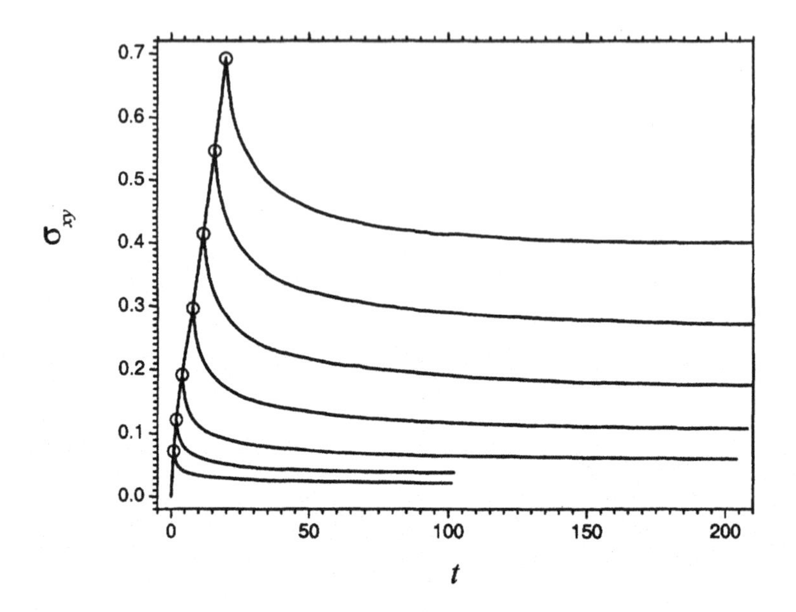

FIG. 12. *Relaxation of the stress $\sigma(t)$ at various fixed values of the strain as a function of time, in the system $n = 8.5$, $\Phi = 0.35$. The shear (upward curve) is performed at $\dot{\gamma} = 0.1$. The time scale in abscissa is in unit of the elementary simulation time scale given by ζ. Circles visualize the starting point of each relaxation experiment. The corresponding γ values are respectively (from top to bottom): 1.876, 1.499, 1.122, 0.745, 0.369, 0.180, 0.086.*

tem to large amplitude shear deformations has been studied by dissipative molecular dynamics. The parameters of the system are the connectivity n and the volume fraction Φ, and the parameters of the applied deformations are the shear rate $\dot{\gamma}$ and the maximum shear $\gamma_{\rm max}$. Note that any tensorial deformation $\Gamma_{\alpha\beta}$ may be implemented in our numerical simulation code. This system exhibits some major effects which correspond to the behavior observed in filled elastomers. First, it reproduces the reinforcement observed in the high temperature regime in [38, 39]. Indeed, we observe that, the higher the filler volume fraction, the higher the shear modulus. Second, our simulations displays the non-affinity of the microscopic displacements, which have been observed in systems such as gels or rubbers. The presence of the fillers tends to enhance the non-affine part of the microscopic displacements. This behavior results from two mechanisms with opposite contribution: the presence of the fillers tends to maintain each unit within a defined neighborhood, which tends to reduce the non-affinity; on the other hand, once a unit has overcome this barrier under the action of the imposed displacement, the presence of the fillers makes it more difficult to drop back in (or very close to) its previous relative position during subsequent relaxation. Our simulations show that this latter effect dominates the former.

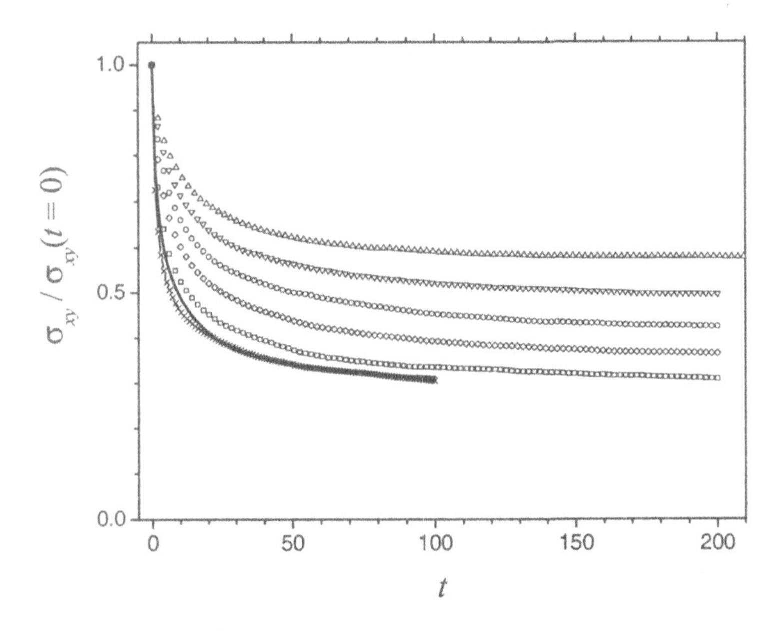

FIG. 13. *Relaxation of the normalized stress $\sigma(t)/\sigma(t=0)$ in the system $n = 8.5$, $\Phi = 0.35$ under different values of the applied strain, from top to bottom curve:* $\triangle$: $\gamma_{max} = 1.88$; ∇: $\gamma_{max} = 1.50$; $\circ$: $\gamma_{max} = 1.12$; $\diamond$: $\gamma_{max} = 0.745$; $\square$: $\gamma_{max} = 0.37$; $\times$: $\gamma_{max} = 0.18$; *plain curve:* $\gamma_{max} = 0.086$.

As mentioned in the introduction, the non-affine nature of the displacement field has also been considered in simulations regarding the case of simple liquids, in the low temperature regime [18, 19]. However, these studies considered wave propagations, on time scales which does not allow the degrees of freedom to change their local environment: the motion of the molecules correspond to vibrations of small amplitudes compared to the molecular diameter, whereas we consider here large scale reorganisations. Note also that these simulations have been performed in 2D only.

The relaxation mechanisms discussed above makes the time evolution of the system very complex. We have also shown that the microscopic state is not a one-to-one function of the macroscopic deformation: the state of this kind of systems depends on their history. In particular, we have shown that the internal energy of a "young" sample can decrease when submitted to deformation cycles. Another major effect is that the complex relaxation mechanisms due to the disorder of the system lead to very long, non-exponential relaxations, with associated time scales much longer than the unit time scale of the system. Therefore, even if the rubber matrix has a well defined relaxation time, these systems should exhibit a broad distribution of relaxation times. The case of a less idealized matrix, with itself a non-exponential relaxation function, could in principle be addressed

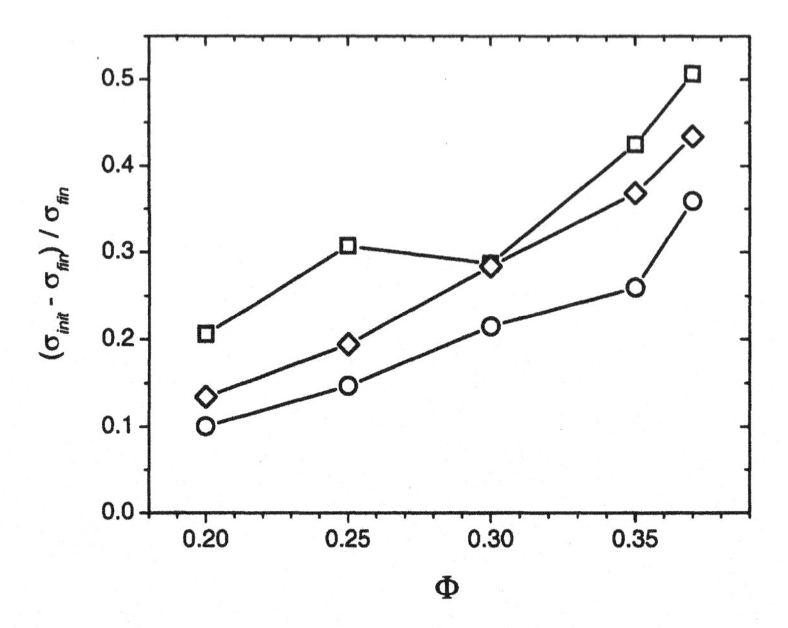

FIG. 14. *The relative percentage of relaxation of the stress expressed as the quantity* $(\sigma(0) - \sigma_{fin})/\sigma(0)$ *at* $\gamma \approx 1.8$ *as a function of* Φ *for different values of* n*:* □: $n = 8.5$; ◇: $n = 9$; ○: $n = 10$.

using our approach, by replacing the friction coefficient ζ^* by a distribution of friction coefficients. These kinds of approaches should be extended to describe in more details the non-linear behavior of gels and rubbers, and also the complex and fascinating behavior of filled elastomers in the low temperature regime, where the glass transition effects lead to dramatic reinforcement and very strong non-linear behaviors [29, 30, 34, 38, 39].

6. Acknowledgments. Both authors are CNRS researchers. D.L. would like to thank the IMA for supporting its stay at the workshop "Modeling of Soft Matter" during which part of this work was completed.

APPENDIX

A. Torque. The torque acting on the system can be written as

$$\vec{T} = \sum_i \vec{R^i} \wedge \vec{F^i} \tag{A.1}$$

where the force $\vec{F^i}$ is the total force acting on particle i. The torque can be written as

$$\vec{T} = \sum_{i,j\text{neighbors}} \vec{R^i} \wedge \vec{F^{i,j}} \tag{A.2}$$

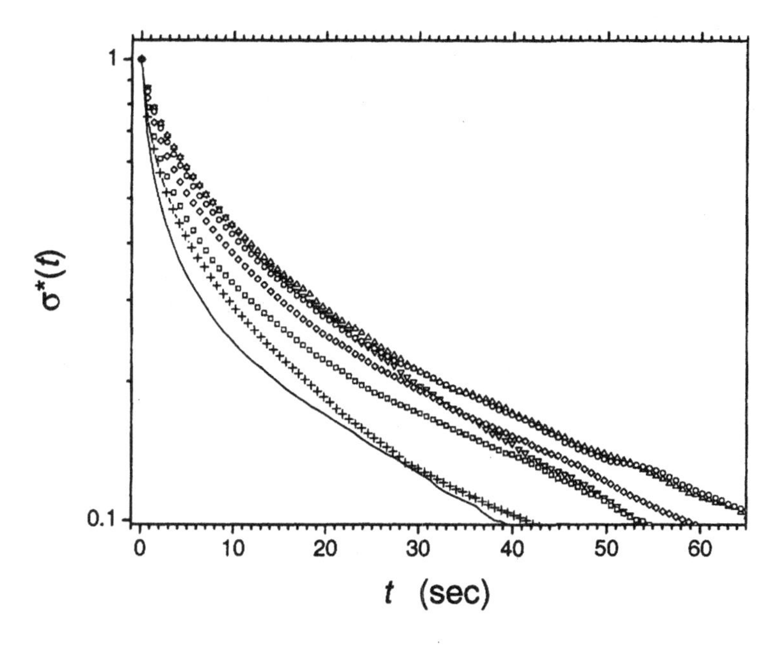

FIG. 15. *The relaxation of the reduced stress* $\sigma^*(t) = (\sigma(t) - \sigma_{fin})/(\sigma(0) - \sigma_{fin})$ *in the system* $n = 8.5$, $\Phi = 0.35$, *for different values of the shear strain* γ_{max}. *From top to bottom curve:* $\triangle$: $\gamma_{max} = 1.88$; ∇: $\gamma_{max} = 1.50$; $\circ$: $\gamma_{max} = 1.12$; $\diamond$: $\gamma_{max} = 0.745$; $\square$: $\gamma_{max} = 0.37$; $\times$: $\gamma_{max} = 0.18$; *plain curve:* $\gamma_{max} = 0.086$.

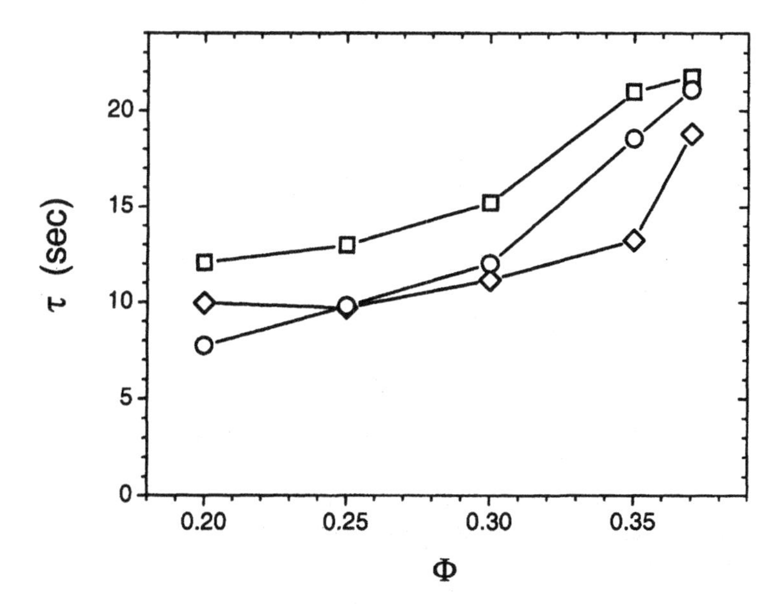

FIG. 16. *The relaxation time* τ *at* $\gamma \approx 1.8$ *as a function of the volume fraction* Φ, *for different values of* n: $\square$: $n = 8.5$; $\diamond$: $n = 9$; $\circ$: $n = 10$.

where the summation runs other all internal forces. $\vec{F}^{i,j}$ is the force that particle j exerts on particle i. The torque can then be written as

$$\vec{T} = \frac{1}{2} \sum_{i,j \text{neighbors}} (\vec{R}^i - \vec{R}^j) \wedge \vec{F}^{i,j} . \tag{A.3}$$

Since all the forces between particles are central forces, this quantity is zero.

B. Deviation of the initial stress tensor. In a macroscopic sample, the stress should be diagonal immediately after the particles have been randomly distributed in the sample, because the sample would be isotropic. On the other hand, because of the finite number of particles in our sample, fluctuations cannot be neglected. After the particles have been distributed, the deviation of non-diagonal elements of the stress tensor can be estimated as:

$$\langle \sigma^2_{\alpha\beta} \rangle = \frac{1}{V^2} \sum_{i,j} \left\langle R^i_\alpha R^j_\alpha F^i_\beta F^j_\beta \right\rangle . \tag{B.1}$$

We consider the case of non-diagonal elements $\alpha \neq \beta$. Then, the previous sum, which contains N^2 terms *a priori*, effectively contains a number of order N of non-negligeable terms only. Indeed

$$\left\langle R^i_\alpha R^j_\alpha F^i_\beta F^j_\beta \right\rangle \approx 0 \tag{B.2}$$

for indices corresponding to particles far apart. Thus

$$\begin{aligned} \langle \sigma^2_{\alpha\beta} \rangle &= \frac{1}{V^2} \sum_{i,j \text{neighbors}} \left\langle R^i_\alpha R^j_\alpha F^i_\beta F^j_\beta \right\rangle \\ &\approx \frac{1}{V^2} \sum_{i,j \text{neighbors}} \langle R^i_\alpha R^j_\alpha \rangle \left\langle F^i_\beta F^j_\beta \right\rangle . \end{aligned} \tag{B.3}$$

When the indices i and j are different, one has, if the particles i and j are neighbors

$$\left\langle F^i_\beta F^j_\beta \right\rangle = \left\langle (\sum_k \vec{F}^{i,k})(\sum_m \vec{F}^{j,m}) \right\rangle = \left\langle \vec{F}^{i,j} \vec{F}^{j,i} \right\rangle (\approx -f^2) \tag{B.4}$$

where f is the typical force exerted by the elastic springs or by the hard core repulsion. If the particles are not neighbors, this average is zero. When the indices i and j are equals, one has

$$\langle F^i_\beta F^i_\beta \rangle = \left\langle \sum_j \vec{F}^{i,j} \vec{F}^{i,j} \right\rangle \tag{B.5}$$

These terms exactly cancel the previous ones, except for the bonds at the boundaries which point out of the box. The contribution of the remaining terms is therefore

$$\langle \sigma^2_{\alpha\beta} \rangle \approx \frac{L^2}{V^2} \sum_{i,B} f_i^2 \approx \frac{L^2}{V^2} f^2 N_{\mathrm{B}} \ . \tag{B.6}$$

The subscript B means that the summation is performed over all particles which interact through the boundary of the box. The number of such particles is N_{B}. Finally:

$$\langle \sigma^2_{\alpha\beta} \rangle \approx \frac{f^2 N_B L^2}{V^2} \propto N^{-2/3} \ . \tag{B.7}$$

Thus, for a box with 10000 particles, the amplitude of non diagonal elements of the stress tensor is of the order 0.1 and is therefore not negligible.

C. Amplitude of the correlation functions. Let us consider an elastic body with a non-homogeneous elastic modulus $\mu = \mu_0 + \delta\mu$. When this system is submitted to an imposed deformation γ, the displacement field $u(\vec{r})$ is such that $\mu(\vec{r})u(\vec{r}) \sim \gamma\mu_0$ is constant. It follows that $\delta u(\vec{r}) \sim \gamma\delta\mu(\vec{r})$, where $\delta u(\vec{r})$ is the local distortion as compared to the macroscopic affine deformation, i.e. the non-affine displacement. Then:

$$\left\langle u(\vec{r_i})u(\vec{r_i} + \vec{R}) \right\rangle \sim \gamma^2 \left\langle \delta\mu(\vec{r_i})\delta\mu(\vec{r_i} + \vec{R}) \right\rangle \ . \tag{C.1}$$

This shows that, for small deformations, the correlation function of the non-affine displacements varies as a function of γ like γ^2.

REFERENCES

[1] S. Alexander, *Amorphous solids: their structure, lattice dynamics and elasticity*, Physics Reports, **296** (1998), pp. 65–236.

[2] J.D. Ferry, *Viscoelastic Properties of Polymers*, Wiley, New York, 1980.

[3] J.H. Weiner, *Statistical Mechanics of Elasticity*, Wiley, New York, 1983.

[4] J. Bastide and L. Leibler, *Large-Scale Heterogeneities in Randomly Cross-Linked Networks*, Macromolecules, **21** (1988), pp. 2647–2649.

[5] J. Bastide, L. Leibler, and J. Prost, *Scattering by Deformed Swollen Gels: Butterfly Isointensity Patterns*, Macromolecules, **23** (1990), pp. 1821–1825.

[6] M. Rubinstein, L. Leibler, and J. Bastide, *Giant Fluctuations of Crosslink Positions in Gels*, Phys. Rev. Lett., **68** (1992), pp. 405–407.

[7] C. Rouf, J. Bastide, J.M. Pujol, F. Schosseler, and J.P. Munch, *Strain Effect on Quasistatic Fluctuations in a Polymer Gels*, Phys. Rev. Lett., **73** (1994), pp. 830–833.

[8] A. Ramzi, F. Zielinski, J. Bastide, and F. Boué, *Butterfly Patterns: Small-Angle Neutron Scattering from Deuterated Mobile Chains in a Randomy Cross-linked Polystyrene Network* Macromolecules, **28** (1995), pp. 3570–3587.

[9] A. Ramzi, A. Hakiki, J. Bastide, and F. Boué, *Uniaxial Extension of End-Linked Polystyrene Networks Containing Deuterated Free Chains Studied by Small-Angle Neutron Scattering: Effect of the Network Chains and the Size of the Free Chains*, Macromolecules, **30** (1997), pp. 2963–2977.

[10] S. Westermann, W. Pyckout-Hintzen, D. Richter, E. Straube, S. Egelhaaf, and R. May, *On the length scale Dependence of Microscopic Strain by SANS*, Macromolecules, **34** (2001), pp. 2186-2194.

[11] P.-G. de Gennes, *On a relation between percolation theory and the elasticity of gels*, J. Physique Lett., **37** (1976), pp. L1–L2.

[12] Stauffer D. and Aharony A., *Introduction to percolation theory*, Taylor and Francis, London, 1994.

[13] S. Feng and P.N. Sen, *Percolation on Elastic Networks: New Exponent and Threshold*, Phys. Rev. Lett., **52** (1984), pp. 216–219.

[14] S. Arbabi and M. Sahimi, *Critical Properties of Viscoelasticity of Gels and Elastic Percolation Networks*, Phys. Rev. Lett., **65** (1990), pp. 725–728.

[15] A.A. Gusev, *Finite Element Mapping for Spring Network Representations of the Mechanics of Solids*, Phys. Rev. Lett., **93** (2004), pp. 0304302–4.

[16] W. Schirmacher, G. Diezemann, and C. Ganter, *Harmonic Vibrational Excitations in Disordered Solids and the "Boson Peak"*, Phys.Rev. Lett., **81** (1998), pp. 136–139.

[17] A.A. Rzepiela, J.H.J. van Opheusden, and T. van Vliet, *Large shear deformation of particle gels studied by Brownian Dynamics simulations*, Computer Physics Communications, **147** (2002), pp. 303–306.

[18] J.P. Wittmer, A. Tanguy, J.-L. Barrat, and L. Lewis, *Vibrations of amorphous, nanometric structures: When does continuum theory apply?*, *Europhysics Letters*, **57** (2002), pp. 423–429.

[19] A. Tanguy, J.P. Wittmer, F. Leonforte, and J.-L. Barrat, *Continuum limit of amorphous elastic bodies: A finite-size study of low-frequency harmonic vibrations*, Phys. Rev. B, **66** (2002), pp. 174205–17.

[20] J. Gao and J.H. Weiner, *Excluded-Volume Effects in Rubber Elasticity. 1. Virial Stress Formulation*, Macromolecules, **20** (1987), pp. 2520–2525.

[21] J. Gao and J. H. Weiner, *Excluded-Volume Effects in Rubber Elasticity. 2. Ideal Chain Assumption*, Macromolecules, **20** (1987), pp. 2525–2531.

[22] C.P. Lusignan, T.H. Mourey, J.C. Wilson, and R.H. Colby, *Viscoleasticity of randomly branched polymers in the vulcanization class*, Phys. Rev. E, **60** (1999), pp. 5657–5669.

[23] E. Gasilova, L. Benyahia, D. Durand, and T. Nicolai, *Influence of entanglements on the Viscoleastic Relaxation of Polyurethane Melts and Gels*, Macromolecules, **35** (2002), pp. 141–150.

[24] M. Pütz, K. Kremer, and R. Everaers, *Self-Similar Chain Conformation in Polymer Gels*, Phys. Rev. Lett., **84** (2000), pp. 298–301.

[25] J. Oberdisse, G. Ianniruberto, F. Greco, and G. Marrucci, *Primitive-chain Brownian simulations of entangled rubbers*, Europhysics Letters, **58** (2002), pp. 530–536.

[26] E.S. Matsuo, M. Orkisz, S.-T. Sun, Y. Li, and T. Tanaka, *Origin of Structural Inhomogeneities in Polymer Gels*, Macromolecules, **27** (1994), pp. 6791–6796.

[27] S. Panyukov and Y. Rabin, *Polymer Gels: Frozen Inhomogeneities and Density Fluctuations*, Macromolecules, **29** (1996), pp. 7960–7975.

[28] C. Rouf-George, J.P. Munch, G. Beinert, F. Isel, A. Pouchelon, J.F. Palierne, F. Boué, and J. Bastide, *About "Defects" in Networks made by End-Linking*, Polymer Gels and Networks, **4** (1996), pp. 435–450.

[29] L.E. Nielsen and R.F. Landel, *Mechanical Properties of Polymers and Composites*, Marcel Dekker, New York, 1994.

[30] A.I. Medalia, *Effect of carbon black on dynamic properties of rubber vulcanizates*, Rubber Chem. Tech., **51** (1978), pp. 437–523.

[31] D.C. Edwards, *Review. Polymer-filler interactions in rubber reinforcement*, J. Mater. Sci. **25** (1990), 4175–4185.

[32] A.R. Payne, *A Note on the Conductivity and Modulus of Carbon Black-Loaded Rubbers*, J. Appl. Polym. Sci., **9** (1965), pp. 1073–1082.

[33] J.A.C. HARWOOD, L. MULLINS, AND A.R. PAYNE, *Stress Softening Effects in Pure Gum and Filler Loaded Rubbers.*, J. Appl. Polym. Sci., **9** (1965), pp. 3011–3021.

[34] G. KRAUS, *Reinforcement of elastomers by carbon black*, Rubber Chem. Tech., **51** (1978), pp. 297–321.

[35] G. KRAUS, *Mechanical Losses in Carbon-Black-Filled Rubbers*, J. Appl. Polym. Sci. Appl. Polym. Symp., **39** (1984), pp. 75–92.

[36] D.W. SCHAEFER, T. RIEKER, M. AGAMALIAN, J.S. LIN, D. FISCHER, S. SUKUMARAN, C. CHEN, G. BEAUCAGE, C. HERD, AND J. IVIE, *Multilevel structure of reinforcing silica and carbon*, J. Appl. Cryst., **33** (2000), pp. 587–591.

[37] S. WESTERMANN, M. KREITSCHMANN, W. PYCKOUT-HINTZEN, D. RICHTER, E. STRAUBE, B. FARAGO, AND G. GOERIGK, *Matrix Chain Deformation in Reinforced Networks: a SANS Approach*, Macromolecules, **32** (1999), pp. 5793–5802.

[38] J. BERRIOT, H. MONTS, F. LEQUEUX, D. LONG, AND P. SOTTA, *Evidence for the shift of the glass transition near the particles in silica-filled elastomers*, Macromolecules, **35** (2002), pp. 9756–9762.

[39] J. BERRIOT, H. MONTS, F. LEQUEUX, D. LONG, AND P. SOTTA, *Gradient of glass transition temperature in filled elastomers*, Europhys. Lett., **64** (2003), pp. 50–56.

[40] L.C.E. STRUIK, *The mechanical and physical aging of semicrystalline polymers: 1.*, Polymer, **28** (1987), pp. 1521–1533.

[41] D. LONG AND F. LEQUEUX, *Heterogeneous dynamics at the glass transition in van der Waals liquids, in the bulk and in thin films*, EPJ E, **4** (2001), pp. 371–387.

[42] Y. KANTOR AND I. WEBMAN, *Elastic Properties of Random Percolating Systems*, Phys. Rev. Lett., **52** (1984), pp. 1891–1894.

[43] M. KELLOMKI, J. ASTRÖM AND J. TIMONEN, *Rigidity and Dynamics of Random Spring Networks*, Phys. Rev. Lett., **77** (1996), pp. 2730–2733.

[44] M. KELLOMKI, J. ALSTRÖM, AND J. TIMONEN, *Rigidity and Dynamics of Random Spring Networks*, Phys. Rev. Lett., **77** (1996), pp. 2730–2733.

[45] C. MOUKARZEL, P.M. DUXBURY, AND P.L. LEATH, *Infinite-Cluster Geometry in Central-Force Networks*, Phys. Rev. Lett., **78** (1997), pp. 1480–1483.

[46] C. MOUKARZEL AND P.M. DUXBURY, *Comparison of rigidity and connectivity percolation in two dimensions*, Phys. Rev. E, **59** (1999), pp. 2614–2622.

[47] O. FARAGO AND Y. KANTOR, *Entropic Elasticity of Two-Dimensional Self-Avoiding Percolation Systems*, Phys. Rev. Lett., **85** (2000), pp. 2533–2537.

[48] O. FARAGO AND Y. KANTOR, *Entropic elasticity of phantom percolation networks*, Europhys. Lett., **52** (2000), pp. 413–419.

[49] M. LATVA-KOKKO, J. MÄKINEN, AND J. TIMONEN, *Rigidity transition in two-dimensional random fiber networks*, Phys. Rev. E, **63** (2001), Art. No. 046113.

[50] O. FARAGO AND Y. KANTOR, *Entropic elasticity at the sol-gel transition*, Europhys. Lett., **57** (2002), pp. 458–463.

[51] M. DOI AND S.F. EDWARDS, *The Theory of Polymers Dynamics* Clarendon Press, Oxford, 1986.

[52] J.G. CURRO AND P. PINCUS, *A theoretical basis for viscoelastic relaxation of elastomers in the long-time limit*, Macromolecules, **16** (1983), pp. 559–562.

[53] L.D. LANDAU AND E.M. LIFSHITZ, *Elasticity*, Pergamon Press, New York, 1993.

[54] L.D. LANDAU AND E.M. LIFSHITZ, *Mechanics*, Pergamon Press, New York, 1989.

STRESS TRANSMISSION AND ISOSTATIC STATES OF NON-RIGID PARTICULATE SYSTEMS

RAPHAEL BLUMENFELD*

Abstract. The isostaticity theory for stress transmission in macroscopic planar particulate assemblies is extended here to non-rigid particles. It is shown that, provided that the mean coordination number in d dimensions is $d+1$, macroscopic systems can be mapped onto equivalent assemblies of perfectly rigid particles that support the same stress field. The error in the stress field that the compliance introduces for finite systems is shown to decay with size as a power law. This leads to the conclusion that the isostatic state is not limited to infinitely rigid particles both in two and in three dimensions, and paves the way to an application of isostaticity theory to more general systems.

Key words. Granular systems; Stress field; Compliance; Isostatic systems.

1. Introduction. Much attention has been given lately to particulate systems both due to their overwhelming technological importance and the fundamental theoretical challenges that they pose [1, 2]. In particular the micro- and macro-mechanics have focused research activity following experimental [3–7] and numerical [8–11] observations of nonuniform stress fields [12]. Specifically, stresses frequently appear to be supported by arch-like regions, termed force chains, that cannot be straightforwardly described by conventional approaches [13–15]. It has been recognized that to understand this phenomenon it is essential to first understand transmission of stresses in 'isostatic' systems [12]. Isostatic states are configurations of particles where the interparticle contact forces are ***statically determinate***, i.e. they can be determined from the mechanical equilibrium conditions of balance of force and torque moments. This means that the interparticle forces can be determined without reference to compliance and hence to stress-strain relations. Isostatic states are characterized by low mean coordination numbers per particle which depend on the dimensionality of the system and on the particles roughness. For rough and infinitely rigid particles in d-dimensional systems ($d = 2, 3$) this number is $z_c = d + 1$, for smooth infinitely rigid particles of arbitrary shape $z_c = d(d + 1)$ [16–19], and for smooth infinitely rigid spheres $z_c = 2d$. Isostatic packings of particles are marginally rigid and such states have been shown to be easy to approach experimentally [19], making them interesting more than only theoretically. Several empirical [20–23] and statistical [12, 18] models have been proposed for the macroscopic stress field equations in these systems, suggesting a linear coupling between the components of the stress tensor. This has been recently established from first principles in the two-dimensional case for systems of infinitely rigid particles [24]. The new

*Biological and Soft Systems, University of Cambridge, Cavendish Laboratory, Madingley Road, Cambridge CB3 0HE, UK (rbb11@phy.cam.ac.uk). I am grateful to Prof. Robin Ball for critical comments.

isostaticity theory (IT) closes the stress field equations with a constitutive relation between the stress tensor $\hat{\sigma}$ and a rank-two symmetric fabric tensor $\hat{p}$ which characterizes the local microstructure:

$$p_{xx}\sigma_{yy} + p_{yy}\sigma_{xx} - 2p_{xy}\sigma_{xy} = 0 \ . \tag{1.1}$$

On the scale of a few particles, this equation is a local manifestation of the torque balance condition beyond the global requirement that $\hat{\sigma} = \hat{\sigma}^T$ [24]. It then transpired that the coarse-graining of Eq. (1.1) is not trivial, but this was eventually resolved, making it applicable for macroscopic systems, albeit with a subtle difference in the interpretation of the constitutive field p_{ij} [30]. This paved the way to several results, most notably it enabled a derivation of the general solution for the stress field in two-dimensional isostatic granular packings [25]. The solution turned out to indeed give rise to force chains and arches. This, not only gave a firm theoretical basis that explains the experimentally observed force chains, but also provided a way to predict the trajectories of individual force chains. Using these predictions made it possible to test the theory by direct comparison with experimental measurements.

However, much controversy surrounds the validity of the new theory. In particular, because it has been developed for infinitely rigid particles, there remained questions concerning its validity to general particulate systems, whose rigidity is unavoidably finite. The clarification of this point is a crucial first step towards bridging between IT and elasticity theory. A detailed examination of this issue, both in two and in three dimensions, is the aim of this paper.

Two dimensions. Consider a polydisperse planar packing of N particles of arbitrary shape and typical area a [26], confined to within a square container of dimensions $L^2 \sim aN$. All the particles are presumed to be made of the same material whose elastic properties are known [27]. The packing is loaded by an infinitesimally small external compressive force F_y on two opposite boundaries, as shown in Figure 1. The load compresses the particles against one another slightly, resulting in contacts between neighboring particles that consist of short lines. The line contacts, rather than point contacts, between particles constitute the main difference between packings of compliant and infinitely rigid particles. The criterion for 'smallness' of F_y is that the contact lines are smaller that the linear size of the corresponding particles [28]. For a system of infinitely rigid particles to be statically determinate in two dimensions the mean coordination number per particle must be $z_c = 3$ up to a boundary-to-bulk correction term. We wish to determine whether stress fields that develop in assemblies of compliant particles that satisfy this condition are also governed by the equations of IT.

An ideal resolution of the issue would be to establish whether *all* the interparticle forces can be determined, at least in principle, from balance

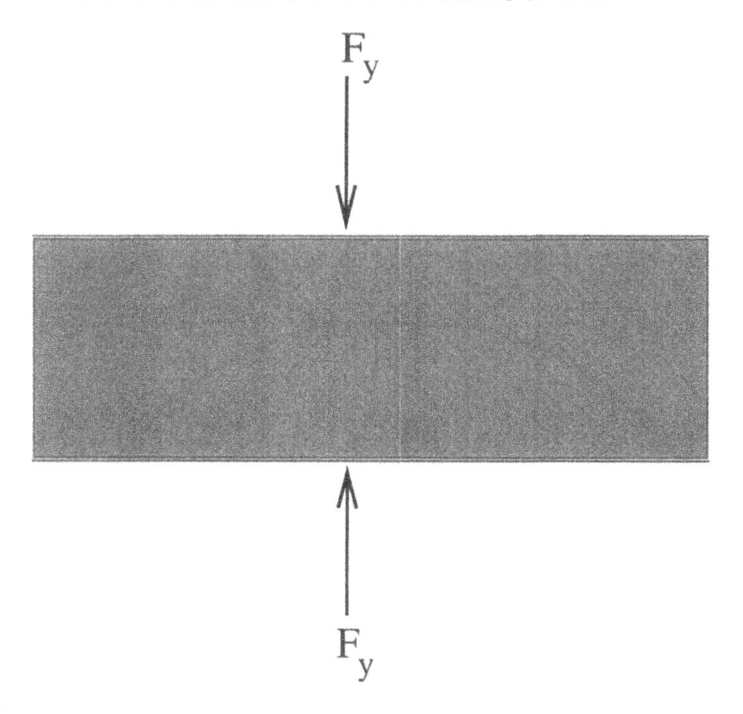

FIG. 1. *The loading on a packing of grains by a force F_y that is distributed evenly on the shown surfaces.*

conditions alone. If this is possible then the system is statically determinate and isostaticity theory must apply. The main difference between the geometries of infinitely rigid and compliant systems is that while in the former the interparticle forces act at a point of contact, in the latter they are continuously distributed along contact lines. Let us examine the contact between two touching particles, g and g'. The particles press on one another with a force that is distributed along the contact line with density $\phi(x)$, where x is a length parameter that varies from 0 to l along the line (see Figure 2). Due to the arbitrary shapes of particles this force density need not be uniform. Both a force and a torque moment are transmitted through the contact and these are given by

$$\vec{f}^{gg'} = \int_0^l \vec{\phi}(x)dx \tag{1.2}$$

and

$$\vec{M}^{gg'} = \int_0^l \vec{\phi}(x) \times \vec{\rho}(x)dx \ . \tag{1.3}$$

Here $\vec{\rho}(x)$ is the position vector from the centroid of the particle (defined as the mean position vector of the contact points of the particle g).

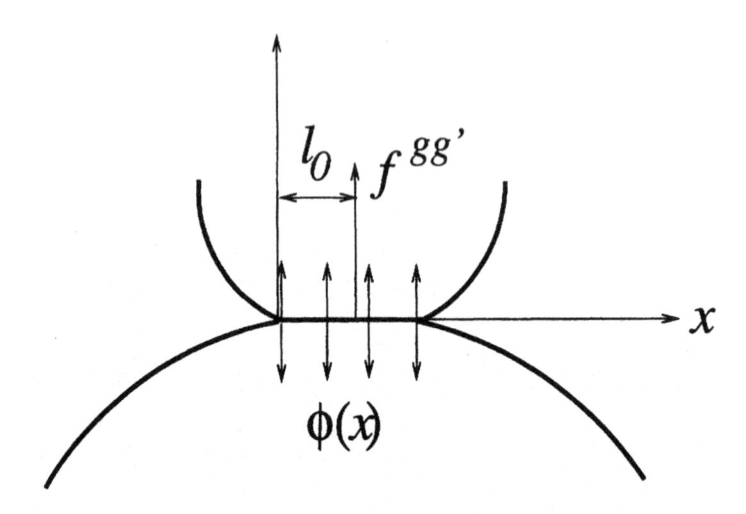

FIG. 2. *The distribution of forces, $\vec{\phi}(x)$ along the contact line between grains g and g'. The contact line is parameterized by $0 < x < l$ from left to right. The mean force is $\vec{f}^{gg'}$ located at a distance $x = l_0$, found from the first and second moments of the force density.*

So, do we need to determine the entire force distributions along the contact lines? Considering relations (1.2) and (1.3), the answer is encouragingly no. The torque moment can be represented by a single force of magnitude $\vec{f}^{gg'}$ acting at a point $x = l_0$ that lies between $x = 0$ and $x = l$ (it is straightforward to see that l_0 cannot be outside this section), and whose location is determined by the relation

$$\vec{f}^{gg'} \times \vec{\rho}(l_0) = \vec{M}^{gg'} . \tag{1.4}$$

Thus, it seems that, at least in principle, we can reduce the problem to find the discrete forces $\vec{f}^{gg'}$. In mechanical equilibrium these interparticle forces balance out [29]

$$\sum_{g'} \vec{f}^{gg'} = 0 . \tag{1.5}$$

The stress field can be defined in terms of the force moments around the particles

$$S^g_{ij} = \sum_{g'} f_i^{gg'} \rho_j^{gg'}(l_0) . \tag{1.6}$$

The torque balance condition for every particle amounts to the requirement that $S^g_{ij} = S^g_{ji}$. With this definition the stress within a given region inside the material is the area average of the force moments over the particles within the region.

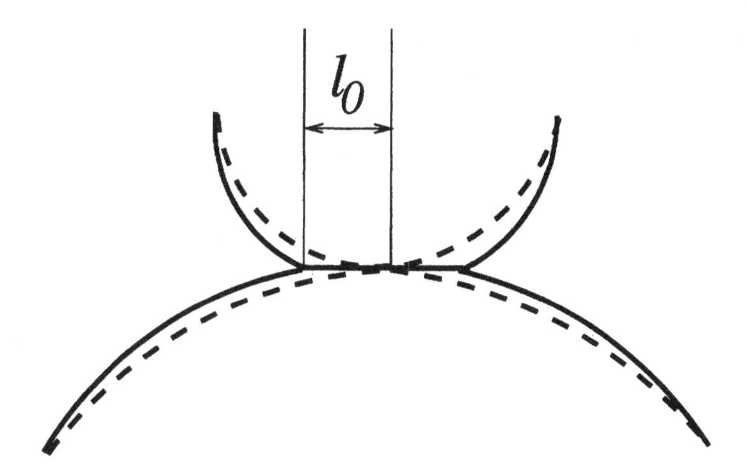

FIG. 3. *The equivalent system of ideally rigid particles (dashed lines) touch at the point $x = l_0$ along the contact line of the original compliant particles (full lines).*

Expressions (1.2), (1.3) and (1.6) suggest that the stress field is determined only by the forces $\vec{f}^{gg'}$, rather than by the entire distributions of the contact forces. It follows that if we knew the locations of the points $l_0^{gg'}$ where the equivalent interparticle forces act then we could map the system of compliant particles onto an equivalent one of infinitely rigid and infinitely rough particles that contact at these points, as illustrated in Figure 3. The equivalent system would have the same mean coordination number and it would transmit the same interparticle forces. Therefore, it would also have the same macroscopic stress field. Since the stress field in the equivalent packing is governed by IT then this would lead to the important conclusion that the original packing of compliant particles is also isostatic and is indeed described by IT.

However, there still remains the issue of the adaptation of the formalism to the structure of compliant systems. In particular, recall that IT relies on the identification of the geometric tensor whose components p_{ij} depend directly on the positions of the contact points. Thus, the question that we are faced with is whether it is possible to identify the points along the contact lines, $l_0^{gg'}$. Here we appear to have a problem. To determine the locations of these points requires using relations (1.3) and (1.4), which in turn require full knowledge of $\vec{\phi}(x)$ at the contact between every two particles. But this is tantamount to a solution of the interparticle forces in the first place. Does this mean that we cannot find the equivalent rigid packing? Have we reached a dead end?

Not necessarily. The conundrum can be resolved as follows. Let us introduce a judiciously chosen approximate equivalent system for which we can use IT to solve for the stress field. The idea is to show that the difference between the approximate and the true fields diminishes as $N^{-\alpha}$ ($\alpha > 0$) when the system size increases and therefore that the approximate solution converges to the true solution for macroscopic systems. The equivalent packing is generated by choosing the forces $\vec{f}^{gg'}$ to act at the centers of the contact lines. This requires only knowledge of the structure, not the force distributions. Let us construct the geometric tensor $\hat{p}$ for the equivalent system, using the definition in [24] and coarse-grain it using the procedure in [30]. Together with the boundary data, we can now determine the stress field using the solution of reference [25].

The deviation of this solution from the 'true' stress field arises from the error made in the position vectors that point from the centroids of the grains to the location of $\vec{\rho}^{gg'}(l_0)$. Defining these error vectors as $\vec{\delta}^{gg'} = \vec{\rho}^{gg'}(l/2) - \vec{\rho}^{gg'}(l_0)$, the error in the stress field around particle g is given by

$$\delta\sigma^g_{ij} = \frac{1}{a^g}\sum_{g'} \delta^{gg'}_i f^{gg'}_j \ , \tag{1.7}$$

where a^g is the area associated with particle g.

Now, a continuous stress field representation is only useful on scales that contain a good number of particles M but where $M \ll N$. For macroscopic description, the system is regarded as a collection of *continuous* such units. The error made in the stress within such a unit region of area $A^\Gamma = \sum_g a^g \sim aM$ is

$$\begin{aligned} \delta\sigma^\Gamma_{ij} &= \delta\left[\frac{1}{A^\Gamma}\sum_g a^g\sigma^g_{ij}\right] \\ &= \frac{1}{A^\Gamma}\sum_g a^g\delta\sigma^g_{ij} = \frac{1}{A^\Gamma}\sum_{g,g'\in\Gamma} \delta^{gg'}_i f^{gg'}_j \ . \end{aligned} \tag{1.8}$$

The error in the stress field is linear in the $\delta^{gg'}$ and its magnitude depends directly on the correlations between these quantities;

$$\left(\delta\sigma^\Gamma_{ij}\right)^2 = \frac{1}{\left(A^\Gamma\right)^2}\sum_{g,g',g'',g'''\in\Gamma} \left(\delta^{gg'}_i\delta^{g''g'''}_i\right) f^{gg'}_j f^{g''g'''}_j \ . \tag{1.9}$$

Let us now assume that in isotropic packings the error vectors $\delta^{gg'}$ are random and uncorrelated. This, of course, may not be the case since correlations may arise from the history of the dynamics that gave rise to the structure, as well as from inhomogeneities in material properties and granular characteristics that lead to nonlinear contact lines. However, in the absence of evidence to the contrary it is plausible that under small loading these effects are negligible. Then the sum in (1.8) can be regarded as a two-dimensional Markovian random walk and it increases as $O(M)$, up to logarithmic terms. But the area also increases as $O(M)$ and hence the entire expression (1.9) decreases as $1/M$. We therefore arrive at the conclusion that

$$\delta\hat{\sigma}^{\Gamma} \sim \frac{1}{\sqrt{M}} . \tag{1.10}$$

This result encouragingly support the idea that the approximate and the true stress fields converge in the macroscopic limit. But we are not finished yet. Having partitioned the system into N/M basic units of M grains each, we now face the acid test of the analysis. We need to determine the size of the discrepancy between the boundary data and the corresponding data derived from the approximate field. Using the same rationale, it is assumed that in isotropic systems the errors in the stresses (1.10) in different basic units are independent. It follows that the error in the stress field at the boundary (which is normalized by the total area) is of order $O(\sqrt{M/N}) \ll 1$. For macroscopic systems this error is indeed negligibly small. We have therefore reached the desired result; in macroscopic packings of compliant grains the approximate and the true stress fields are the same and can be obtained by solving for the isostatic stress in the equivalent infinitely rigid packing.

To make the analysis even more quantitative, let the particles' Young modulus be E and let us assume that their local radius of curvature is typically $R = c\sqrt{a}$. In isotropic systems it is expected that the value of the parameter c would be distributed over particles around $1/\sqrt{\pi}$. For monodisperse circular particles the distribution of c is almost a δ-function around this value. A sensible choice of M is such that there are many particles in a unit region on the boundary that are pressed by the boundary loading. With this choice the fluctuations of the force on the boundary particles can be disregarded and the mean force per particle in the y-direction is $F_y/\sqrt{N}$. Two particles in contact exerting a normal force f_n on one another deform slightly and according to Hertz theory the line contact between them is

$$w = 4\sqrt{\frac{f_n R^e}{\pi E'}} \tag{1.11}$$

long. In this expression R^e is an effective radius, $1/R^e = 1/R^g + 1/R^{g'} \approx 2\sqrt{\pi/a}$, $E' = 2(1-\nu^2)/E$ and ν is Poisson's ratio. Substituting for the compressive force f_n gives that the width of a typical line contact is

$$w \approx \left(\frac{4}{\pi}\right)^{3/4} \sqrt{\frac{F_y}{E'}} \left(\frac{a}{N^{1/2}}\right)^{1/4} . \tag{1.12}$$

The error in the distance between the middle of the line and the true point l_0 is at most $w/2$. Thus, using relations (1.7) and (1.12), and taking into consideration that there are on average three contacts per particle, the typical error made in the computation of the stress around any one particle is bounded by

$$|\delta\sigma^g_{ij}| < \frac{3wf_n}{2a} \approx \frac{3\sqrt{2}}{\pi^{3/4}}\sqrt{\frac{F_y^3}{E'}}\,(aN)^{-3/4} \ . \tag{1.13}$$

This calculation gives the precise power $\alpha = 3/4$ with which the error between the stress fields decays with the number of particles N and completes the proof.

Recalling that theories for ideal rigid particles predict stresses that propagate nonuniformly along arches [12, 20, 25], this explains why such force chains are also observed in packings of compliant particles [3–11]. Moreover, since the trajectories of force chains can be predicted in ideal packings [25] then the above suggests that these predictions can be extended to systems of compliant particles. It would be interesting to compare these predictions with the actual trajectories observed in realistic systems, such as those of references [3–7].

Three dimensions. The discussion of the three-dimensional systems follows the same rationale. The systems considered here consist of compliant and non-slipping particles, slightly compressed under a low external load. The mean coordination number is four per particle and the contacts make small two-dimensional surfaces. We assume, for simplicity, that the particles have homogeneous elastic properties, in which case the contact surfaces are planar. It is straightforward to lift this assumption and extend the results to systems of particles with nonuniform properties. The difference between such a system and one of infinitely rigid particles is that the interparticle forces are distributed across the contact surfaces. The contact surface between particles g and g' can be described by a position vector $\vec{\rho}^{gg'}(x,y)$, where x and y parameterise the surface of the contact. Terming the interparticle force density $\vec{\phi}^{gg'}(x,y)$, the mean force between the two particles is

$$\vec{f}^{gg'} = \int_s \vec{\phi}^{gg'}(x,y)dxdy \ , \tag{1.14}$$

where s stands for an area integration across the contact surface. The mean torque moment on the surface is

$$\vec{M}^{gg'} = \int_s \vec{\phi}^{gg'}(x,y)\vec{\rho}^{gg'}(x,y)dxdy \ . \tag{1.15}$$

From Eqs. (1.14) and (1.15) we now extract a position vector $\vec{\rho}_0^{gg'}$ by using

$$\vec{f}^{gg'} \times \vec{\rho}_0^{gg'} = \vec{M}^{gg'} \ . \tag{1.16}$$

It is straightforward to verify that if the contact surface is planar then $\vec{\rho}_0^{gg'}$ corresponds to a point on the surface $\vec{\rho}_0^{gg'} \equiv \vec{\rho}^{gg'}(x_0,y_0)$. Non-planar surfaces, which may result from non-uniform elastic properties in the particles, do not pose a limitation on this analysis as long as the deviation from the plane is smaller than the size of either of the particles in contact. We can now define the equivalent ideal system by postulating that its infinitely rigid particles make contacts at the points $\vec{\rho}_0^{gg'}$.

This done, we face the same conundrum as in the two-dimensional case; the determination of the contact points of the equivalent system poses the same level of difficulty as the original determination of the contact force distributions. For the resolution of this problem we follow the same logic as before. We construct an approximate system, for which isostaticity theory can be applied to determine the stress field, and then we show that the approximate field converges to the true field as the size of the system increases.

The contact points of the approximate system of infinitely rigid particles are postulated to be at the centroid of the contact surfaces of the true system of compliant particles,

$$\vec{\rho}^{gg'}_{\text{approx}} = \frac{\int_s \vec{\rho}^{gg'}(x,y)dxdy}{\int_s dxdy} . \tag{1.17}$$

These points are well defined from the geometry. The stress field in the approximate ideal packing of rigid particles is

$$\sigma^g_{ij} = \frac{1}{V^g}\sum_{g'} f_i^{gg'} \left(\rho^{gg'}_{\text{approx}}\right)_j , \tag{1.18}$$

where V^g is the volume associated with particle g. There are several ways to define the volume V^g such that $V^{sys} = \sum_g V^g$, but the precise definition is not essential for the present discussion. The error in the stress around particle g originates from the deviations of the true positions of the effective forces from the approximate positions $\delta\vec{\rho}^{gg'} = \vec{\rho}^{gg'}_{\text{approx}} - \vec{\rho}_0^{gg'}$,

$$\delta\sigma^g_{ij} = \frac{1}{V^g}\sum_{g'} f_i^{gg'}\delta\rho_j^{gg'} . \tag{1.19}$$

Consider a region Γ containing $M \ll N$ particles. The volume of the region is of order $V^\Gamma = \sum_g V^g \approx VM$, where V is the typical particle volume. The error in the stress over this region is

$$\begin{aligned} \delta\sigma^\Gamma_{ij} &= \delta\left[\frac{1}{V^\Gamma}\sum_g V^g\sigma^g_{ij}\right] \\ &= \frac{1}{V^\Gamma}\sum_{g,g'\in\Gamma} \delta\rho_i^{gg'} f_j^{gg'} . \end{aligned} \tag{1.20}$$

Following a similar analysis as in two dimensions we end up with a three-dimensional Markovian random walk of the vectors $\delta\vec{\rho}^{gg'}$. Assuming now that in isotropic systems these are uncorrelated (at least above some length scale), we conclude that $| \delta\sigma_{ij}^{\Gamma} |$ increases as $O(M^{1/2})$ and therefore that the error decreases as $O(M^{-1/2})$. By considering many such regions we are then led to the conclusion that the error at the boundary between the true and approximate stress fields decreases again as $O(\sqrt{M/N}) \sim O(L^{-3/2}) \ll 1$.

We can make the analysis again more quantitative by taking into consideration the elasticity of the particles, their typical size and assuming Hertzian interaction. Following the line of reasoning as that leading to Eq. (1.13), we find that according to Hertz theory the diameter of the area of contact between two grains is

$$w = 2\left(\frac{R^e f_n}{K}\right)^{1/3} , \tag{1.21}$$

where f_n is the force pressing them together, $1/K \equiv (3/4)[(1-\mu^2)/E + (1-\mu'^2)/E']$ is an effective elastic constant, and $1/R^e = 1/R^g + 1/R^{g'}$ is an effective curvature. Taking a cube of N particles and pressing on one of its surfaces by a force F_z gives that on average per grain there is a normal force of order $F_z/N^{2/3}$. Recalling that there are on average four contacts per particles, the error in the stress around a particle can be bounded by

$$|\delta\sigma_{ij}^g| < \frac{2wf_n}{V} \approx \frac{3}{\pi\,(R^e)^{8/3}\,K^{1/3}} F_z^{4/3} N^{-8/9} . \tag{1.22}$$

Thus we have demonstrated that, just as in two dimensions, the isostatic solution for the equivalent system of ideally rigid particles converges to the true stress field of the packing of compliant particles as the size of the system increases.

To conclude, it has been shown in this note that isostaticity theory is not limited to packings of infinitely particles. Rather, this theory can be used to describe stress fields in macroscopic systems of compliant particles. The only condition that such systems must satisfy is the same as the one for rigid-particle isostatic systems; that the number of contacts per particle is $z_c = d+1$ in $d = 2,3$ dimensions.

The next step towards bridging between isostaticity and elasticity theories involves considering packings where the mean coordination number is larger than z_c. A suggestion in this direction has been made by this author [25] and a detailed formulation of an isoelasticity theory will be reported elsewhere.

REFERENCES

[1] H.M Jaeger, S.R. Nagel, and R.P. Behringer, *Granular solids, liquids, and gases*, Rev. Mod. Phys. **68** (1996), p. 1259.

[2] P.G. DE GENNES, *Granular matter: a tentative view*, Rev. Mod. Phys. **71** (1999), S379.

[3] C. LIU, S.R. NAGEL, D.A. SCHECTER, S.N. COPPERSMITH, S. MAJUMDAR, O. NARAYAN, AND T.A. WITTEN, *Force fluctuations in bead packs*, Science **269**, (1995), p. 513 and references within.

[4] T. WAKABAYASHI, *Photoelastic method for determination of stress in powdered mass*, Proc. 7th Jpn. Nat. Cong. Appl. Mech. (1957), p. 153.

[5] P. DANTU, *Contribution à l'Étude Mécanique et Géométrique des Milieux Pulvérulents*, Proc. 4th Int. Conf. Soil Mech. and Found. Eng. (1957), p. 144.

[6] D. HOWELL AND R.P. BEHRINGER, *Fluctuations and dynamics for a two-dimensional sheared granular material*, in *Powders and Grains* **97**, eds. R.P. Behringer and J.T. Jenkins (Rotterdam: balkema, 1997), p. 337.

[7] L. VANEL, D. HOWELL, D. CLARK, R.P. BEHRINGER, AND E. CLEMENT, *Memories in sand: Experimental tests of construction history on stress distributions under sandpiles*, Phys. Rev. **E 60** (1999), R5040.

[8] D.F. BAGSTER AND R. KIRK, *Computer generation of a model to simulate granular material behaviour*, J. Powder Bulk Solids Technol. **1**, (1985), p. 19.

[9] K. LIFFMAN, D.Y. CHAN, AND B.D. HUGHES, *Force distribution in a two dimensional sandpile*, Powder Technol. **72** (1992), p. 225.

[10] J.R. MELROSE AND R.C. BALL, *The pathological behavior of sheared hard spheres with hydrodynamic interactions*, Europhys. Lett. **32** (1995), p. 535.

[11] C. THORNTON, *Force transmission in granular media* Kona Powder and Particle **15** (1997), p. 81.

[12] S.F. EDWARDS AND R.B.S. OAKESHOTT, *The transmission of stress in an aggregate*, Physica **D38** (1989), p. 88.

[13] F.H. HUMMEL AND E.J. FINNAN, *The distribution of pressure on surfaces supporting a mass of granular material*, Proc. Inst. Civil Eng. **212** (1920), p. 369.

[14] T. JOTAKI AND R. MORIYAMA, *On the bottom pressure distribution of the bulk material piled with the angle of repose*, J. Soc. Powder Technol. Jpn. **60** (1979), p. 184.

[15] J. SMID, AND J. NOVOSAD, *Pressure distribution under heaped bulk solids*, Proc. of 1981 Powtech Conf., Ind. Chem. Eng. Symp. **63** D3/V (1981), p. 1.

[16] R.C. BALL, *The propagation of stress through static powders*, in *Structures and dynamics of materials in the mesoscopic domain*, eds. M. Lal, R.A. Mashelkar, B.D. Kulkarni, and V.M. Naik (Imperial College Press, London 1999).

[17] R.C. BALL AND D.V. GRINEV, *The Stress Transmission Universality Classes of Periodic Granular Arrays*, cond-mat/9810124.

[18] S.F. EDWARDS AND D.V. GRINEV, *Statistical mechanics of stress transmission in disordered granular arrays*, Phys. Rev. Lett. **82** (1999), p. 5397.

[19] R. BLUMENFELD, S.F. EDWARDS, AND R.C. BALL, *Granular matter and the marginally rigid state*, cond-mat/0105348.

[20] J.P. WITTMER, P. CLAUDIN, M.E. CATES, AND J.-P. BOUCHAUD, *A new approach to stress propagation in sandpiles and silos*, Nature **382** (1996), p. 336.

[21] J.P. WITTMER, P. CLAUDIN, AND M.E. CATES, *Stress propagation and arching in static sandpiles*, J. Phys. I (France) **7** (1997), p. 39.

[22] M.E. CATES, J.P. WITTMER, J.-P. BOUCHAUD, AND P. CLAUDIN, *Jamming, force chains, and fragile matter*, Phys. Rev. Lett. **81** (1998), p. 1841.

[23] A.V. TKACHENKO AND T.A. WITTEN, *Stress propagation through frictionless granular material*, Phys. Rev. E **60** (1999), p. 687.

[24] R.C. BALL AND R. BLUMENFELD, *Stress field in granular systems: Loop forces and potential formulation*, Phys. Rev. Lett. **88** (2002), p. 115505.

[25] R. BLUMENFELD, *Stresses in isostatic granular systems and emergence of force chains*, Phys. Rev. Lett. **93** (2004), p. 108301.

[26] A typical grain size is defined here for convenience only; there is no loss of generality and the discussion is equally valid for packings of particles with any distribution of grain sizes.

[27] The extension of the analysis presented here to packings of particles of different materials is straightforward.

[28] It should be noted that for finite systems it is always possible to find a finite small force F_y that satisfies this criterion. However, the magnitude of this force decreases with system size.

[29] Without loss of generality, the packing is presumed to experience no body forces.

[30] R. BLUMENFELD, *Stress in planar cellular solids and isostatic granular assemblies: Coarse-graining the constitutive equation*, Physica A **336** (2004), p. 361.

LIST OF WORKSHOP PARTICIPANTS

- Douglas N. Arnold, Institute for Mathematics and its Applications, University of Minnesota
- Donald G. Aronson, Institute for Mathematics and its Applications, University of Minnesota
- Gerard Awanou, Institute for Mathematics and its Applications, University of Minnesota
- Peter W. Bates, Department of Mathematics, Michigan State University
- Patricia Bauman, Department of Mathematics, Purdue University
- Martin Z. Bazant, Department of Mathematics, Massachusetts Institute of Technology
- Josef Bemelmans, Institute for Mathematics, Aachen University of Technology
- Yuxing Ben, Department of Mathematics, Massachusetts Institute of Technology
- Daniel E. Bentil, Department of Mathematics and Statistics, University of Vermont
- Jorge Berger, Department of Physics, Technion - Israel Institute of Technology
- Keith Berrier, Department of Computational and Applied Mathematics, Rice University
- Fulvio Bisi, Dipartimento di Matematica, Università di Pavia
- Raphael Blumenfeld, Department of Physics, Cambridge University
- Helmut Brand, Physikalisches Institut, Universität Bayreuth
- Maria-Carme T. Calderer, School of Mathematics, University of Minnesota
- Qianyong Chen, Institute for Mathematics and its Applications, University of Minnesota
- L. Pamela Cook, Department of Mathematical Science, University of Delaware
- Zhenlu Cui, Department of Mathematics, Florida State University
- Antonio DeSimone, Applied Mathematics, International School of Advanced Studies, SISSA-Italy
- Brian DiDonna, Institute for Mathematics and its Applications, University of Minnesota
- Masao Doi, Department of Applied Physics, University of Tokyo
- Georg Dolzmann, Department of Applied Mathematics, University of Maryland
- James J. Feng, Department of Chemical and Biological Engineering, University of British Columbia

- Xiaobing Feng, Department of Mathematics, University of Tennessee
- Eliot Fried, Department of Theoretical and Applied Mechanics, University of Illinois - Urbana-Champaign
- Tim Garoni, Institute for Mathematics and its Applications, University of Minnesota
- Matthias Gobbert, Department of Mathematics and Statistics, University of Maryland - Baltimore County
- Francois Graner, SpectroMetrie Physique Université Joseph Fourier
- Alexander Grosberg, Department of Physics and Astronomy, University of Minnesota
- Colette Guillope, Department of Mathematics, University of Paris XII
- Robert Gulliver, School of Mathematics, University of Minnesota
- Andrei Gusev, Institut fuer Polymere, ETH Zentrum
- Cheng-Cher Huang, School of Physics and Astronomy, University of Minnesota
- Michel E. Jabbour, Department of Mathematics University of Kentucky
- Antal Jakli, Liquid Crystal Institute, Kent State University
- Richard D. James, Aerospace Engineering and Mechanics, University of Minnesota
- Sookyung Joo, Institute for Mathematics and its Applications, University of Minnesota
- Randall D. Kamien, Department of Physics and Astronomy, University of Pennsylvania
- Chiu Yen Kao, Institute for Mathematics and its Applications, University of Minnesota
- David Kinderlehrer, Department of Mathematics, Carnegie Mellon University
- Bernhard Klampfl, Department of Materials Science, Klaiss Inc.
- Isaac Klapper, Department of Mathematical Sciences, Montana State University
- Richard Kollar, Institute of Mathematics and its Applications, University of Minnesota
- Matthias Kurzke, Institute for Mathematics and its Applications, University of Minnesota
- Frederic Legoll, Institute for Mathematics and its Applications, University of Minnesota
- Benedict Leimkuhler, Department of Mathematics and Computer Science, University of Leicester
- Debra Lewis, Institute for Mathematics and its Applications, University of Minnesota

- Xiantao Li, Institute for Mathematics and its Applications, University of Minnesota
- Fanghua Lin, Department of Mathematics, New York University
- Chun Liu, Department of Mathematics, Pennsylvania State University
- Zuhan Liu, Department of Mathematics, Xuzhou Normal University
- Didier Long, Laboratoire de Physique des Solids, Universite Paris Sud
- Mitchell Luskin, School of Mathematics, University of Minnesota
- Govind Menon, Division of Applied Mathematics, Brown University
- Robert Meyer, Martin Fisher School of Physics, Brandeis University
- David Morse, Department of Chemical Engineering, University of Minnesota
- Jinhae Park, School of Mathematics, University of Minnesota
- Robert Pelcovits, Department of Physics, Brown University
- Peter Philip, Institute for Mathematics and its Application, University of Minnesota
- Petr Plechac, Mathematics Institute, University of Warwick
- Harald Pleiner, Max Planck Institute for Polymer Research
- Lea Popovic, Institute for Mathematics and its Applications, University of Minnesota
- Yitzhak Rabin, Department of Physics, Bar-Ilan University
- Amit Ranjan, Department of Chemical Engineering and Material Sciences, University of Minnesota
- Riccardo Rosso, Dipartimento di Matematica, Università di Pavia
- Rolf Ryham, Department of Mathematics, Pennsylvania State University
- Arnd Scheel, Institute for Mathematics and its Applications, University of Minnesota
- Jonathan V. Selinger, Center for Bio/Molecular Science and Engineering, Naval Research Laboratory
- George R. Sell, School of Mathematics, University of Minnesota
- Tien-Tsan Shieh, Department of Mathematics, Indiana University
- André M. Sonnet, Department of Mathematics, University of Strathclyde
- Peter J. Sternberg, Department of Mathematics, Indiana University
- Vladimir Sverak, Department of Mathematics, University of Minnesota
- Ellad Tadmor, Department of Mechanical Engineering, Technion - Israel Institute of Technology

- Eugene M. Terentjev, Cavendish Laboratory, Cambridge University
- Robert T. Tranquillo, Department of Chemical Engineering and Material Sciences, University of Minnesota
- Erkan Tuzel, Department of Physics and Astronomy, University of Minnesota
- Jon Fredric Urban, Department of Biomedical Engineering, University of Minnesota
- Epifanio G. Virga, Dipartimento di Matematica, Universita di Pavia
- Karl Voss, Department of Mathematics, Bucknell University
- Jimmy Wang, Aerospace Engineering and Mechanics, University of Minnesota
- Xiaoqiang Wang, Department of Mathematics, Pennsylvania State University
- Zhi-Qiang Wang, Department of Mathematics and Statistics, Utah State University
- Stephen J. Watson, ESAM, Northwestern University
- Jon A. Wilkening, Courant Institute of Mathematical Sciences, New York University
- Baisheng Yan, Department of Mathematics, Michigan State University
- Toshio Yoshikawa, Liu Bie Ju Centre for Mathematical Sciences, City University of Hong Kong

1999–2000 Reactive Flows and Transport Phenomena
2000–2001 Mathematics in Multimedia
2001–2002 Mathematics in the Geosciences
2002–2003 Optimization
2003–2004 Probability and Statistics in Complex Systems: Genomics, Networks, and Financial Engineering
2004–2005 Mathematics of Materials and Macromolecules: Multiple Scales, Disorder, and Singularities
2005-2006 Imaging
2006-2007 Applications of Algebraic Geometry
2007-2008 Mathematics of Molecular and Cellular Biology

IMA SUMMER PROGRAMS

1987 Robotics
1988 Signal Processing
1989 Robust Statistics and Diagnostics
1990 Radar and Sonar (June 18–29)
New Directions in Time Series Analysis (July 2–27)
1991 Semiconductors
1992 Environmental Studies: Mathematical, Computational, and Statistical Analysis
1993 Modeling, Mesh Generation, and Adaptive Numerical Methods for Partial Differential Equations
1994 Molecular Biology
1995 Large Scale Optimizations with Applications to Inverse Problems, Optimal Control and Design, and Molecular and Structural Optimization
1996 Emerging Applications of Number Theory (July 15–26)
Theory of Random Sets (August 22–24)
1997 Statistics in the Health Sciences
1998 Coding and Cryptography (July 6–18)
Mathematical Modeling in Industry (July 22–31)
1999 Codes, Systems, and Graphical Models (August 2–13, 1999)
2000 Mathematical Modeling in Industry: A Workshop for Graduate Students (July 19–28)
2001 Geometric Methods in Inverse Problems and PDE Control (July 16–27)
2002 Special Functions in the Digital Age (July 22–August 2)
2003 Probability and Partial Differential Equations in Modern Applied Mathematics (July 21–August 1)
2004 n-Categories: Foundations and Applications (June 7–18)
2005 Wireless Communications (June 22–July 1)
2006 Symmetries and Overdetermined Systems of Partial Differential Equations (July 17 - August 4)

IMA "HOT TOPICS" WORKSHOPS

- Challenges and Opportunities in Genomics: Production, Storage, Mining and Use, April 24–27, 1999
- Decision Making Under Uncertainty: Energy and Environmental Models, July 20–24, 1999
- Analysis and Modeling of Optical Devices, September 9–10, 1999
- Decision Making under Uncertainty: Assessment of the Reliability of Mathematical Models, September 16–17, 1999
- Scaling Phenomena in Communication Networks, October 22–24, 1999
- Text Mining, April 17–18, 2000
- Mathematical Challenges in Global Positioning Systems (GPS), August 16–18, 2000
- Modeling and Analysis of Noise in Integrated Circuits and Systems, August 29–30, 2000
- Mathematics of the Internet: E-Auction and Markets, December 3–5, 2000
- Analysis and Modeling of Industrial Jetting Processes, January 10–13, 2001
- Special Workshop: Mathematical Opportunities in Large-Scale Network Dynamics, August 6–7, 2001
- Wireless Networks, August 8–10 2001
- Numerical Relativity, June 24–29, 2002
- Operational Modeling and Biodefense: Problems, Techniques, and Opportunities, September 28, 2002
- Data-driven Control and Optimization, December 4–6, 2002
- Agent Based Modeling and Simulation, November 3–6, 2003
- Enhancing the Search of Mathematics, April 26-27, 2004
- Compatible Spatial Discretizations for Partial Differential Equations, May 11-15, 2004
- Adaptive Sensing and Multimode Data Inversion, June 27–30, 2004
- Mixed Integer Programming, July 25–29, 2005
- New Directions in Probability Theory, August 5–6, 2005

SPRINGER LECTURE NOTES FROM THE IMA:

The Mathematics and Physics of Disordered Media
Editors: Barry Hughes and Barry Ninham
(Lecture Notes in Math., Volume 1035, 1983)

Orienting Polymers
Editor: J.L. Ericksen
(Lecture Notes in Math., Volume 1063, 1984)

New Perspectives in Thermodynamics
Editor: James Serrin
(Springer-Verlag, 1986)

Models of Economic Dynamics
Editor: Hugo Sonnenschein
(Lecture Notes in Econ., Volume 264, 1986)

Current Volumes:

24 **Mathematics in Industrial Problems, Part 2** A. Friedman
25 **Solitons in Physics, Mathematics, and Nonlinear Optics**
P.J. Olver and D.H. Sattinger (eds.)
26 **Two Phase Flows and Waves**
D.D. Joseph and D.G. Schaeffer (eds.)
27 **Nonlinear Evolution Equations that Change Type**
B.L. Keyfitz and M. Shearer (eds.)
28 **Computer Aided Proofs in Analysis**
K. Meyer and D. Schmidt (eds.)
29 **Multidimensional Hyperbolic Problems and Computations**
A. Majda and J. Glimm (eds.)
30 **Microlocal Analysis and Nonlinear Waves**
M. Beals, R. Melrose, and J. Rauch (eds.)
31 **Mathematics in Industrial Problems, Part 3** A. Friedman
32 **Radar and Sonar, Part I**
R. Blahut, W. Miller, Jr., and C. Wilcox
33 **Directions in Robust Statistics and Diagnostics: Part I**
W.A. Stahel and S. Weisberg (eds.)
34 **Directions in Robust Statistics and Diagnostics: Part II**
W.A. Stahel and S. Weisberg (eds.)
35 **Dynamical Issues in Combustion Theory**
P. Fife, A. Liñán, and F.A. Williams (eds.)
36 **Computing and Graphics in Statistics**
A. Buja and P. Tukey (eds.)
37 **Patterns and Dynamics in Reactive Media**
H. Swinney, G. Aris, and D. Aronson (eds.)
38 **Mathematics in Industrial Problems, Part 4** A. Friedman
39 **Radar and Sonar, Part II**
F.A. Grünbaum, M. Bernfeld, and R.E. Blahut (eds.)
40 **Nonlinear Phenomena in Atmospheric and Oceanic Sciences**
G.F. Carnevale and R.T. Pierrehumbert (eds.)
41 **Chaotic Processes in the Geological Sciences** D.A. Yuen (ed.)
42 **Partial Differential Equations with Minimal Smoothness and Applications** B. Dahlberg, E. Fabes, R. Fefferman, D. Jerison, C. Kenig, and J. Pipher (eds.)
43 **On the Evolution of Phase Boundaries**
M.E. Gurtin and G.B. McFadden
44 **Twist Mappings and Their Applications**
R. McGehee and K.R. Meyer (eds.)
45 **New Directions in Time Series Analysis, Part I**
D. Brillinger, P. Caines, J. Geweke, E. Parzen, M. Rosenblatt, and M.S. Taqqu (eds.)
46 **New Directions in Time Series Analysis, Part II**
D. Brillinger, P. Caines, J. Geweke, E. Parzen, M. Rosenblatt, and M.S. Taqqu (eds.)
47 **Degenerate Diffusions**
W.-M. Ni, L.A. Peletier, and J.-L. Vazquez (eds.)
48 **Linear Algebra, Markov Chains, and Queueing Models**
C.D. Meyer and R.J. Plemmons (eds.)

49 **Mathematics in Industrial Problems, Part 5** A. Friedman
50 **Combinatorial and Graph-Theoretic Problems in Linear Algebra**
R.A. Brualdi, S. Friedland, and V. Klee (eds.)
51 **Statistical Thermodynamics and Differential Geometry of Microstructured Materials**
H.T. Davis and J.C.C. Nitsche (eds.)
52 **Shock Induced Transitions and Phase Structures in General Media** J.E. Dunn, R. Fosdick, and M. Slemrod (eds.)
53 **Variational and Free Boundary Problems**
A. Friedman and J. Spruck (eds.)
54 **Microstructure and Phase Transitions**
D. Kinderlehrer, R. James, M. Luskin, and J.L. Ericksen (eds.)
55 **Turbulence in Fluid Flows: A Dynamical Systems Approach**
G.R. Sell, C. Foias, and R. Temam (eds.)
56 **Graph Theory and Sparse Matrix Computation**
A. George, J.R. Gilbert, and J.W.H. Liu (eds.)
57 **Mathematics in Industrial Problems, Part 6** A. Friedman
58 **Semiconductors, Part I**
W.M. Coughran, Jr., J. Cole, P. Lloyd, and J. White (eds.)
59 **Semiconductors, Part II**
W.M. Coughran, Jr., J. Cole, P. Lloyd, and J. White (eds.)
60 **Recent Advances in Iterative Methods**
G. Golub, A. Greenbaum, and M. Luskin (eds.)
61 **Free Boundaries in Viscous Flows**
R.A. Brown and S.H. Davis (eds.)
62 **Linear Algebra for Control Theory**
P. Van Dooren and B. Wyman (eds.)
63 **Hamiltonian Dynamical Systems: History, Theory, and Applications**
H.S. Dumas, K.R. Meyer, and D.S. Schmidt (eds.)
64 **Systems and Control Theory for Power Systems**
J.H. Chow, P.V. Kokotovic, R.J. Thomas (eds.)
65 **Mathematical Finance**
M.H.A. Davis, D. Duffie, W.H. Fleming, and S.E. Shreve (eds.)
66 **Robust Control Theory** B.A. Francis and P.P. Khargonekar (eds.)
67 **Mathematics in Industrial Problems, Part 7** A. Friedman
68 **Flow Control** M.D. Gunzburger (ed.)
69 **Linear Algebra for Signal Processing**
A. Bojanczyk and G. Cybenko (eds.)
70 **Control and Optimal Design of Distributed Parameter Systems**
J.E. Lagnese, D.L. Russell, and L.W. White (eds.)
71 **Stochastic Networks** F.P. Kelly and R.J. Williams (eds.)
72 **Discrete Probability and Algorithms**
D. Aldous, P. Diaconis, J. Spencer, and J.M. Steele (eds.)
73 **Discrete Event Systems, Manufacturing Systems, and Communication Networks**
P.R. Kumar and P.P. Varaiya (eds.)
74 **Adaptive Control, Filtering, and Signal Processing**
K.J. Åström, G.C. Goodwin, and P.R. Kumar (eds.)

75 **Modeling, Mesh Generation, and Adaptive Numerical Methods for Partial Differential Equations** I. Babuska, J.E. Flaherty, W.D. Henshaw, J.E. Hopcroft, J.E. Oliger, and T. Tezduyar (eds.)

76 **Random Discrete Structures** D. Aldous and R. Pemantle (eds.)

77 **Nonlinear Stochastic PDEs: Hydrodynamic Limit and Burgers' Turbulence** T. Funaki and W.A. Woyczynski (eds.)

78 **Nonsmooth Analysis and Geometric Methods in Deterministic Optimal Control** B.S. Mordukhovich and H.J. Sussmann (eds.)

79 **Environmental Studies: Mathematical, Computational, and Statistical Analysis** M.F. Wheeler (ed.)

80 **Image Models (and their Speech Model Cousins)** S.E. Levinson and L. Shepp (eds.)

81 **Genetic Mapping and DNA Sequencing** T. Speed and M.S. Waterman (eds.)

82 **Mathematical Approaches to Biomolecular Structure and Dynamics** J.P. Mesirov, K. Schulten, and D. Sumners (eds.)

83 **Mathematics in Industrial Problems, Part 8** A. Friedman

84 **Classical and Modern Branching Processes** K.B. Athreya and P. Jagers (eds.)

85 **Stochastic Models in Geosystems** S.A. Molchanov and W.A. Woyczynski (eds.)

86 **Computational Wave Propagation** B. Engquist and G.A. Kriegsmann (eds.)

87 **Progress in Population Genetics and Human Evolution** P. Donnelly and S. Tavaré (eds.)

88 **Mathematics in Industrial Problems, Part 9** A. Friedman

89 **Multiparticle Quantum Scattering With Applications to Nuclear, Atomic and Molecular Physics** D.G. Truhlar and B. Simon (eds.)

90 **Inverse Problems in Wave Propagation** G. Chavent, G. Papanicolau, P. Sacks, and W.W. Symes (eds.)

91 **Singularities and Oscillations** J. Rauch and M. Taylor (eds.)

92 **Large-Scale Optimization with Applications, Part I: Optimization in Inverse Problems and Design** L.T. Biegler, T.F. Coleman, A.R. Conn, and F. Santosa (eds.)

93 **Large-Scale Optimization with Applications, Part II: Optimal Design and Control** L.T. Biegler, T.F. Coleman, A.R. Conn, and F. Santosa (eds.)

94 **Large-Scale Optimization with Applications, Part III: Molecular Structure and Optimization** L.T. Biegler, T.F. Coleman, A.R. Conn, and F. Santosa (eds.)

95 **Quasiclassical Methods** J. Rauch and B. Simon (eds.)

96 **Wave Propagation in Complex Media** G. Papanicolaou (ed.)

97 **Random Sets: Theory and Applications** J. Goutsias, R.P.S. Mahler, and H.T. Nguyen (eds.)

98 **Particulate Flows: Processing and Rheology** D.A. Drew, D.D. Joseph, and S.L. Passman (eds.)

99 **Mathematics of Multiscale Materials** K.M. Golden, G.R. Grimmett, R.D. James, G.W. Milton, and P.N. Sen (eds.)
100 **Mathematics in Industrial Problems, Part 10** A. Friedman
101 **Nonlinear Optical Materials** J.V. Moloney (ed.)
102 **Numerical Methods for Polymeric Systems** S.G. Whittington (ed.)
103 **Topology and Geometry in Polymer Science** S.G. Whittington, D. Sumners, and T. Lodge (eds.)
104 **Essays on Mathematical Robotics** J. Baillieul, S.S. Sastry, and H.J. Sussmann (eds.)
105 **Algorithms For Parallel Processing** M.T. Heath, A. Ranade, and R.S. Schreiber (eds.)
106 **Parallel Processing of Discrete Problems** P.M. Pardalos (ed.)
107 **The Mathematics of Information Coding, Extraction, and Distribution** G. Cybenko, D.P. O'Leary, and J. Rissanen (eds.)
108 **Rational Drug Design** D.G. Truhlar, W. Howe, A.J. Hopfinger, J. Blaney, and R.A. Dammkoehler (eds.)
109 **Emerging Applications of Number Theory** D.A. Hejhal, J. Friedman, M.C. Gutzwiller, and A.M. Odlyzko (eds.)
110 **Computational Radiology and Imaging: Therapy and Diagnostics** C. Börgers and F. Natterer (eds.)
111 **Evolutionary Algorithms** L.D. Davis, K. De Jong, M.D. Vose, and L.D. Whitley (eds.)
112 **Statistics in Genetics** M.E. Halloran and S. Geisser (eds.)
113 **Grid Generation and Adaptive Algorithms** M.W. Bern, J.E. Flaherty, and M. Luskin (eds.)
114 **Diagnosis and Prediction** S. Geisser (ed.)
115 **Pattern Formation in Continuous and Coupled Systems: A Survey Volume** M. Golubitsky, D. Luss, and S.H. Strogatz (eds.)
116 **Statistical Models in Epidemiology, the Environment, and Clinical Trials** M.E. Halloran and D. Berry (eds.)
117 **Structured Adaptive Mesh Refinement (SAMR) Grid Methods** S.B. Baden, N.P. Chrisochoides, D.B. Gannon, and M.L. Norman (eds.)
118 **Dynamics of Algorithms** R. de la Llave, L.R. Petzold, and J. Lorenz (eds.)
119 **Numerical Methods for Bifurcation Problems and Large-Scale Dynamical Systems** E. Doedel and L.S. Tuckerman (eds.)
120 **Parallel Solution of Partial Differential Equations** P. Bjørstad and M. Luskin (eds.)
121 **Mathematical Models for Biological Pattern Formation** P.K. Maini and H.G. Othmer (eds.)
122 **Multiple-Time-Scale Dynamical Systems** C.K.R.T. Jones and A. Khibnik (eds.)
123 **Codes, Systems, and Graphical Models** B. Marcus and J. Rosenthal (eds.)
124 **Computational Modeling in Biological Fluid Dynamics** L.J. Fauci and S. Gueron (eds.)
125 **Mathematical Approaches for Emerging and Reemerging Infectious Diseases: An Introduction** C. Castillo-Chavez with S. Blower, P. van den Driessche, D. Kirschner and A.A. Yakubu (eds.)

126 **Mathematical Approaches for Emerging and Reemerging Infectious Diseases: Models, Methods, and Theory** C. Castillo-Chavez with S. Blower, P. van den Driessche, D. Kirschner, and A.A. Yakubu (eds.)

127 **Mathematics of the Internet: E-Auction and Markets** B. Dietrich and R.V. Vohra (eds.)

128 **Decision Making Under Uncertainty: Energy and Power** C. Greengard and A. Ruszczynski (eds.)

129 **Membrane Transport and Renal Physiology** H. Layton and A.M. Weinstein (eds.)

130 **Atmospheric Modeling** D.P. Chock and G.R. Carmichael (eds.)

131 **Resource Recovery, Confinement, and Remediation of Environmental Hazards** J. Chadam, A. Cunningham, R.E. Ewing, P. Ortoleva, and M.F. Wheeler (eds.)

132 **Fractals in Multimedia** M.F. Barnsley, D. Saupe, and E.R. Vrscay (eds.)

133 **Mathematical Methods in Computer Vision** P.J. Olver and A. Tannenbaum (eds.)

134 **Mathematical Systems Theory in Biology, Communications, Computation, and Finance** J. Rosenthal and D.S. Gilliam (eds.)

135 **Transport in Transition Regimes** N. Ben Abdallah, A. Arnold, P. Degond, I. Gamba, R. Glassey, C.D. Lawrence, and C. Ringhofer (eds.)

136 **Dispersive Transport Equations and Multiscale Methods** N. Ben Abdallah, A. Arnold, P. Degond, I. Gamba, R. Glassey, C.D. Lawrence, and C. Ringhofer (eds.)

137 **Geometric Methods in Inverse Problems and PDE Control** C.B. Cooke, I. Lasiecka, G. Uhlmann, and M.S. Vogelius (eds.)

138 **Mathematical Foundations of Speech and Language Processing** M. Johnson, S. Khudanpur, M. Ostendorf, and R. Rosenfeld (eds.)

139 **Time Series Analysis and Applications to Geophysical Systems** D.R. Brillinger, E.A. Robinson, and F.P. Schoenberg (eds.)

140 **Probability and Partial Differential Equations in Modern Applied Mathematics** E.C. Waymire and J. Duan (eds.)

141 **Modeling of Soft Matter** Maria-Carme T. Calderer and Eugene M. Terentjev (eds.)

142 **Compatible Spatial Discretizations** Douglas N. Arnold, Pavel B. Bochev, Richard B. Lehoucq, Roy A. Nicolaides, and Mikhail Shashkov (eds.)

Printed in the United States
By Bookmasters